AF803281

INDUSTRIAL HYGIENE SCIENCE SERIES

ADVANCES IN AIR SAMPLING

American Conference of Governmental Industrial Hygienists

CRC Press
Taylor & Francis Group
Boca Raton London New York

CRC Press is an imprint of the
Taylor & Francis Group, an **informa** business

First published 1988 by CRC Press
Taylor & Francis Group
6000 Broken Sound Parkway NW, Suite
300 Boca Raton, FL 33487-2742

Reissued 2018 by CRC Press

© 1988 by Taylor & Francis
CRC Press is an imprint of Taylor & Francis Group, an Informa business

No claim to original U.S. Government works

This book contains information obtained from authentic and highly regarded sources. Reasonable efforts have been made to publish reliable data and information, but the author and publisher cannot assume responsibility for the validity of all materials or the consequences of their use. The authors and publishers have attempted to trace the copyright holders of all material reproduced in this publication and apologize to copyright holders if permission to publish in this form has not been obtained. If any copyright material has not been acknowledged please write and let us know so we may rectify in any future reprint.

Except as permitted under U.S. Copyright Law, no part of this book may be reprinted, reproduced, transmitted, or utilized in any form by any electronic, mechanical, or other means, now known or hereafter invented, including photocopying, microfilming, and recording, or in any information storage or retrieval system, without written permission from the publishers.

For permission to photocopy or use material electronically from this work, please access www.copyright.com (http://www.copyright.com/) or contact the Copyright Clearance Center, Inc. (CCC), 222 Rosewood Drive, Danvers, MA 01923, 978-750-8400. CCC is a not-for-profit organiza-tion that provides licenses and registration for a variety of users. For organizations that have been granted a photocopy license by the CCC, a separate system of payment has been arranged.

Trademark Notice: Product or corporate names may be trademarks or registered trademarks, and are used only for identification and explanation without intent to infringe.

A Library of Congress record exists under LC control number: 88011736

Publisher's Note
The publisher has gone to great lengths to ensure the quality of this reprint but points out that some imperfections in the original copies may be apparent.

Disclaimer
The publisher has made every effort to trace copyright holders and welcomes correspondence from those they have been unable to contact.

ISBN 13: 978-1-138-50605-3 (hbk)
ISBN 13: 978-1-315-14983-7 (ebk)

Visit the Taylor & Francis Web site at http://www.taylorandfrancis.com and the CRC Press Web site at http://www.crcpress.com

PREFACE

The ACGIH Symposium, Advances in Air Sampling, was held February 16–18, 1987 at the Asilomar Conference Center, Pacific Grove, California. Organized by the ACGIH Air Sampling Procedures Committee, the symposium was convened to discuss the establishment of Threshold Limit Values (TLVs) for particulate substances based on size-selective sampling, to review new developments in techniques for the sampling of workplace and community atmospheres, and to stimulate the exchange of ideas and information on the sampling of gases and vapors. Session titles included Particle Size-Selective Sampling, Sampling Gases and Vapors for Analysis, Special Topics, Real-Time Samplers, Sampling Strategy, and Posters.

In 1985, the ACGIH Air Sampling Procedures Committee recommended particle size-selective criteria for sampling in the workplace and defined inspirable, thoracic, and respirable particle mass fractions. The subsequent adoption of the criteria by the ACGIH made urgent the task of applying the criteria to the TLVs of specific substances in order to optimize worker protection. At the Symposium, case studies were presented for beryllium, wood dust, and sulfuric acid aerosols.

An unusual feature of the symposium was the participation not only of workers from the industrial hygiene community, but also by those concerned with nonoccupational air exposures, affording the cross-fertilization of ideas and techniques for the sampling of pollutants. For example, it was pointed out that the U.S. Environmental Protection Agency's program to develop a new particle standard for ambient air, designated PM-10, has provided samplers which are also useful for thoracic mass sampling, as specified by ACGIH. Likewise, diffusion denuders for the separation of gases from particles, developed for ambient air sampling, may also find applications in the workplace. Several new sampling devices for gases and vapors in ambient indoor and outdoor atmospheres were presented.

Several chapters are concerned with the all-important subject of sampling strategies. It was emphasized that sampling for compliance was not very effective in preventing acute exposure and, moreover, the current need is for prospective surveillance to correlate with and to prevent chronic disease. Such surveillance, involving the sampling of many substances and at low concentrations, requires a completely new strategy for sampling. Two presentations stressed that the sampling should be designed according to biological considerations involving toxicokinetics.

In many cases the chapters herein give a more complete account of their topics than was possible at the symposium. Although papers were peer reviewed, the authors are responsible for their content.

Program Committee:

Walter John, Ph.D., Chair
Robert F. Phalen, Ph.D., Chair, Air Sampling Procedures Committee
Joseph D. Bowman, Ph.D.
Robert G. Lewis, Ph.D.
Paul J. Lioy, Ph.D.
Otto G. Raabe, Ph.D.
Ronald S. Ratney, Ph.D.
Bruce O. Stuart, Ph.D.

CONTENTS

SECTION I
PARTICLE SIZE-SELECTIVE SAMPLING

SECTION II
SAMPLING GASES AND VAPORS FOR ANALYSIS

SECTION III
SPECIAL TOPICS

SECTION IV
REAL-TIME AEROSOL SAMPLERS

SECTION V
SAMPLING STRATEGY

SECTION I

Particle Size-Selective Sampling

CHAPTER 1

Rationale for and Implications of Particle Size-Selective Sampling

ROBERT F. PHALEN,[A] BRUCE O. STUART[B] and PAUL J. LIOY[C]

[A]University of California, Irvine; [B]Wright-Patterson AFB, Ohio; [C]Rutgers Medical School, New Jersey

INTRODUCTION

It is difficult to imagine a material that, when inhaled in a liquid or solid aerosol form, does not have the potential for producing lung disease. Thus, most commonly encountered substances are now associated with air contamination criteria for protecting people from overexposure. Most air contamination criteria relate to the total airborne mass per unit volume of air. Unfortunately, the accurate measurement of such a total airborne mass concentration is usually impractical but, even if successful, it would be inadequate for predicting the pathological effects of inhaling most aerosols. The reason for this is that the human respiratory tract does not uniformly sample airborne particles of various sizes. The application of information on how the aerodynamic size of aerosols determines the inhalability (the fraction of airborne mass that actually enters the nose or mouth during inhalation) and the regional deposition of particles within the respiratory tract can lead to acquiring size-selective samples that more closely relate to aerosol inhalation hazards. Size-selective aerosol samples may be defined as reliably collected aerosol fractions which are expected to be available for deposition in the various subregions of the respiratory tract.

Historically, the British Medical Research Council, the U.S. Atomic Energy Commission, the American Conference of Governmental Industrial Hygienists (ACGIH), the U.S. Occupational Safety and Health Administration, the U.S. Environmental Protection Agency, and the International Standards Organization have recognized the importance of particle size factors in inhalation-related risks. They have each recommended various forms of particle size-selective sampling for specific aerosols including pneumoconiosis-producing dusts, insoluble radioactive particles, quartz dust, coal dust, and inhalable air pollutant particles. Despite the efforts of these agencies, there has not been, until recently, any consistent rational approach to defining the size fractions of interest.

The accrual of a reliable data base on the aerosol sampling and collection efficiencies of the human respiratory tract has led to recommendations by the ACGIH Air Sampling Procedures Committee for the collection efficiencies of workplace air sampling instruments which have been listed in the 1986–87 issue of the Threshold Limit Values (TLVs) booklet published by the ACGIH. Three sets of collection efficiencies (Figure 1) are recommended for obtaining three mass fractions: inspirable particulate mass (IPM) for materials which may be hazardous when they reach any region of the respiratory tract; thoracic particulate mass (TPM) for materials which may be hazardous when they reach the thoracic airways or the gas exchange region; and respirable particulate mass (RPM) for materials which may be hazardous when they reach the gas exchange region of the lung. Each mass fraction is described by a collection efficiency curve and an associated tolerance band of ± 10 percent collection efficiency. The tolerance band recognizes uncertainties in the particle sampling and collection characteristics of the human respiratory tract, and it allows for practical differences in the performance of various sampling instruments.

In order for such size-selective sampling strategies to become realized in the workplace several things must occur:

1. Appropriate aerosol fractions that relate to aerosol exposure of the relevant sites in the respiratory tract must be defined.
2. Instruments that can reliably collect the specified samples must be available.
3. Acceptable criteria for workplace aerosol concentrations must be developed that relate to the relevant aerosol mass fractions rather than to total airborne mass; such criteria include TLVs and Maximum Permissible Concentrations (MPCs).

The remainder of this chapter will focus on each of these three items.

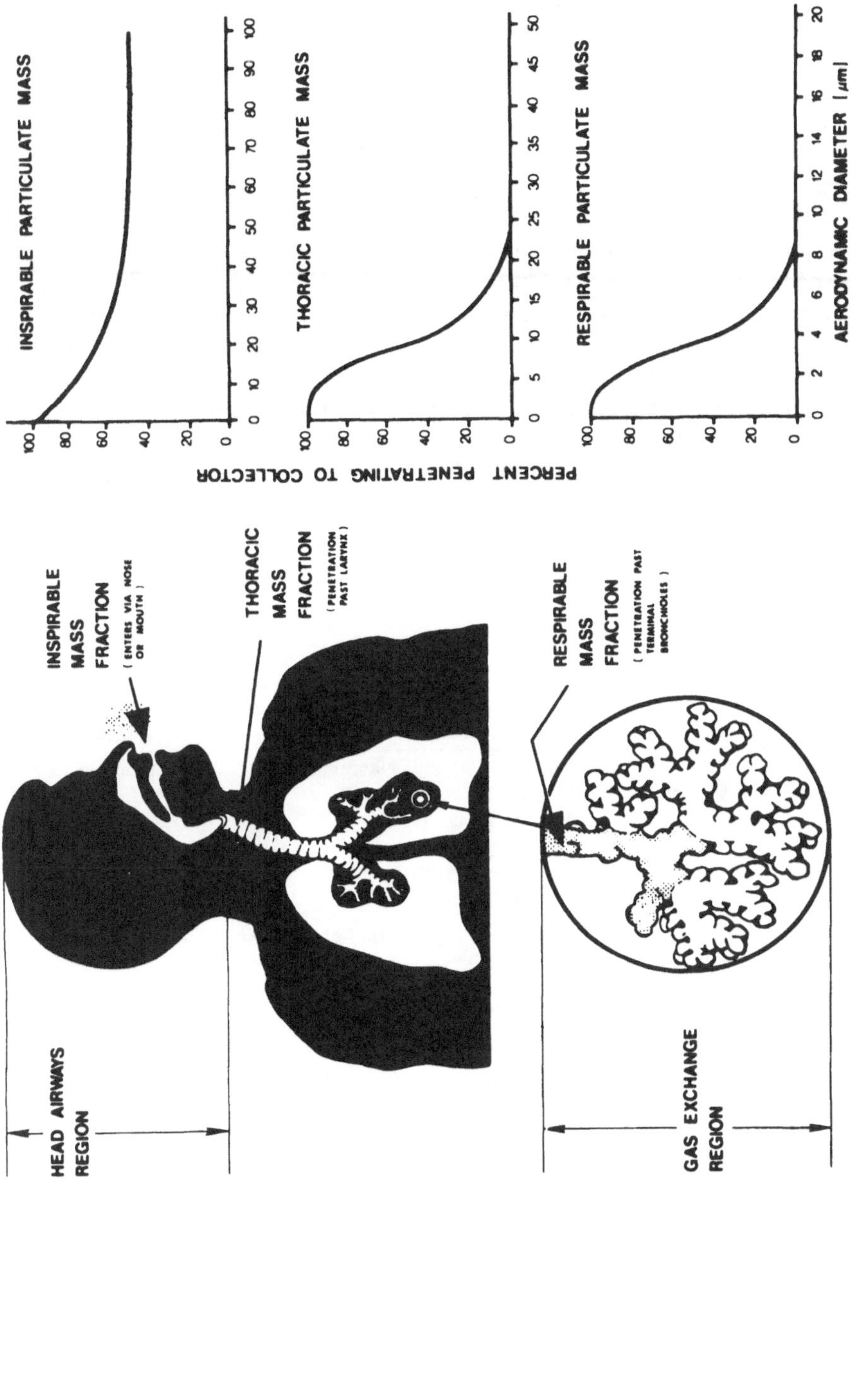

Figure 1. The three aerosol mass fractions recommended for particle size-selective sampling.

APPROPRIATE MASS FRACTIONS

Relevant Airway Structures

By following the path of an inhaled particle, it is possible to identify over 20 separate airway structures that both influence aerosol particle deposition and can be affected by inhaled irritants or toxics. However, for the purpose of simplification it is useful to focus on larger airway regions that include collections of structures that are similar with respect to particle deposition mechanisms, particle clearance phenomena, and their potential diseases. After reviewing several regional models, the Air Sampling Procedures Committee of ACGIH selected a three-region model of the respiratory tract (Table I).[1] Region 1, the head airways region (HAR), includes the nose and mouth, the nasopharynx, the oropharynx, the laryngopharynx, and the larynx. Region 2, the tracheobronchial region (TBR), includes the trachea, bronchi, and bronchioles. Region 3, the gas exchange region (GER), includes the respiratory bronchioles, alveolar ducts, alveolar sacs, and all of the associated alveoli. These three regions are similar to those previously used by the Task

TABLE I. Respiratory Tract Regions

Region	Anatomic Structures	Task Group Region	Region
1. HAR (Head Airways Region)	Nose Mouth Nasopharynx Oropharynx Laryngopharynx Larynx	Nasopharynx (NP)	Extrathoracic (E)
2. TBR (Tracheobronchial Region)	Trachea Bronchi Bronchioles (to terminal bronchioles)	Tracheobronchial (TB)	Tracheobronchial (B)
3. GER (Gas Exchange Region)	Respiratory bronchioles Alveolar ducts Alveolar sacs Alveoli	Pulmonary (P)	Alveolar (A)

Group on Lung Dynamics of the International Commission on Radiological Protection[2] and the Ad Hoc Working Group to Technical Committee TC 146, Air Quality, of the International Standards Organization.[3,4] These three regions have been relatively well-studied in adults with respect to collection efficiencies of aerosol particles over a wide range of aerodynamic sizes. These regions are the anatomical basis for identifying the relevant aerosol mass fractions, criteria for sampling instruments, and size-selective TLVs.

Inspirable Mass Fraction

Some particulate air contaminants are hazardous when they deposit in the airways of the head or anywhere else within the respiratory system. For them, an air sample that includes all of the inspirable mass is useful. The inspirable particulate mass fraction of an aerosol is that portion of the total ambient aerosol that can actually enter either the nose or mouth during inhalation. It is also known as the aspiration efficiency of the HAR. Because the head is not an isokinetic sampler, inspirability will vary with aerosol aerodynamic size as well as wind direction and speed, body position and ventilatory pattern. Considering the published information on measurements of aerosol inspirability on humans and mannequins, Soderholm[5] recommended the following function for describing percent inspirability E as a function of particulate aerodynamic diameter d_a.

$$E = 50\,(1 + \exp[-0.06\, d_a]) \pm 10 \qquad (1)$$

for $0 < d_a \leq 100$ micrometers
E is unknown for $d_a > 100$ micrometers

This function, with a ± 10 percent band (Figure 2), has become known as the ACGIH recommended inspirable particulate mass (IPM) size-selective criteria.

Thoracic and Respirable Mass Fractions

Some aerosols must penetrate beyond the airways of the head before they can affect human health. For such materials a sample that represents the exposure of the thoracic airways (tracheobronchial and gas exchange regions) is recommended.

The thoracic particulate mass (TPM) fraction refers to that portion of the airborne particles that may penetrate the head airways and enter the

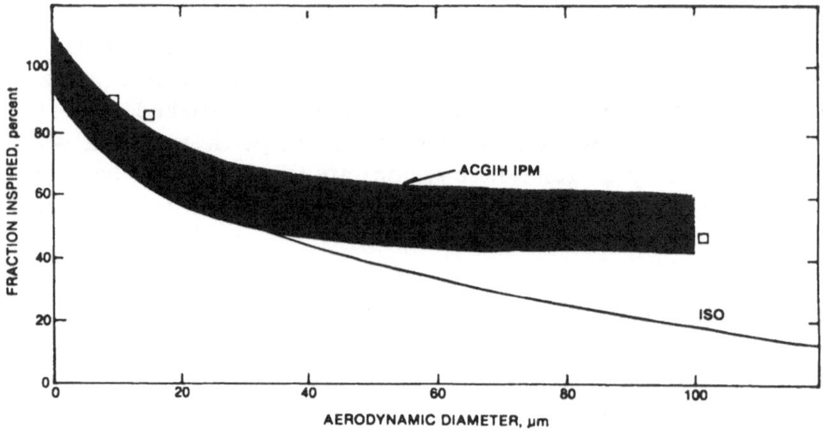

Figure 2. ACGIH recommended inspirable particulate mass (IPM) size-selective criteria compared to the ISO curve and to recently available data averaged over orientation and wind speed.

tracheobronchial airways region (TBR). The respirable particulate mass (RPM) fraction refers to that portion of the airborne particles that may penetrate the head and tracheobronchial airways and thereby expose the gas exchange region of the lung. A respirable mass fraction is most appropriate for materials that are hazardous only when they reach the gas exchange region.

Clearly, aerosol deposition in the tracheobronchial region and gas exchange region is greater during mouth breathing than it is during nose breathing.[6] This suggests that the thoracic sample should be based upon those particles that penetrate the head airways region during mouth breathing. An estimate of the fraction of particles available for deposition in the thoracic airways region therefore requires an accurate estimate of the deposition of particles in the head airways region during mouth breathing.

Based upon a recent review of the data, Raabe[7] recommended a thoracic particulate mass (TPM) size-selective sampling criterion as a tolerance band consisting of those particles that penetrate a separator whose size collection efficiency is described by a cumulative lognormal function with median aerodynamic diameter (d_{50}) of 10 ± 1 μm and with geometric standard deviation (σ_g) of 1.5 ± 0.1. The recommended tolerance band is shown in Figure 3. This band is seen to be conservative when compared to data corrected to the average air flow (Q = 43.5 lpm) of the mouth-breathing reference worker in that the actual penetration to the

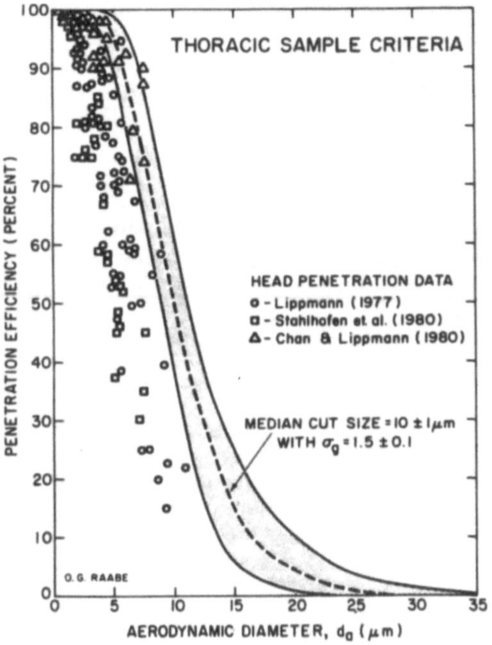

Figure 3. Thoracic particulate mass (TPM) tolerance band given as penetration efficiency to a sample collector. Also shown are selected values for observed human head penetration during inhalation by mouth.

lung tends to be generally less than the TPM fraction size-selective sample. The TPM criterion therefore tends to overestimate the amount of lung exposure and, correspondingly, to provide a reasonable level of protection when used as the basis for determining compliance with TLVs.

For situations where the risks due to aerosol inhalation exposure are primarily associated with the deposition of particles in the deep lung, the ACGIH has previously recommended a respirable dust standard. In this context, the word "respirable" has been used to describe that portion of an aerosol that is available to the gas exchange region. Using the ACGIH respirable dust sample as the basis, the new recommended respirable particulate mass (RPM) size-selective sampling criterion is a tolerance band consisting of those particles that penetrate a separator whose size collection efficiency is described by a cumulative lognormal function with median aerodynamic diameter (d_{50}) of 3.5 ± 0.3 μm and with geometric standard deviation (σ_g) of 1.5 ± 0.1.[7] This tolerance band is shown in Figure 4 along with the points representing the ACGIH respirable dust criteria. The RPM size-selective sampling criteria incorporate and clarify the ACGIH respirable dust criteria.

IMPLICATIONS FOR SAMPLERS

The specification of aerosol mass fractions that expose the respiratory tract and its three sub-regions permits one to design new aerosol samplers for industrial hygiene applications. Such samplers should collect airborne particles for analysis according to the criteria for IPM, TPM, and RPM.

It is desirable that IPM sampling eventually replace the present method of total dust sampling using open-face filter holders. So-called total dust samplers, such as open-face filter cassettes, do not measure total dust and are unsuitable for most monitoring of airborne particles

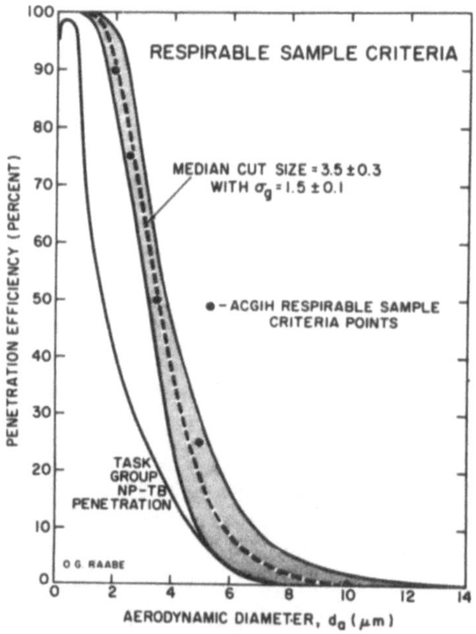

Figure 4. Respirable particulate mass (RPM) tolerance band given as penetration efficiency to a sample collector. Also shown are the assumed penetration values of head and tracheobronchial airways based upon the recommendations of the ICRP Task Group on Lung Dynamics.

larger than a few micrometers because their sampling efficiency for large particles is sensitive to wind velocity and direction.[8] Implementation of IPM sampling will require the development and testing of suitable sampling instruments. The current state of development of IPM samplers has been reviewed and discussed by Hinds[9] and by Mark and Vincent.[10] It appears that simple, inexpensive, reliable IPM samplers are not only feasible but are close to becoming practical.

Several types of aerosol samplers are suitable for collecting TPM and RPM samples. These include filters, particle size-selective inlets followed by filters, inertial impactors (including virtual impactors such as the dichotomous sampler), cyclone samplers, elutriators, and combinations of these devices. Relatively recent reviews of TPM and RPM sampling have been prepared by John.[11,12]

All such size-selective samplers must be able to collect or to analyze *in situ* sufficient airborne material to accurately define the particular mass fraction of interest. This sets limits on the sampling rate, permissible physical artifacts (such as particle bounce and loading effects), and degree of conformance to the recommended sampling criteria curves. Consideration of such factors in the design, testing, and operation of such samplers is a highly sophisticated and rapidly-evolving topic.[13,14] Solutions to problems in this area will lead to a new generation of improved sampling instruments.

IMPLICATIONS FOR THRESHOLD LIMIT VALUES

Analyses of specific air contaminants and the potential diseases associated with different regions of the respiratory tract indicate that size-selective sampling is necessary for a meaningful evaluation of the inhalation hazard to the worker. Different particle size distributions of the same contaminant will cause major changes in deposition in the various regions of the respiratory tract, altering not only the relative quantity of material deposited within the region but also the probability and nature of the associated disease processes. Thus, the development of Particle Size-Selective Sampling-TLVs (PSS-TLVs) is an important and necessary step toward the improvement of air-contamination standards established for the protection of workers.

The first step in deriving a PSS-TLV is the identification of the chemical substance that constitutes a potential air pollutant. This involves an examination of the physicochemical properties related to its airborne and biological behavior. Concomitantly, the literatures of epidemiology, industrial hygiene, and toxicology should be analyzed to identify diseases that may be associated with the chemical substance affecting specific regions of the respiratory tract or systemic organ systems.

If no potential diseases related to the chemical substance are found, then the evaluation can be terminated or a nuisance dust-type standard recommended. If a disease potential exists, but the physicochemical nature of the chemical substance is such that no airborne particle phase can be produced, the procedure can revert to the traditional procedures for protecting exposed persons. However, if the physicochemical properties of the chemical substance suggest that it may become airborne as an aerosol and have the potential for producing disease upon inhalation, the analysis proceeds. At this stage, the physical and chemical properties of the substance are carefully evaluated for the conditions likely to be encountered by people.

The aerodynamic particle size distribution will usually determine the mass fraction of the aerosol that will enter the head airways, tracheobronchial, or gas exchange regions of the respiratory tract. The particle size-selective sampling criteria can be used to estimate the actual quantity of chemical substance that will be presented to the three principal regions of the respiratory tract. Thus, the mass of the substance presented to each region will be estimated for the purposes of continuing the process of evaluating the airborne hazards.

In addition to considering lung disease, extrapulmonary sites of action should also be evaluated. Subsequently, it will be determined whether the incorporated doses of the substance are likely to cause disease. Once the particle size and particulate mass fraction are determined and the hazard analyses are completed, critical mass concentrations can be determined for each size fraction and the most restrictive PSS-TLV applied to a given exposure situation. This review will result in formal recommendations for PSS-TLVs. A more complete discussion of the development of PSS-TLVs has been recently published.[15] Analyses of some specific airborne chemical substances are presented in some of the chapters that follow.

SOME IMPORTANT LIMITATIONS

Epidemiological Applications

Because the criteria for TPM and RPM are designed to overestimate the actual inhalation exposure of the average worker, they have built-in protection factors. Although this conservatism is desirable from the industrial hygienist's point of view, it does not necessarily serve the purposes of the epidemiologist. In an epidemiological study of hazardous aerosols, it is often desirable to determine the effect of aerosol particle size on risk. Ideally, the actual size distribution of the airborne material or the actual deposition in the exposed individuals is directly determined or otherwise estimated. However, being able to collect TPM and/or RPM in an epidemiological investigation could reduce the uncertainties in estimating the actual particulate exposures.

Nonoccupational Environments

The criteria for IPM, TPM, and RPM relate to exposures of healthy adult workers. Newborns, infants, children, adolescents, people with respiratory tract diseases, and other segments of the population may have aerosol sampling and deposition characteristics that significantly deviate from those of typical workers. Unfortunately, little relevant information exists on the general population. Thus, extreme caution is necessary in applying particle size-selective considerations beyond the workplace.

Nonideal Particles

Several classes of aerosol particles do not behave in the simple aerodynamic manner that has been assumed in recommending particle size-selective sampling procedures. Examples are aerosols that are highly electrically charged, aerosol particles that may be either rapidly growing (hygroscopic) or rapidly decreasing (volatile) in size during inhalation, and aerosol particles that have extreme shapes, e.g., fibers and thin plates. Aerosols of these types must be considered as special cases in hazard evaluation, and the applicability of the particle size-selective sampling recommendations must be evaluated on a case-by-case basis.

SUMMARY

The inspirability and the principal deposition sites (within the human respiratory tract) of inhaled aerosols, and hence their hazards, are greatly influenced by their aerodynamic sizes. Realization of these facts has led the ACGIH Air Sampling Procedures Committee to recommend the sampling of three specific mass fractions for industrial hygiene purposes. The mass fractions are: inhalable particulate mass (IPM) for aerosols that are hazardous when they deposit anywhere in the respiratory tract; thoracic particulate mass (TPM) for aerosols that are hazardous when they deposit in the tracheobronchial or alveolarized airways; and respirable particulate mass (RPM) for aerosols that are hazardous only when they deposit in the gas exchange region of the lungs.

These mass fractions can be reliably collected only by using sampling instruments that conform to the criteria that define each mass fraction. In some cases, new instruments will be required to obtain valid samples.

PSS-TLVs should be developed for several substances in recognition of the effects of regional deposition of aerosols on their potential for producing disease. These new TLVs will require the gathering of new information on aerosol aerodynamic characteristics and on the various diseases associated with each material.

Caution must be applied in attempting to use the recommended sampling procedures in several situations including epidemiological studies, nonoccupational environments, and exposures involving aerosols that do not exhibit simple aerodynamic behavior when inhaled. However, the successful institution of appropriate particle size-selective samples is anticipated to be a major step in the effective protection of workers from diseases related to aerosol inhalation.

ACKNOWLEDGMENT

The American Conference of Governmental Industrial Hygienists supported the activities of the Air Sampling Procedures Committee which developed the particle size-selective sampling recommendations. The committee members were W.C. Hinds, W. John, P.J. Lioy, M. Lippmann, M.A. McCawley, R.F. Phalen, O.G. Raabe, S.C. Soderholm, and B.O. Stuart. The authors thank the Southern Occupational Health Center and the Electric Power Research Institute (Contract No. RP 1962-1) for their supports. The manuscript was prepared by Sonia Usdansky.

REFERENCES

1. Phalen, R.F., W.C. Hinds, W. John et al.: Rationale and Recommendations for Particle Size-Selective Sampling in the Workplace. *Appl. Ind. Hyg.* *1*:3–14 (1985).
2. Morrow, P.E., D.V. Bates, B.R. Fish et al.: Deposition and Retention Models for Internal Dosimetry of the Human Respiratory Tract. (Report of the International Commission on Radiological Protection: ICRP: Task Group on Lung Dynamics.) *Health Phys.* *12*:173–207 (1966).
3. Ad hoc Working Group to Technical Committee TC-146, Air Quality, Inter-

national Standards Organization: Size Definitions for Particle Sampling. *Am. Ind. Hyg. Assoc. J. 42*(5):A64–6B (1981).

4. International Standards Organization: *Air Quality-Particle Size Fraction Definitions for Health Related Sampling*. ISO/TR 7708–1983 (E). Geneva (1983).
5. Soderholm, S.C.: Size-Selective Sampling Criteria for Inspirable Mass Fraction. *Ann. Am. Conf. Govt. Ind. Hyg. 11*:47–52 (1984).
6. Lippmann, M.: Regional Deposition of Particles in the Human Respiratory Tract. *Handbook of Physiology*, Section 9, *Reactions to Environmental Agents*, pp. 213–232. D.H.K. Lee, H.L. Falk and S.D. Murphy, Eds. American Physiological Society, Bethesda, MD (1977).
7. Raabe, O.G.: Size-Selective Sampling Criteria for Thoracic and Respirable Mass Fractions. *Ann. Am. Conf. Govt. Ind. Hyg. 11*:53–65 (1984).
8. Buchan, R.M., S.C. Soderholm and M.I. Tillery: Aerosol Sampling Efficiency of 37 mm Filter Cassettes. *Am. Ind. Hyg. Assoc. J. 47*:825–831 (1986).
9. Hinds, W.C.: Sampler Efficiencies: Inspirable Mass Fraction. *Ann. Am. Conf. Govt. Ind. Hyg. 11*:67–74 (1984).
10. Mark, D. and J.H. Vincent: A New Personal Sampler for Airborne Total Dust in Workplaces. *Ann. Occup. Hyg. 30*:89–102 (1986).
11. John, W.: Sampler Efficiencies: Thoracic Mass Fraction. *Ann. Am. Conf. Govt. Ind. Hyg. 11*:75–79 (1984).
12. John, W.: Sampler Efficiencies: Respirable Mass Fraction. *Ann. Am. Conf. Govt. Ind. Hyg. 11*:81–84 (1984).
13. McCawley, M.A.: Performance Considerations for Size-Selective Samplers. *Ann. Am. Conf. Govt. Ind. Hyg. 11*:97–100 (1984).
14. Barley, D.L. and L.J. Doemeny: Critique of 1985 ACGIH Report on Particle Size-Selective Sampling in the Workplace. *Am. Ind. Hyg. Assoc. J. 47*:443–447 (1986).
15. Stuart, B.O., P.J. Lioy and R.F. Phalen: Particle Size-Selective Sampling in Establishing Threshold Limit Values. *Appl. Ind. Hyg. 1*:138–144 (1986).

CHAPTER 2

Sampling for Total Aerosol as Defined by the Inspirability Criterion

JAMES H. VINCENT

Institute of Occupational Medicine, 8 Roxburgh Place, Edinburgh EH8 9SU, United Kingdom

BACKGROUND

Sampling for total aerosol is carried out by hygienists or occupational epidemiologists when it is suspected that airborne particles of all sizes present a potential risk to health. Unfortunately, in the past, total aerosol was not explicitly defined; available sampling instruments displayed a wide range of particle size-selection characteristics. Therefore, total aerosol measurements and the Threshold Limit Values (TLVs) with which they were associated have been of limited relevance to health. In recent years, however, recommendations have been made to define total aerosol quantitatively in terms of the fraction of airborne particles inspired by humans during breathing.[1,2] This common sense approach is now leading to a new generation of total aerosol samplers—both static (or area) and personal—having performance characteristics based on the Inspirable Particulate Mass (IPM) criterion. The details of the evolution of such instruments are given in publications which have already appeared (see references). This chapter will review the body of relevant work and the current status of sampler development. The bibliography should provide a useful foundation on which to base further study.

REVIEW OF RESEARCH LEADING TO A NEW GENERATION OF TOTAL AEROSOL SAMPLERS BASED ON THE INSPIRABILITY CRITERION

The long history of aerosol sampler research and development at the Institute of Occupational Medicine (IOM) in Edinburgh derived initially from the hygiene and epidemiological needs of the United Kingdom coal mining industry. More recently, however, it has been stimulated increasingly by the aerosol sampling needs of other industries, especially those where total aerosol is thought to be of health-related interest. The body of work falls into three main categories: 1) the generation of sampling criteria based on health-risk considerations, 2) the construction of a scientific basis for describing the performance characteristics of samplers, and from the preceding, 3) the development and engineering of new practical sampling instruments.

Inspirability Criteria

Inspirability criteria have been generated from the results of wind tunnel experiments to investigate the efficiency with which airborne particles entered the nose and/or mouth of a life-size mannequin during simulated breathing. In the early stages, such particles were referred to as "inhalable" (see References); the adoption of the term "inspirable" occurred only later in order to avoid confusion with terminology used elsewhere.

The first experiments were carried out at Edinburgh[3] in a small wind tunnel using a head and shoulders mannequin with wind speeds (U, describing the relative motion between the worker and the surrounding air) from about 0.75 to 2.75 m/s and particle aerodynamic diameter (d_{ae}) up to about 30 μm. These limited data were used to construct a single curve relating the efficiency of sampling to d_{ae}, and this was then used as the basis of the first quantitative definition of inspirability in the form recommended by the International Standards Organization (ISO).[1] Later, installation of a large new wind tunnel enabled the investigation of the inspirability characteristics of a mannequin with head and full torso for the extended ranges $1 < U < 4$ m/s and d_{ae} up to 100 μm.[4] Meanwhile, at Bergbau-Forschunginstitut in Essen, similar experiments were taking place, using a head-only mannequin with U in the range 1 to 8 m/s and d_{ae} up to 60 μm.[5] When the data from all three studies were compared, they were found to be in excellent agreement. They were therefore combined by Vincent and Armbruster[6] and later recommended by the

ACGIH Air Sampling Procedures Committee as the latest definition for IPM.[2] This version effectively supersedes the earlier ISO one (which, it is hoped, will be updated in due course).

Before proceeding to use the IPM criterion as the basis of TLVs and new sampling devices, it is important to recognize that it embodies certain idealizations and simplifications.

1. It essentially represents a single set of breathing conditions corresponding approximately to typical human subjects carrying out "moderately hard" work.
2. It represents the situation where the aerosol is uniformly mixed in the vicinity of the worker and is moving horizontally and unidirectionally.
3. It represents a range of wind speeds which may be uncharacteristic of some workplaces; the lower limit in particular was imposed by experimental difficulties involved in conducting aerosol sampling studies for relatively coarse particles with U lower than about 0.5 m/s.

These factors have been discussed at length, and it is generally agreed that the idealizing assumptions embodied in the IPM criterion have been justified in order to enable resolution of the total aerosol problem.

Scientific Framework for Describing Sampler Performance

Most previous knowledge about the basic performance characteristics of aerosol samplers has been based on experimental and theoretical investigations of cylindrical thin-walled probes, usually facing the wind. In order to assist in the development of samplers for workplace applications, the generation of relevant basic information about such samplers facing away from the wind and for the more general case of *blunt* samplers (i.e., samplers which present substantial geometrical blockage to the movement of air in their vicinity) was considered necessary. To this end, a program of research into some of the basic physical aspects of sampler performance has been conducted. This work has added to knowledge of the behavior of thin-walled samplers both facing and at yaw orientations to the wind,[7] the airflow patterns near blunt samplers of simple shape,[4,8] their aspiration efficiency characteristics,[4,9] the effects of turbulence on the performances of thin-walled and blunt samplers,[10] generalization of aspiration theory to a variety of sampler configurations,[11] and particle bounce or blow-off from external surfaces.[12,13]

Although basic sampler theory has not yet reached the stage where practical samplers for individual applications or meeting given criteria can be designed from first principles, the good qualitative picture of functional behavior which has emerged is useful in the first stages of practical sampler development. For example, it can guide assessment of the broad effects of changing particle size distribution, wind speed, sampling flow rate, sampler body size and sampler entry size. However, at present the final stages of convergence toward a practical sampler for IPM must be achieved by a combination of trial-and-modification. In view of the importance of a firm basic understanding of sampler theory to the future development of practical aerosol sampling devices, it is essential that further such studies be carried out.

ASSESSMENT OF THE PERFORMANCES OF PREVIOUS SAMPLERS AGAINST THE INSPIRABILITY CRITERION

The performances of sampling devices for "total" aerosol already in existence need to be considered before any search for new samplers is initiated. A certain amount of experimental data (from wind tunnel experiments) on which to base such assessment is contained in the literature.[14]

For a range of typical static (or area) samplers of the type used in industrial hygiene in Europe, available data suggest that none satisfactorily collects the inspirable fraction.[14-16] The most common adverse feature of performance is that aspiration efficiency is dependent on wind speed and direction to an unacceptably high degree.

For personal samplers, less data are available. Furthermore, any such data must be inspected critically since many performance tests reported in the literature have been carried out for the samplers in question mounted in isolation. This is unsatisfactory since, when a personal sampler is used in the intended manner, mounted on the breast or lapel, the effect of the aerodynamically bluff torso is to substantially alter the nature of the airflow in its vicinity. The aspiration efficiency of the sampler must be correspondingly affected. Therefore, reliable performance data may only be obtained for such samplers mounted on life-size torsos, e.g., on mannequins. This implies the use of large wind tunnels in order that blockage effects may be kept low; this limits the amount of reliable information which is available. Two independent studies have been carried out in the United Kingdom. The first suggested that, for d_{ae} up to 18 μm, the performance of the "seven-hole" personal sampler recommended for total aerosol by the UK Health and Safety Executive (HSE)[17] satisfactorily matches the inspirability criterion.[18] However, the second study, carried out in our much larger wind tunnel in Edinburgh, indicated that the seven-hole sampler substantially undersamples with respect to inspirability for larger particle sizes.[19] The performance of the "single-hole" sampler recommended by the HSE for the sampling of total lead aerosol[20] was found to undersample even more strongly. As in the case of the various static samplers for which data are available, the magnitude of wind speed effects are unacceptable in the performances of these personal samplers.

It would appear that the only previous sampler which was developed specifically to match inspirability data is the IOM 2 lpm "ORB" static instrument.[21] Its performance was found to agree quite well with the earlier inspirability data limited to d_{ae} up to about 30 μm.[3] However, later tests showed that aspiration efficiency for this sampler became very erratic for particles with d_{ae} exceeding about 25 μm.[22]

Based on the available information, it may be concluded that, while certain of the previously-available samplers meet the IPM criterion under some conditions (e.g., small particle sizes, low wind speeds), their general applicability is limited; therefore, they cannot be unreservedly recommended. The need for a new generation of improved, inspirable aerosol samplers is clearly indicated.

NEW SAMPLERS FOR INSPIRABLE AEROSOL

The search for new sampling instruments for total aerosol consistent with the new IPM criterion has begun, and progress has been achieved in a number of areas.

The first practical devices were developed with the coal industry in mind at a time when it was felt that particles coarser than respirable might be associated with some respiratory ill-health (e.g., bronchitis) among workers in that industry. More recently, support for that view—certainly as far as total aerosol is concerned—has receded. However, applications in other industries may be more appropriate.

The first practical instrument was the IOM 3-lpm *static* inspirable aerosol sampler.[23] It features a single, slot-shaped, slightly protruding, sampling orifice in a cylindrical sampling head which rotates slowly about a vertical axis. Thus it is ensured that 1) sampling takes place with no preferred orientation with respect to the local air movement, 2) effects associated with changes in wind speed or sampling flow rate are reduced, and 3) effects of coarse particle blow-off from external surfaces of the sampling head are eliminated. The air to be sampled is drawn through the entry orifice and into a metal cassette (weighing about 7 g) which houses the filter. In order to assess the collected aerosol gravimetrically, the whole cassette is weighed before and after sampling; in each case it is allowed to stabilize in the balance room overnight prior to weighing. All that is aspirated is assessed and so there are no errors associated with internal wall losses. The sampling head itself is mounted on a cylindrical casing which houses the pump, motor and gearbox (for driving the rotating head), battery, and controls. The whole instrument stands 25 cm high and weighs 2.5 kg. Tests in our large wind tunnel have shown that its performance closely matches the IPM curve for particles with d_{ae} up to 100 μm and for conditions of air movement representative of many of those found in workplaces. More recently, static sampling heads capable of selecting the inspirable fraction have been developed for sampling flowrates of 10 and 30 lpm,[24,25] intended as entry systems both for an aerosol spectrometer (i.e., cascade impactor) and a sampler for atmospheric suspended particulate matter (SPM) in the range of d_{ae} up to about 30 μm, respectively.

The next development was the IOM 2 lpm *personal* inspirable aerosol sampler.[19] The main feature of the sampler itself is the single, circular sampling orifice of 15 mm diameter which faces outward when the device is worn on the breast or lapel. As for the static sampler described above — and for the same reasons—this entry orifice also forms an integral part of an aerosol collection cassette. The sampler may be used in conjunction with any suitable personal sampling pump. Wind tunnel tests with it mounted on the torso of a life-size tailor's mannequin have shown that its performance closely matches the IPM curve for particles with d_{ae} at least up to 100 μm for wind conditions (reflecting the relative motion between the worker and his surrounding air) representative of many workplaces.

In addition to the family of samplers described above for total aerosol based on the IPM criterion, a range of particle size-selective samplers is being developed in which aerosol reflecting particle deposition in given regions of the respiratory tract is collected. This is achieved by first sampling the inspirable fraction and then selecting from it the particular *sub*fraction of interest. This two-stage process—one taking place *outside* the sampler and the other taking place *inside* it—is consistent with what happens in the case of human exposure.

The new instruments have been, and are being, tested in the occupational setting during the course of industrial hygiene and epidemiological studies. From the experience gained so far, they are looked on favorably by industrial hygienists. These studies are continuing. Meanwhile, the instruments are available commercially from Rotheroe and Mitchell (Aylesbury, UK).

ACKNOWLEDGMENTS

The author wishes to thank British Coal and the Commission of European Communities for their financial support of a large proportion of the work reviewed in this paper; also the many colleagues at the Institute of Occupational Medicine and elsewhere who have participated in the research and development which are described.

REFERENCES

1. International Standards Organization: *Air Quality—Particle Size Fraction Definitions for Health-related Sampling*. Technical Report ISO/RE7708-1983. Geneva (1983).
2. American Conference of Governmental Industrial Hygienists: *Particle Size-selective Sampling in the Workplace*, 80 pp. Cincinnati, OH (1985).
3. Ogden, T.L. and J.L. Birkett: The Human Head as a Dust Sampler. *Inhaled Particles IV*, pp. 93–105. W.H. Walton, Ed. Pergamon Press, Oxford (1977).
4. Vincent, J.H. and D. Mark: Application of Blunt Sampler Theory to the Definition and Measurement of Inhalable Dust. *Inhaled Particles V*, pp. 3–19. W.H. Walton, Ed. Pergamon Press, Oxford (1982).
5. Armbruster, L. and H. Breuer: Investigations into Defining Inhalable Dust. *Inhaled Particles V*, pp. 21–32. W.H. Walton, Ed. Pergamon Press, Oxford (1982).
6. Vincent, J.H. and L. Armbruster: On the Quantitative Definition of the Inhalability of Airborne Dust. *Ann. Occup. Hyg. 24*:245–248 (1981).
7. Vincent, J.H., D.C. Stevens, D. Mark and M. Marshall: On the Entry Characteristics of Large-diameter, Thin-walled Aerosol Sampling Probes at Yaw Orientations with Respect to the Wind. *J. Aerosol Sci. 17*:211–224 (1986).
8. Vincent, J.H., D. Hutson and D. Mark: The Nature of Air Flow Near the Inlets of Blunt Dust Sampling Probes. *Atmos. Environ. 16*:1243–1249 (1982).
9. Vincent, J.H.: A Comparison Between Models for Predicting the Performances of Blunt Dust Samplers. *Atmos. Environ. 18*:1033–1035 (1984).
10. Vincent, J.H., P.C. Emmett and D. Mark: The Effects of Turbulence on the Entry of Airborne Particles into a Blunt Dust Sampler. *Aerosol Sci. Technol. 4*:17–29 (1985).
11. Vincent, J.H.: Recent Advances in Aspiration Theory for Thin-walled and Blunt Aerosol Sampling Probes. *J. Aerosol Sci. 18* (in press).
12. Vincent, J.H. and H. Gibson: Sampling Errors in Blunt Samplers Arising from External Wall Losses. *Atmos. Environ. 15*:703–712 (1981).
13. Mark, D., J.H. Vincent and W. Witherspoon: Particle Blow-off: A Source of Error in Blunt Dust Samplers. *Aerosol Sci. Technol. 1*:463–469 (1982).

14. Barrett, C.F., M.O. Ralph and S.L. Upton: *Wind Tunnel Measurements of the Inlet Efficiency of Four Samplers of Suspended Particulate Matter*. Final Report on CEC Contract 6612/10/2. Commission of European Communities, Luxembourg (1984).
15. Armbruster, L., H. Breuer, J.H. Vincent and D. Mark: *The Definition and Measurement of Inhalable Dust. Aerosols in the Mining and Industrial Work Environment*, pp. 205–217. V.A. Marple and B.Y.H. Liu, Eds. Ann Arbor Science, Ann Arbor, MI (1983).
16. Mark, D., J.H. Vincent, D.C. Stevens and M. Marshall: Investigations of the Entry Characteristics of Dust Samplers of the Type Used in the British Nuclear Industry. *Atmos. Environ. 20*:2389–2396 (1986).
17. Health and Safety Executive: *General Methods for the Gravimetric Determination of Respirable and Total Inhalable Dust*. M.D.H.S. 14, London, UK (1983).
18. Chung, K.Y.K., T.L. Ogden and N.P. Vaughan: Wind Effect on Personal Dust Samplers. *J. Aerosol Sci. 18* (in press).
19. Mark, D. and J.H. Vincent: A New Personal Sampler for Airborne Total Dust in Workplaces. *Ann. Occup. Hyg. 30*:89–102 (1986).
20. Health and Safety Executive: *Control of Lead: Air Sampling Techniques and Strategies*. Guidance Note EH28. London, UK (1981).
21. Ogden, T.L. and J.L. Birkett: An Inhalable Dust Sampler for Measuring the Hazard from Total Airborne Particulate. *Ann. Occup. Hyg. 21*:41–50 (1978).
22. Vincent, J.H., D. Mark, H. Gibson et al.: *Measurement of Inhalable Dust in Wind Conditions Pertaining to Mines*. Final Report on CEC Contract 7256–21/029/08, Institute of Occupational Medicine Report TM/83/7. Edinburgh (1983).
23. Mark, D., J.H. Vincent, H. Gibson and G. Lynch: A New Static Sampler for Airborne Total Dust in Workplaces. *Am. Ind. Hyg. Assoc. J. 46*:127–133 (1985).
24. Aitken, R.J., H. Gibson, G. Lynch et al.: *Development of a Static Sampler for the Measurement of Suspended Particulates in the Ambient Atmosphere*. Final Report on CEC Study Contract 85-B-6600–11–045–11-N, Institute of Occupational Medicine Report TM/87/02. Edinburgh (1987).
25. Vincent, J.H.: Work not yet published.

CHAPTER 3

Thoracic and Respirable Particulate Mass Samplers: Current Status and Future Needs

WALTER JOHN

Air and Industrial Hygiene Laboratory, California Department of Health Services, 2151 Berkeley Way, Berkeley, California 94704

SAMPLER CRITERIA: TESTING PROCEDURES

Recommended size-selective sampling criteria for thoracic and respirable mass fractions were published in Chapter 4.B of the 1985 Report of the American Conference of Governmental Industrial Hygienists (ACGIH) Technical Committee on Air Sampling Procedures, *Particle Size-Selective Sampling in the Workplace.*[1] The general requirements for an ideal sampler satisfying the criteria and a review of currently available samplers were presented in Chapters 5.B and 5.C for thoracic and respirable samplers, respectively.

There are advantages to specifying a generic rather than a reference sampler. Within the general particle size-selective requirements, a sampler may be chosen with characteristics appropriate to the application. The desirable flow rate can be determined from the expected aerosol concentration and the time available for sampling. The collection substrate can be made compatible with the chemical species to be sampled and the subsequent analysis. The sampler can be sized to fit space constraints or need for portability. Perhaps most importantly, improvements in samplers and innovations in sampling technology are encouraged.

The generic sampler approach requires criteria which define not only the ideal sampler's efficiency as a function of particle size, but also the limits to the allowable deviation from the ideal performance. The ACGIH Committee put an envelope around the ideal curve, as shown in Figure 1, for a thoracic mass sampler. Recently, the U.S. Environmental Protection Agency (EPA) proposed a new primary standard for particulate matter in ambient air measured as PM-10 and an associated Federal Reference Method.[2] The PM-10 sampler is to have 50 percent sampling efficiency for particles of 10 μm aerodynamic diameter as in the case of the ACGIH thoracic mass sampler. However, the allowable shape of the efficiency curve for the PM-10 sampler is to be determined by calculating the mass which would be sampled, assuming a specified ambient particle size distribution. The calculated mass must be within ten percent of the mass which would be sampled by an ideal sampler having a specified efficiency curve based on lung deposition. This procedure would, for

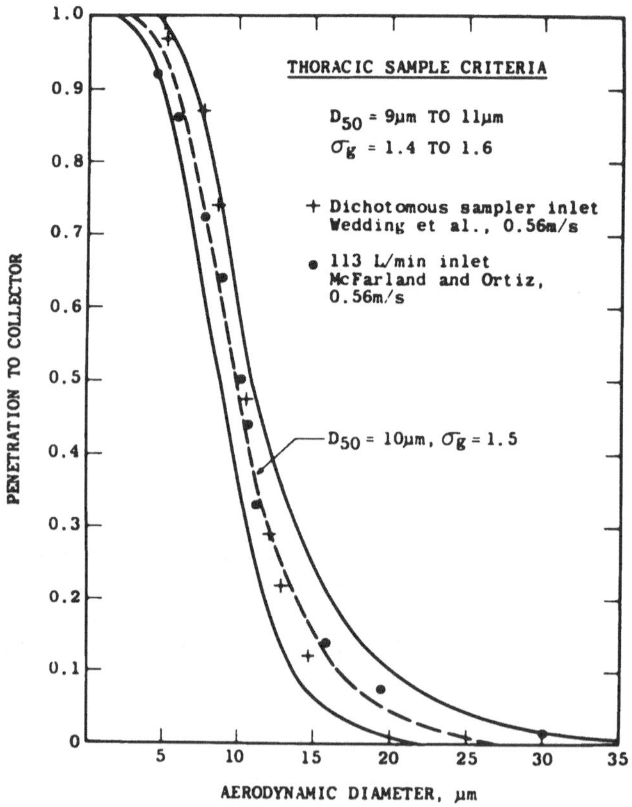

Figure 1. The ACGIH thoracic mass sampling criteria with data for two samplers.

example, allow a sampler to oversample (compared to the ideal curve) large particles if this were compensated by an undersampling of small particles.

A similar approach for respirable mass sampling criteria has been suggested by Bartley and Doemeny,[3] who used coal mine particle size distributions to illustrate inaccuracies associated with the ACGIH criteria envelope. The ACGIH Committee considered the sampled mass approach but rejected it because of concern that the wide variability of size distributions in the workplace would introduce unacceptable deviations from ideal sampling. The full implications of either approach to sampling criteria are not known at this time. This is an important subject which merits additional investigation. One of the consequences of the differing criteria is uncertainty as to the interchangeability of thoracic and PM-10 samplers. Fortunately, most of the current PM-10 samplers have efficiency curves which satisfy the ACGIH criteria (examples are shown in Figure 1).

Testing for Particle Bounce

For most samplers which segregate particles according to aerodynamic diameter, there is an inherent problem due to the bouncing of particles from sampler surfaces, which may lead to excessive sampling of large particles. Accordingly, the EPA imposed an additional criterion for PM-10 samplers; namely, that the sampling effectiveness for solid, 20 μm aerodynamic diameter particles is to be no more than five percent absolute above that for liquid particles of the same size. However, two candidate PM-10 high-volume samplers which have satisfied the EPA criteria in laboratory tests, the Sierra Andersen SA 321A (Figure 2) and the Wedding PM-10 Hivol (slightly modified version of the GMW-9000, Figure 3), have shown differences in the field with the SA 321A yielding higher mass concentrations. Recently, the EPA conducted a field study in Phoenix[4] under conditions affording large wind-blown particles. An array of samplers included the SA 321A, the Wedding PM-10 Hivol, the Sierra Andersen Dichotomous Sampler (SA 246B, Figure 4) and the Wedding Dichotomous Sampler (GMW-9200, Figure 5).

The results of the Phoenix study are summarized in Figure 6. In the lower half of the figure, the concentrations obtained from the two dichotomous samplers show excellent agreement. In the upper half of the figure, the average of the concentrations from the dichotomous samplers is used as a reference and the concentrations from the other samplers are plotted in terms of the difference from the reference. The SA 321A is

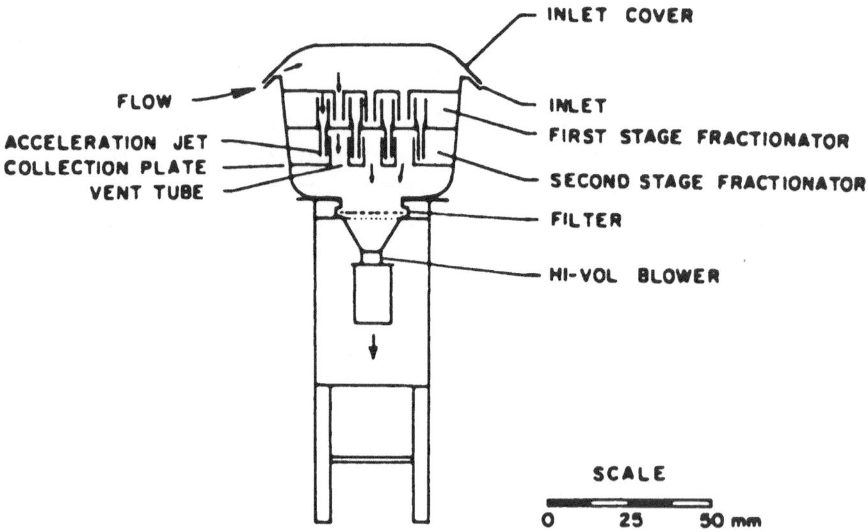

Figure 2. The Sierra Andersen SA 321A high-volume PM-10 sampler.

consistently higher than the dichotomous reference and the Wedding PM-10 Hivol is consistently lower. There is some tendency for the differences to anticorrelate. A second SA 321A with oiled impaction surfaces can be seen to differ from the dichotomous sampler reference by less than ten percent, suggesting that the oil suppresses particle bounce. A second Wedding sampler which received a light cleaning between runs shows much less difference from the reference. The downward trend may be due to incomplete cleaning. The lowered efficiency of the Wedding sampler with loading has been interpreted to be the result of increased particle losses on the particle deposits which protrude into the air stream. This problem can be minimized by a maintenance procedure. The latest version of the Sierra Andersen sampler incorporates an oiled impaction surface to suppress particle bounce. However, the results of the field study imply a need for revised criteria for laboratory testing. An encouraging result is the good performance of the dichotomous samplers. They agreed closely with each other and the application of remedial measures to the other samplers brought them into near agreement with the dichotomous samplers.

In experiments with a test impactor, Wang and John[5] found that the fraction of particles which bounced (FOB) and penetrated the impactor was a function mainly of the kinetic energy of the particles. Figure 7

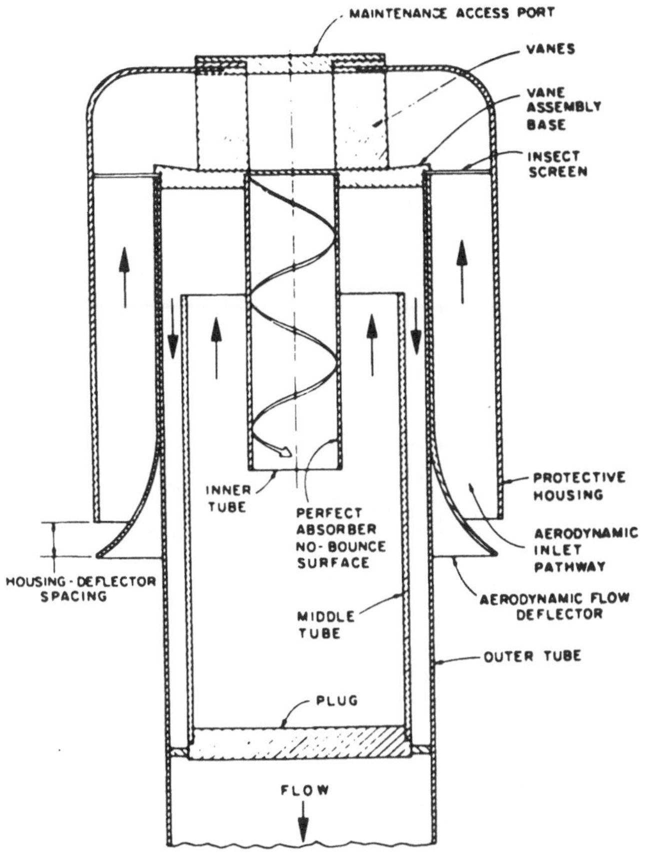

Figure 3. The Wedding PM-10 inlet for the high-volume sampler.

shows the fit of an empirical function to the data. This function was used to predict the FOB for several bounce criteria. Figure 8 shows that the predicted FOB increases extremely rapidly beyond a certain particle size, depending on the criteria. Calculations combining the FOB with typical ambient particle size distributions indicate that the currently allowed five percent bounce of 20-μm particles could permit 20 to 30 percent oversampling. A stricter criterion, five percent at 50 μm, reduces the estimated oversampling to less than ten percent. These estimates require verification in actual sampler testing.

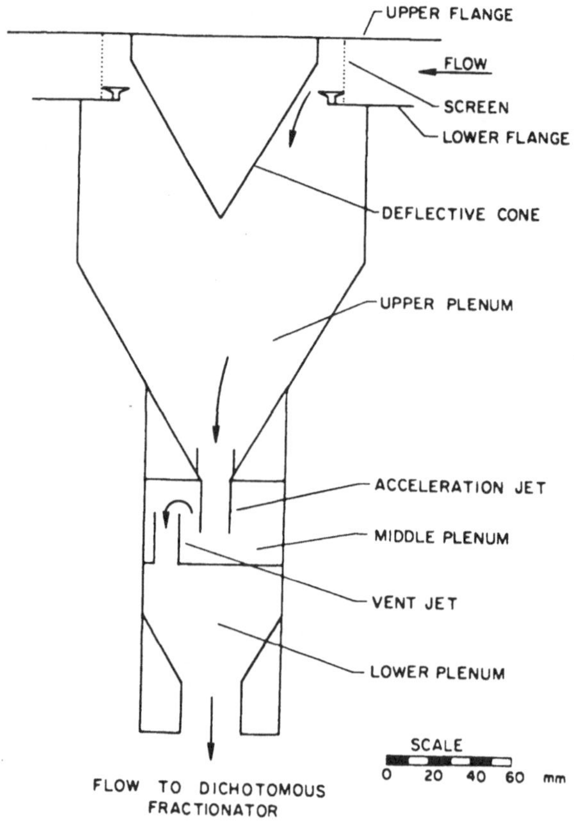

Figure 4. The Sierra Andersen SA 246B PM-10 inlet for the dichotomous sampler.

CURRENT SAMPLER NEEDS

Respirable Mass Sampler

The 10-mm nylon cyclone is the most widely used size-selective sampler in the workplace. In spite of decades of experimentation, some calibration problems remain to be resolved. It is convenient to separate examination of the shape of the sampling efficiency curve from the question of the proper flow rate. This can be done as shown in Figure 9 where the ratio of the particle diameter to the diameter at the 50 percent cutpoint (D_{50}) is plotted vs. penetration on a log-probability plot. The data points are from recent measurements by Baron[6] using polydisperse

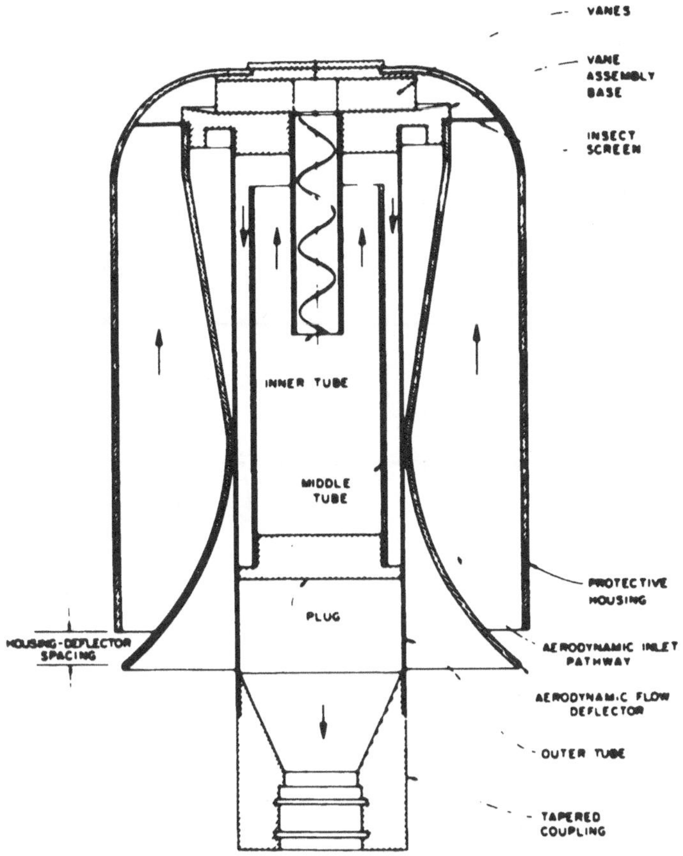

Figure 5. The Wedding PM-10 inlet (GMW-9200) for the dichotomous sampler.

aluminum oxide particles. The particle size distributions before and after penetration of the cyclone were measured with an aerodynamic particle sizer (APS). The solid line is a curve which was drawn through the data of Blachman and Lippmann[7] by Saltzman.[8] The curve appears to be a good fit to the data points, i.e., the two sets of data obtained by quite different techniques are in good agreement. Also shown in Figure 9 are the ACGIH criteria for respirable mass sampling. The nylon cyclone cuts more sharply than the ideal sampler, but just fits within the envelope.

A more serious problem concerns the determination of the flow rate required to produce the specified 50 percent cutpoint at 3.5 μm aerodynamic diameter. In Figure 10, D_{50} is plotted vs. flow rate for the same

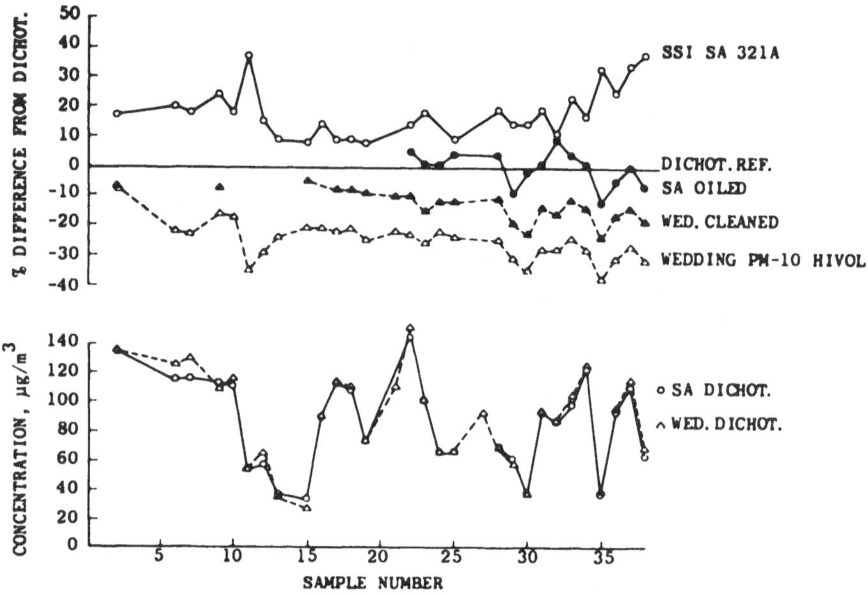

Figure 6. Data from an EPA field study comparing PM-10 samplers.[4]

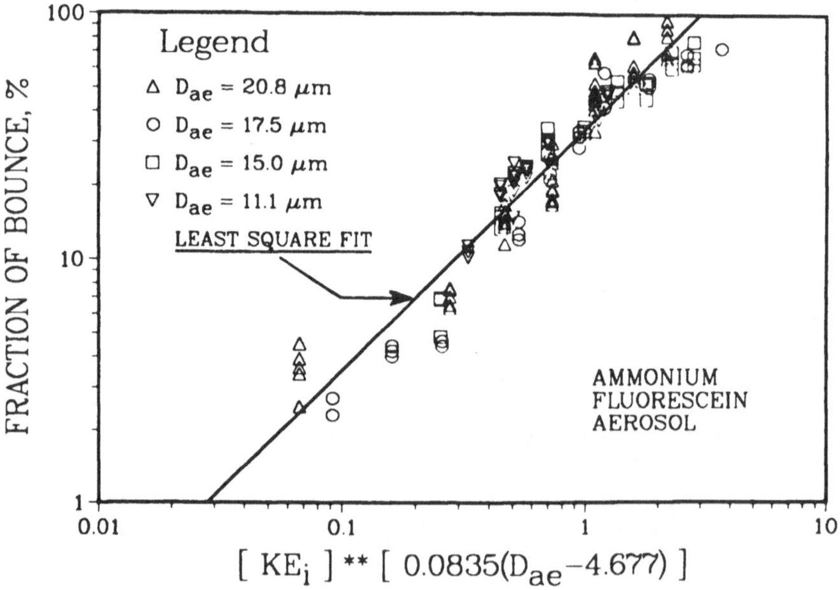

Figure 7. Correlation of the fraction of particles bouncing and penetrating a test impactor with an empirical function of the particle kinetic energy and aerodynamic diameter.[5]

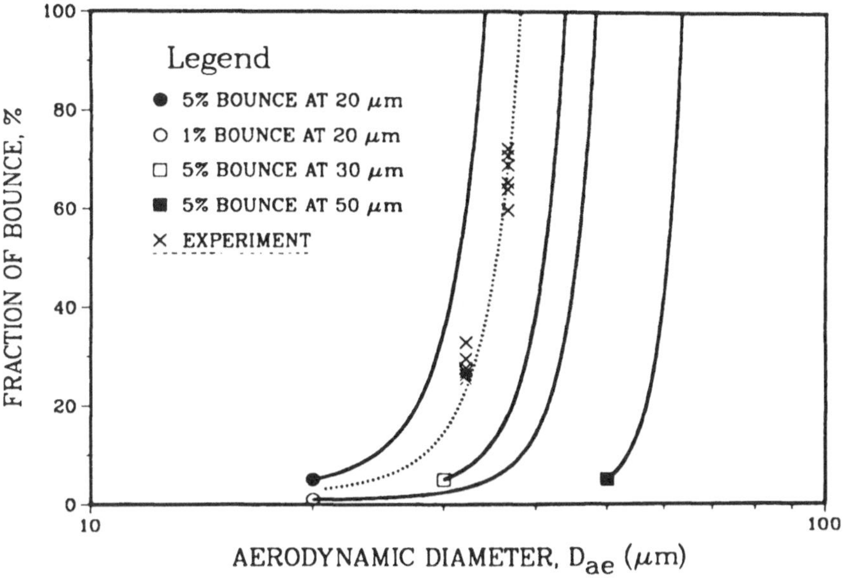

Figure 8. The fraction of particles bouncing and penetrating a PM-10 sampler as a function of particle aerodynamic diameter. The lines are calculated from the empirical function in Figure 7 for several bounce criteria. The crosses are experimental measurements for ammonium fluorescein particles in a test impactor with a 10 μm cutpoint.

data sets discussed above. Although each set of data fit a straight line on the log-log plot, the slopes are different and the flow rates corresponding to 3.5 μm differ by about 15 percent. It has been shown that the calibration of the APS should be corrected for the density of the particle material.[9] The line for the density-corrected data of Baron is shifted even farther to smaller flow rate. This may be an over-correction because the effective density may have been lower than bulk density due to particle agglomeration.[10] Other data from the literature give various lines between the extremes shown in Figure 10. There is a need for definitive measurements on this important sampler. This is well within the capabilities of aerosol technology. Measurements of D_{50} vs. flow rate should be coupled with measurements of the pressure drop just before the filter on the cyclone outlet since D_{50} is a function of this pressure drop.[11] Also, the couplings and filter configuration at the outlet should be documented because it has been suspected that these might influence cyclone perform-

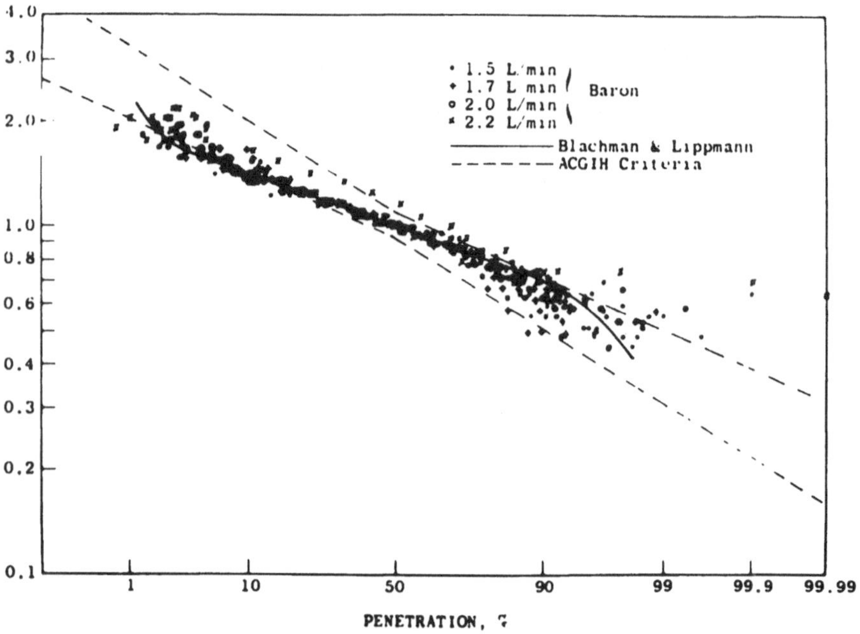

Figure 9. Data for particle penetration of the 10-mm nylon cyclone plotted against normalized particle diameter. The ACGIH respirable mass sampling criteria are shown for comparison.

ance.[12] An investigation of these effects would increase confidence in the general applicability of the cyclone calibration.

Thoracic Mass Sampler

There is no personal thoracic mass sampler available. This is clearly a major need for the workplace. It would be desirable that the sampler be capable of operating on the pumps and flow controllers which were developed for the respirable mass samplers.

FUTURE SAMPLING NEEDS

An important need is for the capability to sample particles without artifacts from the absorption or reaction of gases and vapors with the deposit or the filter substrate. For some species it has been shown that diffusion denuders can eliminate artifactual errors from gases.[13] It can be

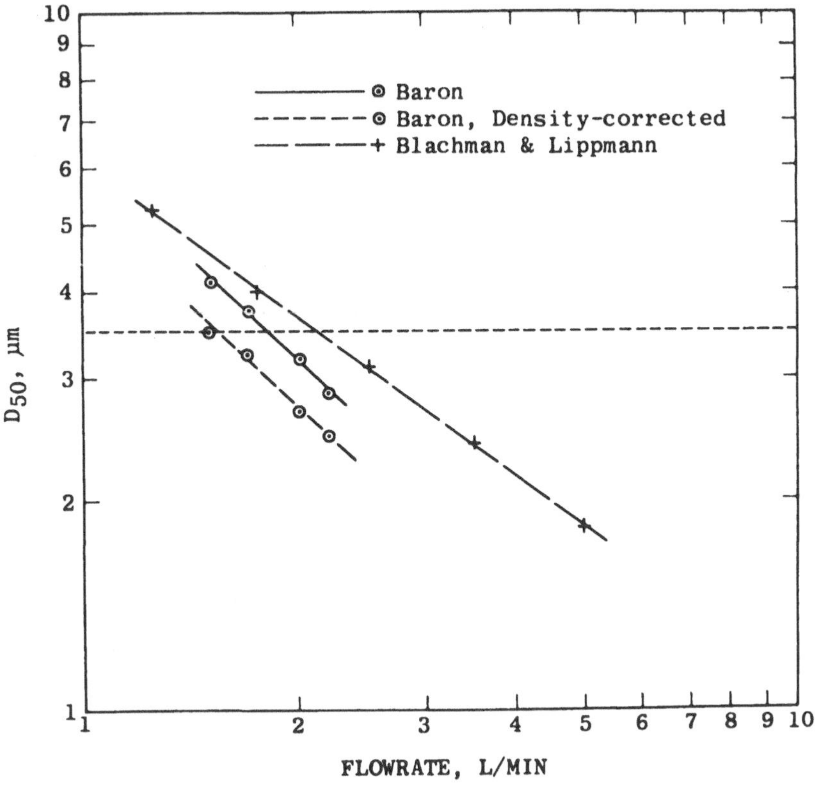

Figure 10. Plot of the 50 percent cutpoint of the 10-mm nylon cyclone vs. flow rate.

anticipated that there will be increasing use of denuders for this purpose. In some cases it may be desirable to include the denuder in the design of the inlet of the sampler. An example is the sampling of particulate nitrate in the presence of gaseous nitric acid. John et al.[14] have shown that the dichotomous sampler has a fortuitous denuding action on nitric acid due to the oxidized aluminum surfaces of the sampler's inlet. Figure 11 is a plot of the fine nitrate particulate sampled with a cyclone preceded by a conventional denuder[15] vs. that sampled with a parallel dichotomous sampler; excellent agreement between the two samplers is shown.

A complimentary theme is the avoidance of negative artifacts due to the loss of volatile species from the particle deposit in a sampler. One way to prevent such loss is the use of a substrate which reacts with the gas, binding it to the substrate. An example is the use of a nylon filter to

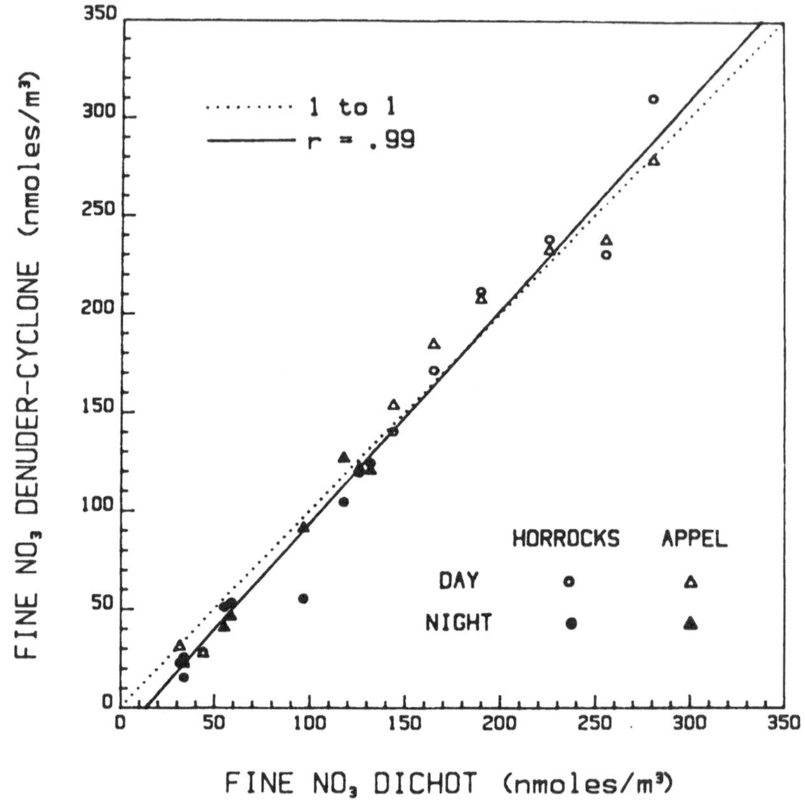

Figure 11. Fine particulate nitrate from a diffusion denuder-cyclone sampler vs. that from a dichotomous sampler with a SA 246B inlet, showing that the dichotomous sampler denudes nitric acid quantitatively. Data taken in during a methods comparison study in Claremont, CA.[15]

sample ammonium nitrate, which volatizes with the release of nitric acid which then binds to the nylon. The use of nylon membrane filters in a dichotomous sampler allows the artifact-free sampling of particulate nitrate in ambient air.[14]

The loss of semivolatile material has been found to be less when sampled by a cascade impactor than by a filter sampler. Wang and John[16] measured less than ten percent loss of submicron particulate ammonium nitrate from a Berner impactor under conditions producing up to 95 percent losses from Teflon filters. This was ascribed to the lower surface-to-volume ratio of the impactor deposits and to the thicker boundary layer over the impactor deposit. This result is remarkable in that the submicron stages of the impactor operate substantially below atmospheric pressure. This result also implies that cascade impactors would be useful for the sampling of particulate semivolatile organic compounds. The foregoing discussion emphasizes the need to take into account more than just particle mechanics in sampler design to ensure the quantitative sampling of specific chemical species.

While filter samplers will probably continue to be the most widely used, real-time or semireal-time samplers with aerodynamic particle separation are needed for many applications. Real-time mass measurement is already feasible, but real-time chemical analysis is still to be developed. With continuing advances in computer technology, more detailed numerical calculations of fluid flow and particle trajectories can be made to reduce the empiricism in sampler design. The adoption of particle size-selective criteria by the ACGIH and the EPA will undoubtedly accelerate progress in sampler technology.

REFERENCES

1. Air Sampling Procedures Committee: *Particle Size-Selective Sampling in the Workplace*. American Conference of Governmental Industrial Hygienists, Cincinnati, OH (1985).
2. *Federal Register 49*:10408–10462 (March 20, 1984).
3. Bartley, D. and Doemeny, L.J.: Critique of 1985 ACGIH Report on Particle Size-Selective Sampling in the Workplace. *Am. Ind. Hyg. Assoc. J. 47*:443–447 (1986).
4. Purdue, L.J., C.E. Rodes, K.A. Rehme et al.: Intercomparison of High-Volume PM-10 Samplers at a Site with High Particulate Concentrations. *J. Air Poll. Control Assoc. 36*:917–920 (1986).
5. Wang, H.C. and W. John: Comparative Bounce Properties of Particle Materials. Submitted to *Aerosol Sci. Technol.* (1986).
6. Baron, P.A.: Sampler Evaluation with an Aerodynamic Particle Sizer. *Aerosols in the Mining and Industrial Work Environments*, Vol. 3, *Instrumentation*, pp. 861–877. V.A. Marple and B.Y.H. Liu, Ed. Butterworth, Stoneham, MA (1983).
7. Blachman, M.W. and M. Lippmann: Performance Characteristics of the Multicylone Aerosol Sampler. *Am. Ind. Hyg. Assoc. J. 38*:311–326 (1974).
8. Saltzman, B.E.: Generalized Performance Characteristics of Miniature Cyclones for Atmospheric Particulate Sampling. *Am. Ind. Hyg. Assoc. J. 45*:671–680 (1984).
9. Wang, H.C. and W. John: Particle Density Correction for the Aerodynamic Particle Sizer. *Aerosol Sci. Technol. 6*:191–198 (1987).
10. Baron, P.A.: Private communication.

11. Saltzman, B.E. and J.M. Hochstrasser: Design and Performance of Miniature Cyclones for Respirable Aerosol Sampling. *Environ. Sci. Technol.* *17*:418–424 (1983).
12. Lippmann, M.: Cyclone Sampler Performance. *Staub-Reinhalt. Luft* *39*:7–11 (1979).
13. Forrest, J., D.J. Spandau, R.L. Tanner and L. Newman: Determination of Atmospheric Nitrate and Nitric Acid Employing a Diffusion Denuder with a Filter Pack. *Atmos. Environ.* *16*:1473–1485 (1982).
14. John, W., S.M. Wall and J.L. Ondo: A New Method for Nitric Acid and Nitrate Aerosol Measurement Using the Dichotomous Sampler. Submitted to *Atmos. Environ.* (July 1986).
15. Hering, S. V. et al.: The Nitric Acid Shootout: Field Comparison of Measurement Methods. Submitted to *Atmos. Environ.* (December 1986).
16. Wang, H.C. and W. John: Characteristics of the Berner Impactor for Sampling Inorganic Ions. Submitted to *Aerosol Sci. Technol.* (1986).

CHAPTER 4

Basis for Particle Size-Selective Sampling for Beryllium

OTTO G. RAABE

Department of Veterinary Pharmacology and Toxicology, and Department of Civil Engineering, University of California, Davis, California 95616

CHARACTERISTICS OF BERYLLIUM

Beryllium (Be) is a light alkaline earth metallic element of atomic number 4 having one natural isotope with atomic mass number 9 associated with five neutrons (atomic weight: 9.0122). A silver-white metal of low density (1.85 g/cm^3), beryllium is the lightest of the structural metals. It has the following natural properties: melting point, 1283°C; boiling point, 2970°C; covalent radius, 0.9 angstroms; ionic radius, 0.31 angstroms; oxidation number, 2; hexagonal crystal structure; and ionization energies, 9.32 ev and 18.21 ev. The principal sources of beryllium are the natural silicate ores: beryl, bertrandite, and phenacite; the most important is the mineral beryl, $Be_2Al_2(SiO_3)_6$. The white, crystalline beryllium oxide is obtained by combustion of beryllium or its compounds in air. It resembles aluminum oxide in being refractory (m.p. 2570°C) and in having polymorphs. The high temperature forms (> 800°C) are exceedingly inert and insoluble. Because of the beryllium atom's small size (ionic radius 0.34 angstroms) and high ionization potential (9.32 ev), its lattice and hydration energies are not sufficient to result in complete charge separation to form Be^{2+} ions, and all beryllium compounds exhibit covalent bonding of Be. Samples containing beryllium can be accurately assayed by dissolution and atomic absorption spectroscopy measurement to 0.01 mg/L.[1-3]

Metallic beryllium is light-weight, strong, rigid, and dimensionally stable. It has high electrical and thermal conductivity, resists corrosion and oxidation, and is both antimagnetic and antisparking. When used as a constituent in alloys with other metals (usually less than 4% by weight), beryllium contributes to improved tensile strength, yield strength, fatigue resistance, hardness, and corrosion resistance. These alloys have good casting and other desirable mechanical properties and are non-sparking.[2-4]

BERYLLIUM INDUSTRIAL USES AND OCCUPATIONAL EXPOSURE

Because of its desirable physical and chemical properties, beryllium metal is widely used in metallic components of various types in aerospace applications, space flight hardware, bearings, gears, optical devices, space mirrors, inertial systems, and nuclear power components. Its low atomic number and density make beryllium metal the best choice for thin window radiation detection instruments designed to measure low energy photons.[3,4]

Beryllium metal alloys of copper (usually less than 4% beryllium by weight), nickel or aluminum have important uses for aerospace applications, automotive parts, computers, communications, electronics, tools, springs, switches, and power equipment.[3,4] Beryllium oxide is used in electronics components, nuclear power applications, and in special plastics. Other beryllium compounds are used in chemical reagents, organic chemicals, and fluorescent phosphors. Because of the hazard associated with the use of beryllium in regular household fluorescent lights, this practice was discontinued in 1949.[2-4]

Overall, the industrial usage of beryllium in the United States is about 600 tons/year, of which 65 percent is in the form of various copper-bronze alloys. About half of the uses are associated with aerospace and nuclear applications. Occupational exposures to beryllium can occur in mining and milling of beryllium minerals, extraction of beryllium, alloy manufacturing and metallurgy, beryllium ceramics preparation, electronic equipment manufacturing, aerospace equipment manufacturing, non-ferrous foundry operations, tool and die manufacturing, preparation and handling of special chemicals, and during machining, welding, and fabrication with beryllium and beryllium alloys.[2-4]

BERYLLIUM TOXICITY

Beryllium is a potent systemic poison that is tenaciously retained by many body organs including lung, liver, and spleen.[4] However, its only ready route of entry into the body is as inhaled aerosols via the airways of the lung. The ingestion of beryllium even in soluble forms results in systemic uptake of only about 0.2 percent of the total entering the stomach.[5] A schematic illustration of the regions of the respiratory airways is shown in Figure 1. The three main functional regions are the head airways region (HAR), including the nose (NP) section, the oral cavity and airways down to and including the larynx; the tracheobronchial (TB)

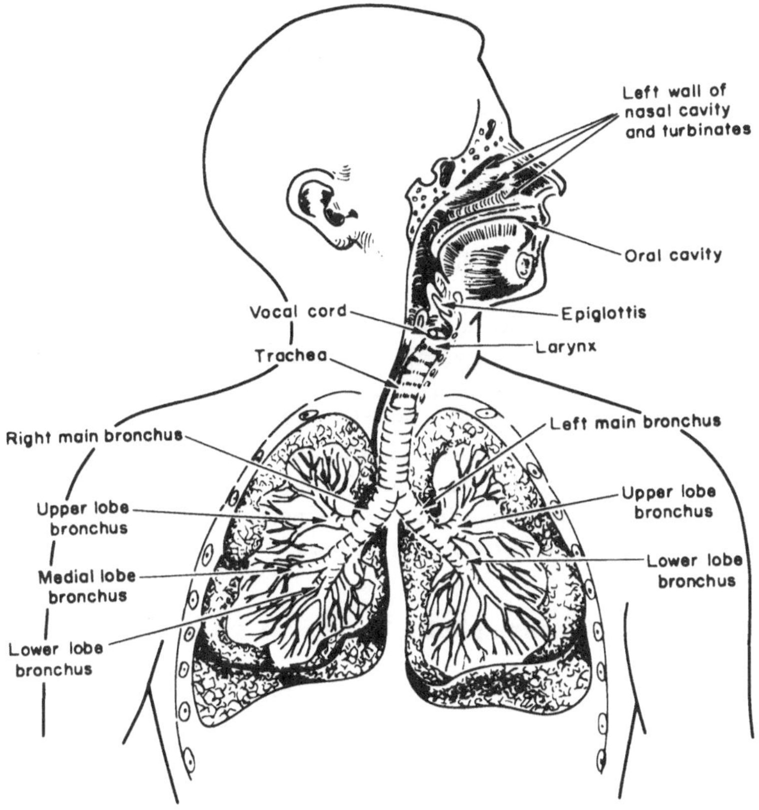

Figure 1. The Respiratory Tract: Schematic illustration of the human respiratory airways.[10]

conductive airways of the lung; and the pulmonary, parenchymal, alveolar or gas exchange region (GER) of the lung.[6]

Particles of inhaled beryllium aerosols that lodge in the pulmonary or parenchymal gas exchange region of the lung may be retained for hundreds of days or years depending upon the solubility of the inhaled chemical forms and particle size. Slow systemic uptake from the lung into the blood with transport to bone, liver, kidneys, and lymph nodes can lead to a condition of chronic poisoning with severe long-term effects and even death. The lung-retained material may adversely affect lung tissue with severe injury to alveolar cells resulting in granulomatous lesions. Insoluble forms of beryllium that deposit in the HAR and tracheobronchial region (TBR) are cleared to the throat within several hours and are swallowed or expectorated with little biological consequence. Certain soluble forms of beryllium can evoke acute responses in all regions of the respiratory tract and can cause skin eruptions and dermatitis upon contact, but these responses are not found with the pure metal, its alloys, or other insoluble forms of beryllium. A type of beryllium-antigen sensitivity has been observed for soluble forms of beryllium leading to acute anaphylactic reactions to exposures to beryllium aerosols usually at concentrations that are well above 0.025 mg/m^3.[2-4,7,8]

Acute Beryllium Disease (ABD)

Even brief inhalation exposure to soluble forms of beryllium at concentrations well above 0.025 mg/m^3 may result in serious tracheal and bronchial airway congestion, unproductive cough, shortness of breath, tightness of chest, loss of appetite, loss of weight, malaise, pneumonitis, pulmonary edema, respiratory distress, and possible death. Examples of these soluble forms include beryllium fluoride, beryllium carbonate, beryllium sulfate, metalorganics (e.g., diethylberyllium), beryllium chloride, beryllium acetate, and beryllium oxalate.[2,4,7,8]

Chronic Beryllium Disease (CBD)

Repeated inhalation exposure to either soluble or insoluble forms of beryllium with associated lung burdens of the element may lead to chronic beryllium disease with up to 20 years latency. This disease is characterized by weakness; malaise; weight loss; unproductive cough; dyspnea; chills; fever; sarcoid-like granulomata of lung, skin, kidney, liver, and spleen; pneumothorax; pulmonary and/or myocardial failure; and death.[2,4,7,8]

Experiments with laboratory animals have shown various forms of beryllium to be carcinogenic, yielding bone sarcomas and lung carcinomas under certain exposure conditions. However, human exposures for over 40 years have not clearly indicated any carcinogenic potential although there were many cases of both acute and chronic beryllium disease before the application of modern industrial hygiene standards. Industrial physicians tend to give a low priority to concerns about potential carcinogenesis in the maintenance of appropriate occupational hygiene practice with beryllium. In any case, the possibility of lung, liver and/or bone cancer associated with chronic beryllium disease should not be completely discounted.[2,7,8]

CURRENT STANDARDS FOR AIRBORNE BERYLLIUM

The current safety standards and the threshold limit value (TLV) for airborne beryllium in the workplace are based upon toxicity predictions stemming from the Saranac Symposium of 1947.[3,4] These predictions involved assumptions about beryllium toxicity based on known toxicity of other highly toxic metals but scaled to beryllium. The Atomic Energy Commission (AEC) first promulgated these standards which stipulated an eight-hour average exposure of 0.002 mg-Be/m^3 of air based upon high volume sampling of total airborne particulate material. This time-weighted average (TWA) concentration was not to be exceeded during occupational exposure to beryllium aerosols. A peak concentration (measured over 30 minutes) was also stipulated at a maximum of 0.025 mg-Be/m^3 to protect against acute responses. An environmental limit for exposure of the general public in the vicinity of beryllium industries was set at 0.00001 mg/m^3. The current National Institute of Occupational Safety and Health (NIOSH) standards and the American Conference of Governmental Industrial Hygienists (ACGIH) TLV are based upon these values.[3,4,9]

RESPIRABLE PARTICULATE MASS THRESHOLD LIMIT VALUE (RPM-TLV)

The current ACGIH TLV-TWA for occupational exposure to beryllium aerosols is 0.002 mg-Be/m^3 of air based upon filter samples of total airborne particulate material of all sizes.[9] This TLV concentration of airborne beryllium is currently applied to all forms including the most soluble and reactive compounds as well as the more insoluble and inert forms; it also encompasses all sizes of aerosol particles including those too large to be readily inhaled. Since the gastrointestinal uptake of beryllium is only about 0.2 percent of that ingested, the risks associated with inhaling beryllium are primarily associated with deposition in the respiratory tract. In addition, the insoluble particles deposited in the conductive tracheobronchial airways of the lung, the nasopharyngeal region, and other parts of the HAR are cleared rapidly to the throat and swallowed or expectorated. Hence, the hazard associated with the inhalation of insoluble inert forms of beryllium, such as metallic beryllium, alloys, and high-fired beryllium oxide, is limited to those inhaled particles that lodge in the pulmonary GER of the deep lung where their long retention leads to slow systemic uptake and gradual cellular reactions in the lung.

The deposition of inhaled particles in the regions of the respiratory tract during normal breathing via the nose is illustrated in Figure 2 as a fraction of the concentration entering the nose with respect to particle aerodynamic equivalent diameter.[10] Particles that are most likely to deposit in the pulmonary GER are those with aerodynamic equivalent diameters well below 10 μm. In fact, less than 50 percent of particles larger than about 3.5 μm are available to enter the pulmonary GER

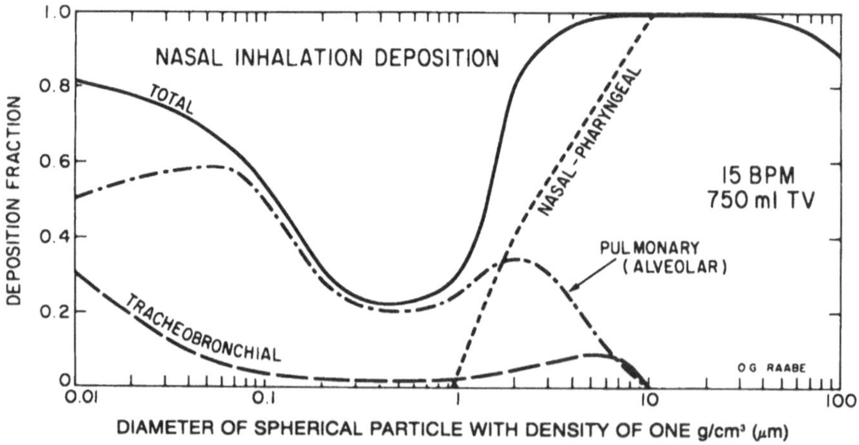

Figure 2. Nasal Inhalation Deposition: Total and regional deposition fractions for aerosols entering the nose for various sizes of inhaled airborne spherical particles with physical density of one gram per cubic centimeter in the human respiratory tract as calculated by the International Commission on Radiological Protection (ICRP) Task Group on Lung Dynamics[13] for nasal breathing at a rate of 15 breaths per minute (BPM) and tidal volume (TV) of 750 ml.

because they are deposited in the head or tracheobronchial airways first. Thus the respirable particulate mass particle size-selective sampler, with collection efficiencies shown in Figure 3, provides a sample of those particles available for deposition in the deep lung.[11] For these reasons a respirable particulate mass particle size-selective threshold limit value (RPM-TLV) of 0.002 mg/m^3 averaged over eight hours is appropriate to limit occupational exposures to aerosols of insoluble forms of beryllium. In these cases larger particles pose minimal risks. This suggested RPM-TLV is equivalent to the current TLV when aerosols consist only of particles with aerodynamic sizes less than about 3.5 μm.

Restated, this is the suggested RPM-TLV which should not be exceeded during occupational inhalation exposure to insoluble forms of beryllium (averaged over eight hours): RPM-TLV = 0.002 mg-Be/m^3. The RPM-TLV is only applicable to relatively insoluble and inert forms of beryllium, since only the pulmonary GER of the deep lung is at risk. Examples of RPM-TLV forms of beryllium include beryllium metal, beryllium oxide (high fired above 1100°C), beryllium silicate ($BeSiO_4$), bertrandite ore [about 4% Be as $Be_4(OH)_2Si_2O_7 \cdot H_2O$], beryl ore (about 4% Be as $3BeO \cdot Al_2O_3 \cdot 6SiO_2$), phenacite ore (about 1.5% Be as $BeSiO_4$), beryllium-copper alloy (less than 4% Be), beryllium-aluminum alloy (about 62% Be), beryllium-nickel alloy (about 2.2% Be), beryllium-copper-cobalt alloy (less than 2.4% Be), tantalum beryllide (37.2% Be as $TaBe_{12}$), vanadium beryllide (68% Be as VBe_{12}), niobium beryllide (53.6% Be as $NbBe_{12}$), and titanium beryllide (69.3% Be as $TiBe_{12}$)

INSPIRABLE PARTICULATE MASS THRESHOLD LIMIT VALUE (IPM-TLV)

In sharp contrast, all inspired soluble and reactive forms of beryllium, such as beryllium fluoride and beryllium sulfate, may yield adverse response at various locations in the respiratory tract including the nose and other parts of the HAR. Particles of all sizes that enter the nose may lead to adverse responses. All inspirable particles need to be considered in establishing safe limits.

The current TLV-TWA for beryllium is based upon the total airborne aerosol beryllium concentration including particles that are too large to be readily inhaled. Also, open filter samples, even those collected with high volume samplers, probably do not actually collect all airborne particles for sizes larger than 10 μm in aerodynamic equivalent diameter. It would, therefore, be most appropriate to substitute the inspirable partic-

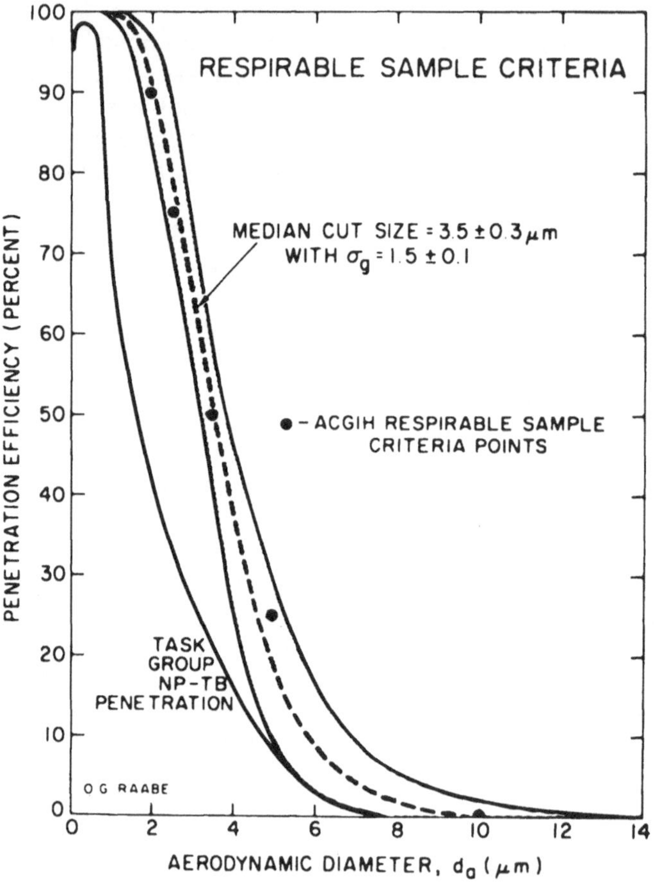

Figure 3. Respirable Particulate Mass Sampling Criteria: Respirable particulate mass (RPM) and respirable particulate mass fraction (RPMF) criteria tolerance band given as penetration efficiency to sample collector for those particles that penetrate a separator whose size collection efficiency is described by a cumulative lognormal function with median cut size of 3.5 $\pm$ 0.3 μm in aerodynamic equivalent diameter and with geometric standard deviation, σ_g, of 1.5 $\pm$ 0.1 as given by Raabe.[11] Also shown are the calculated penetration through the human nasopharyngeal (NP), head, and the tracheobronchial (TB) airways representing particles available to the pulmonary gas exchange region (GER) for a typical active worker breathing with tidal volume (TV) of 1450 ml at 15 breaths per minute (BPM) based upon the recommendations of the Task Group on Lung Dynamics.[13]

ulate mass particle size-selective sample for total airborne particulate material in the TLV for soluble beryllium. Hence, it is suggested that the current TLV should be used as the IPM-TLV for occupational exposure to aerosols of soluble forms of beryllium as measured with an inspirable particulate mass particle size-selective sampler.[12]

Restated, this is the suggested IPM-TLV which should not be exceeded during occupational inhalation exposure to beryllium (averaged over eight hours): IPM-TLV = 0.002 mg-Be/m^3. The IPM-TLV is applicable to beryllium aerosols of all physical and chemical forms including soluble and reactive forms of beryllium, since all parts of the respiratory tract are at risk. Examples of IPM-TLV forms of beryllium include beryllium fluoride (BeF_2), potassium beryllium fluoride (K_2BeF_4), beryllium sulfate ($BeSO_4 \cdot 4H_2O$), beryllium nitrate [$Be(NO_3)_2$], beryllium oxalate [$Be_4O(CH_3COO)_6$], beryllium carbonate ($BeCO_3$), beryllium chloride ($BeCl_2$) and beryllium metalorganics (such as diethylberyllium).

THORACIC PARTICULATE MASS THRESHOLD LIMIT VALUE (TPM-TLV)

For aerosols of other less soluble forms of beryllium, consideration needs to be given to potential adverse responses in the tracheobronchial airways as well as in the pulmonary GER of the lung. This region is especially at risk during mouth breathing since the protective collection of particles in the nasopharyngeal region of the head is prevented. Deposition of these low solubility particles in the head airways results in their eventually being swallowed or expectorated with little consequence, however.

The deposition of inhaled particles in the human respiratory tract during normal breathing via the mouth is illustrated in Figure 4 as a fraction of the concentration entering the mouth with respect to particle aerodynamic equivalent diameter. Particles that are most likely to deposit in the lung are those with aerodynamic equivalent diameters well below 20 μm; in fact, less than 50 percent of particles larger than about 10 μm are available to enter the lung region because they are deposited in the mouth or other parts of the head airways first. Thus the thoracic particulate mass particle size-selective sampler, with collection efficiencies shown in Figure 5, provides a sample of those particles available for deposition anywhere in the lung.[11]

For these reasons a TPM-TLV of 0.002 mg/m^3 averaged over eight hours would be appropriate for exposures to aerosols of insoluble forms of beryllium. In these cases larger particles pose minimal risks. This

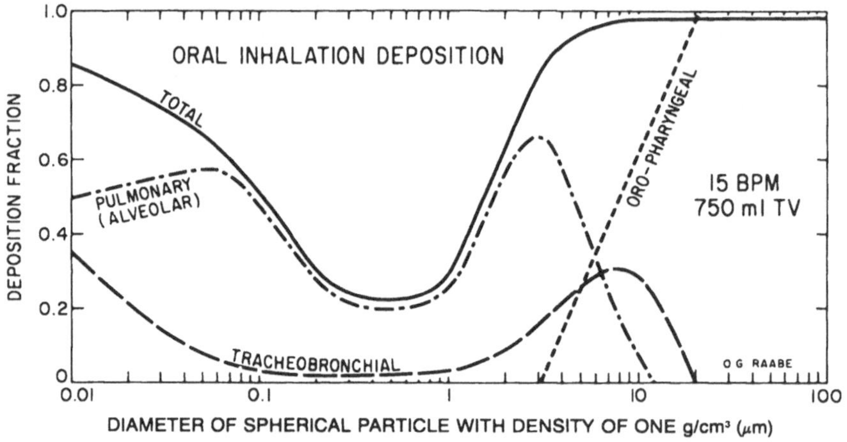

Figure 4. Oral Inhalation Deposition: Total and regional deposition fractions for aerosols entering the mouth for various sizes of inhaled airborne spherical particles with physical density of one gram per cubic centimeter in the human respiratory tract as calculated by the International Commission on Radiological Protection (ICRP) Task Group on Lung Dynamics[13] with the head airway deposition function given by Raabe[11] for oral breathing at a rate of 15 breaths per minute (BPM) and tidal volume (TV) of 750 ml.

suggested TPM-TLV is equivalent to the current TLV when aerosols consist only of particles with aerodynamic sizes less than about 10 μm. For aerosols of beryllium of less soluble forms such as low-fired beryllium oxide or beryllium hydroxide, industrial hygiene controls can utilize a TPM-TLV of 0.002 mg/m³. For convenience the TPM-TLV can be applied to all but the most soluble forms of beryllium.

Restated, this is the suggested TPM-TLV which should not be exceeded during occupational inhalation exposure to relatively insoluble forms of beryllium (averaged over eight hours): TPM-TLV = 0.002 mg-Be/m³. The TPM-TLV is applicable to all but the most soluble and reactive forms of beryllium since only the lung is at risk. Examples of TPM-TLV forms of beryllium include beryllium hydroxide [$Be(OH)_2$], hydrated beryllium oxide ($BeO \cdot H_2O$), beryllium oxide (BeO), and beryllium phosphors (oxides of Be, Mg, Mn, and Zn with SiO_4); in addition, all of the insoluble forms that are applicable to the RPM-TLV may also be sampled using a thoracic particulate mass particle size-selective sampler and the TPM-TLV.

Since the Environmental Protection Agency is promoting the use of particle size-selective sampling equivalent to collection of the thoracic

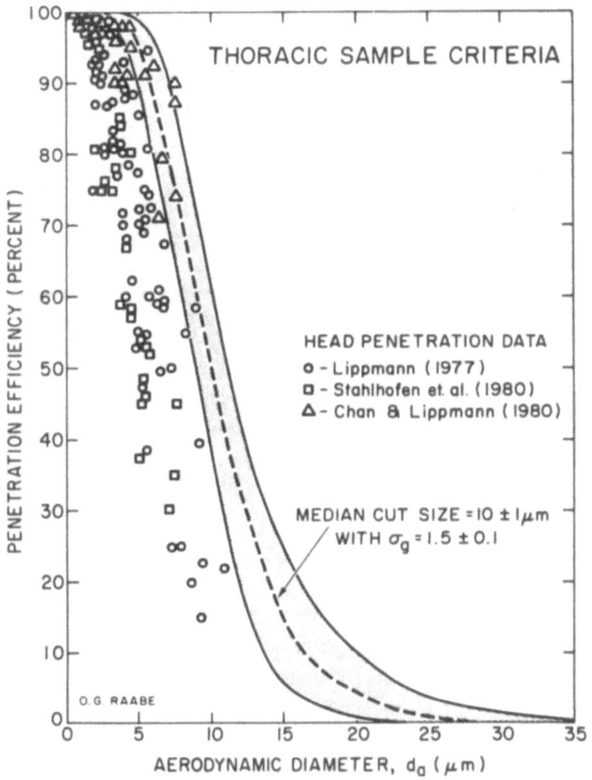

Figure 5. Thoracic Particulate Mass Sampling Criteria: Thoracic particulate mass (TPM) and thoracic particulate mass fraction (TPMF) criteria tolerance band given as penetration efficiency to sample collector for those particles that penetrate a separator whose size collection efficiency is described by a cumulative lognormal function with median cut size of 10 ± 1 μm in aerodynamic equivalent diameter and with geometric standard deviation, σ_g, of 1.5 ± 0.1 as given by Raabe.[11] Also shown are selected human data for head penetration during inhalation via the mouth corrected to the appropriate aerodynamic diameters for a typical active worker breathing with a tidal volume (TV) of 1450 ml at 15 breaths per minute (BPM).[11]

particulate fraction (the so-called PM-10 for airborne particulate matter) for sampling of ambient environmental aerosols, high volume samplers outfitted with 10 μm cut-size inlets are available for use for occupational sampling of beryllium. The current use of high-volume sampling for measurement of airborne beryllium in the workplace can, therefore, be readily converted to thoracic sampling as long as no soluble forms of beryllium are being sampled.

SUMMARY

The current TLV-TWA recommended by ACGIH for occupational exposure to beryllium aerosols is 0.002 mg-Be/m^3 of air based upon filter samples of total airborne particulate material of all sizes. This TLV concentration of airborne beryllium is currently applied to all forms including the most soluble and reactive compounds as well as the more insoluble and inert forms. Since the gastrointestinal uptake of beryllium is only about 0.2 percent of that ingested, the risks associated with inhaling beryllium are primarily associated with deposition in the respiratory tract and especially in the GER of the lung.

A RPM-TLV of 0.002 mg-Be/m^3 is suggested for occupational exposures to aerosols of insoluble forms of beryllium, such as beryllium metal and alloys, since only particles available to the deep lung are effectively hazardous. For soluble and reactive forms of beryllium, such as beryllium fluoride and beryllium sulfate, an IPM-TLV of 0.002 mg-Be/m^3 is suggested since all regions of the respiratory tract may be affected. For other relatively insoluble forms whose deposition anywhere in the lung is to be avoided, such as beryllium oxide or beryllium hydroxide, industrial hygiene controls can utilize a TPM-TLV of 0.002 mg/m^3 since only particles available to the lung are effectively hazardous. For convenience the TPM-TLV can be applied to all but the most soluble forms of beryllium.

REFERENCES

1. Cotton, F.A. and G. Wilkinson: *Advanced Inorganic Chemistry*. Interscience Publishers, John Wiley & Sons, New York (1972).
2. Wilber, C.G.: *Beryllium—A Potential Environmental Contaminant*. C.C. Thomas, Springfield, IL (1980).
3. Stokinger, H.E., Ed.: *Beryllium—Its Industrial Hygiene Aspects*. Academic Press, New York (1966).
4. Tabershaw, I.R., Ed.: *The Toxicology of Beryllium*. DHEW (NIOSH) Pub. No. 2173 (1972).
5. International Commission on Radiological Protection: *Report of Committee II on Permissible Dose for Internal Radiation*. Pergamon Press, New York (1959).
6. Raabe, O.G.: Deposition and Clearance of Inhaled Particles. *Occupational Lung Diseases*, pp. 1–37. J.B.L. Gee, W.K.C. Morgan and S.M. Brooks, Eds. Raven Press, New York (1984).
7. Cullen, M.R., M.G. Cherniack and J.R. Kominsky: Chronic Beryllium Disease in the United States. *Semin. Resp. Med.* 7:203–209 (1986).

8. Hardy, H.L.: Beryllium Disease: A Clinical Perspective. *Environ. Res. 21*:1–9 (1980).
9. American Conference of Governmental Industrial Hygienists: *Documentation of Threshold Limit Values*, 4th ed. Cincinnati, OH (1980).
10. Raabe, O.G.: Deposition and Clearance of Inhaled Aerosols. *Mechanisms in Respiratory Toxicology*, Chap. 2, pp. 27–76. H.R. Witschi and P. Nettesheim, Eds. CRC Press, Inc., West Palm Beach, FL (1982).
11. Raabe, O.G.: Size-selective Sampling Criteria for Thoracic and Respirable Mass Fractions. *Particle Size-selective Sampling in the Workplace*, Chap. 4.B. American Conference of Governmental Industrial Hygienists, Cincinnati, OH (1985).
12. Soderholm, S.C.: Size-selective Sampling Criteria for Inspirable Mass Fraction. *Particle Size-selective Sampling in the Workplace*, Chap. 4.A. American Conference of Governmental Industrial Hygienists, Cincinnati, OH (1985).
13. Morrow, P.E., D.V. Bates, B.R. Fish et al.: Deposition and Retention Models for Internal Dosimetry of the Human Respiratory Tract (Task Group on Lung Dynamics). *Health Phys. 12*:173–207 (1966).

CHAPTER 5

Basis for Particle Size-Selective Sampling for Wood Dust

WILLIAM C. HINDS

University of California Southern Occupational Health Center, UCLA School of Public Health, Los Angeles, California 90024

INTRODUCTION

The wood products industry is divided into the primary industry group consisting of sawmills, pulp mills, and plywood manufacturing, where relatively wet or green wood is used; the secondary industry group, primarily mill work; and the tertiary industry group, which includes furniture, cabinet, and pattern making. Most of the health problems occur in the secondary and tertiary groups, where the wood is much drier and produces more dust. These two groups represented 72 percent of a total wood products industry work force of 760,000 in the U.S. in 1982. Dust concentrations in the tertiary industry group are quite variable and can range up to 200 mg/m^3; earlier measurements averaged 10 mg/m^3 or more while more recent measurements average less than 5 mg/m^3.

There are three basic wood working operations that produce dust: sawing, milling, and sanding. Sawing, the most important and most widely used, is common in all three industry groups. For all three operations both shattering of the lignified wood cells and breaking out of whole cells or groups of cells (chips) occur. The more cell shattering that occurs, the finer the dust particles that are produced. Sawing and milling are mixed shattering and chip forming processes, whereas sanding is exclusively cell shattering. In general, the harder the wood the more tightly bound are the cells and the more shattering takes place, causing more dust formation. Similarly in drier wood the cells are less plastic and are more likely to be shattered, leading to dust formation. Since wood cell dimensions are on the order of 1 mm, *airborne* dust concentrations depend primarily on the extent of cell shattering rather than the size or extent of chip formation.

Wood dust is composed of cellulose, hemicellulose, lignin, and hundreds of compounds collectively known as "wood extractives." These are high and low molecular weight organic compounds that can be extracted, primarily from the heartwood boundary, and include terpenes, polyphenolic compounds, tropolones, glycocides, sugars, fatty acids, and inorganic compounds.

HEALTH EFFECTS OF WOOD DUST

Since 1980 several reviews and evaluations of the health effects of wood dust have been published.[1-7] The carcinogenic aspects of exposure to wood dust are reviewed by Darcy[7] and in greater detail in an International Agency for Research on Cancer (IARC) monograph on wood dust.[8]

The primary health effects associated with wood dust are summarized in Table I. Omitted from this table are the health effects associated with toxic woods such as satinwood, boxwood, oleander, and yew. Skin and eye irritation are believed to result primarily from irritating chemicals in the sapwood and bark of certain tropical and domestic trees. A fungus that develops in cut sapwood, called sap stain or blue stain fungus (Ascomycetes), is aerosolized along with the sawdust which frequently causes irritation and may cause sensitization. Sensitizing chemicals are often found in the "extractives" compounds. The most common respiratory sensitizer is Western Red Cedar. An evaluation by Whitehead et al.[9] found pulmonary function changes consistent with an obstructive effect in a population of workers exposed to maple and pine dust.

The most serious health concern associated with wood dust is nasal cancer. The highest risk group are workers exposed to hardwood dust especially in the furniture and cabinet-making industries. These workers were found to have a 10- to 20-fold increased risk of nasal cancer and a 100- to 500-fold increased risk of nasal adenocarcinoma, a rare tumor which occurs at a rate of less than one per million in the general population.[5,6,10] Of the tumors found in these workers, 75 to 90 percent were nasal adenocarcinomas. These tumors have a long latency period (average about 40 years) and have been produced by exposures as brief as five

TABLE I. Health Effects Associated with Exposure to Wood Dust

Health Effects	Type of Wood	Respiratory Region
Skin and eye irritation	Douglas fir, white cedar, beech, oak, tropical woods, fungal spores	—
Allergic skin response	Heartwood, fungal spores, Douglas fir, white cedar, beech, oak, tropical woods	—
Rhinitis, nasal dryness	Hardwood	Head airways
Pulmonary function changes	Hardwood, softwood	Lung airways and gas exchange
Allergic respiratory response (asthma)	Western Red Cedar, oak, beech, mahogany, "exotic" woods, fungal spores	Lung airways (and gas exchange)
Cancer of nasal cavity and paranasal sinus	Hardwood (and softwood)	Head airways

years.[6] An Association of Schools of Public Health/National Institute for Occupational Safety and Health (ASPH/NIOSH) publication identifies wood dust as associated with nasal cancer with a relative risk of 500.[10] Furniture workers have also been found to have impaired nasal mucociliary clearance. A wood dust concentration of 2.2 mg/m^3 produced mucostasis in 11 percent of exposed subjects and 25.5 mg/m^3 produced it in 63 percent of exposed subjects. These effects are suspected of playing a role in development of nasal cancer.

The association between wood dust exposure in the furniture industry and nasal cancer was first reported in England in the 1960s.[8] The increase in the incidence of this tumor during the period 1955 to 1975 corresponds to an increase in mechanization in the furniture manufacturing industry that occurred after World War I. Since 1950 local exhaust ventilation has been used increasingly in furniture and cabinet making plants in England and this may be responsible for a decrease in the incidence of nasal cancer among these wood workers.[6] This and other factors such as the type of wood used, moisture content, and type of cutting may explain differences in nasal cancer rate between countries.

The furniture woods involved are oak, beech, mahogany, maple, walnut, teak, and birch. Carpenters in the building industry and others exposed to softwood dust appear to have an increased risk of nasal cancer, but not as great as those exposed to hardwood dust.[4,5] Wood dust is designated as a suspected carcinogen in Sweden and West Germany.[5] The *Documentation of the Threshold Limit Values and Biological Exposure Indices* cites three studies to support the idea that it is the wood dust itself that causes these health effects and not varnishes, polishes, or lacquers used in the furniture industry.[6] Other studies have evaluated the health effects of wood dust separately from the health effects of finishes and preservatives used in the wood products industry.[5,8]

SIZE DISTRIBUTION OF WOOD DUST

Several investigators have reported size distributions for wood dust.[5-9,11-13] McCammon et al. measured size distributions in automotive model shops with an Andersen Non-Viable Cascade Impactor at 29 lpm and found mass median aerodynamic diameter (MMAD) that ranged from 5 to 10 μm and geometric standard deviations (GSD) from 1.1 to 2.8[11] They found respirable fractions that ranged from 19 to 38 percent. The respirable fraction was approximately inversely proportional to total dust loading over the range 0.3 to 0.9 mg/m^3 (total dust). Darcy reported that rough mill operations have a bimodal size distribution with a large mode having a MMAD of 30 μm with 25 percent less than 15 μm.[7,12] The MMAD for the small mode is 0.8 μm with 95 percent less than 2 μm. Softwood sanding operations produce a somewhat finer dust with a large mode MMAD of 20 μm and 40 percent less than 15 μm and a small mode with a MMAD of 1.3 μm and 90 percent less than 5 μm. Hardwood sanding is similar with a large mode MMAD of 24 μm and 35 percent less than 15 μm. For all of these operations the mass in the small mode represents a small percentage of the total mass of the bimodal distribution. Whitehead and colleagues give size distribution information for softwood and hardwood dust obtained with a six-stage Sierra Model 216 cascade impactor.[13] For 14 of 15 samples the percentage of mass greater than 22.5 μm is in the range of 64 to 85 percent and less than 4 percent of the mass is contributed by particles less than 1.2 μm. Representative size distributions for hard and softwood sanding dust are shown in Figure 1.[7] These results support the idea that the major portion of the wood dust mass is contributed by particles larger than 10 μm, a size range for which use of inspirable mass sampling is necessary and important.

DEPOSITION SITE OF WOOD DUST

It seems likely that the increased risk of nasal cancer is a result of direct deposition of wood dust in the nose. Work rates in cabinet-making

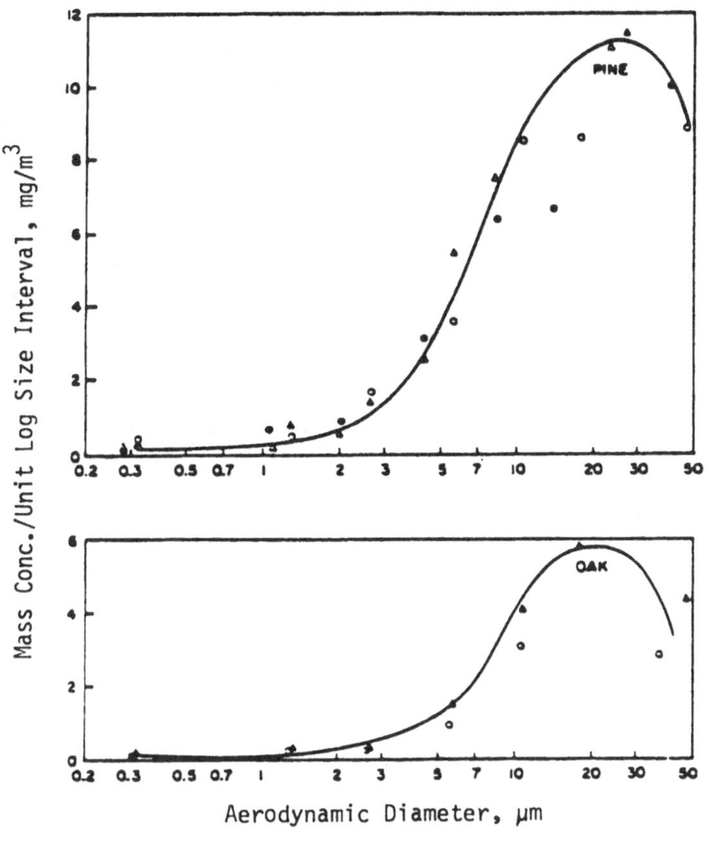

Figure 1. Mass size distribution for wood dust from typical sanding operations.[7]

and furniture industries are mostly light to moderate, and most people would use nose breathing. Both Whitehead and Darcy report that, except for some sanding operations, more than 80 percent of the dust is contained in particles greater than 10 µm.[12,13] Such particles will deposit in the head airways (primarily in the nose for nose breathing) with high efficiency (approximately 100%). For sanding dust the majority of particulate mass is contained in particles greater 10 µm. There is no evidence on whether the smaller particles that do deposit in the nose are more dangerous because of their higher surface-to-mass ratio. Some of the earlier work indicated that it was the "fine" hardwood dust that was responsible for the nasal cancer. In this context the term "fine" is used in a qualitative sense reflecting the proportion of chips and dust and likely refers to suspended particles in the 1–50 μm range. Asthma and pulmonary effects are presumably caused by the thoracic or respirable fraction.

SAMPLING AIRBORNE WOOD DUST

In light of the above, it is appropriate that Inspirable Particulate Mass (IPM) sampling be used to assess general wood dust concentration in furniture, cabinet making, and other industries where hardwood dust is present. To do otherwise would exclude from the sample the major source of particles that would enter the nose or mouth in a typical work situation. To a large degree this reflects the particles that deposit in the nose; consequently, IPM sampling will be the environmental measurement that is most closely related to the risk of developing nasal cancer.

As explained below, traditional "total dust" sampling with open-face or in-line 37-mm filter cassettes is an unreliable method for assessing the concentration of particles in the 10–100 μm size range that are of primary concern for nasal cancer. Even though IPM samples include the thoracic particulate mass (TPM), separate sampling for TPM or respirable particulate mass (RPM) should be used when the health concern is occupational asthma, as it might be at a plant processing Western Red Cedar.

Figure 2[14] shows the performance of 37-mm cassette samplers as a function of particle size for various configurations. Isolated open-face and in-line cassette data are from reference 15 and torso-mounted cassette data are for a Casella Model T13037 cassette with a 4-mm diameter inlet.[16] While the shape of the latter is equivalent to an in-line 37-mm cassette, it is mounted on a frame so that its inlet is facing forward (axis horizontal) at all times. The difference in performance when sampling with an isolated cassette and sampling with a cassette mounted on a torso is dramatic for particles larger than 15 μm.

Similar data for 37-mm cassettes mounted on a full-torso mannequin facing the wind are given by Buchan et al.[17] They evaluated sampling performance for open-face and in-line 37-mm cassettes on a torso facing a wind of 1.0 m/s. Cassettes were either hanging free with their inlet axes aligned vertically downward or attached to a mannequin with their inlet axes 46 degrees below the horizontal. In all cases the downward facing sampler showed lower sampling efficiency than the 46-degree-angled sampler. For in-line cassettes the ratio of down to 46-degree sampling efficiency ranged from 0.73 for 2.4-μm to 0.23 for 24-μm particles. This underscores the sensitivity of sampler performance to the orientation of the inlet and whether or not it is mounted on a torso.

Researchers at the Institute of Occupational Medicine in Edinburgh,

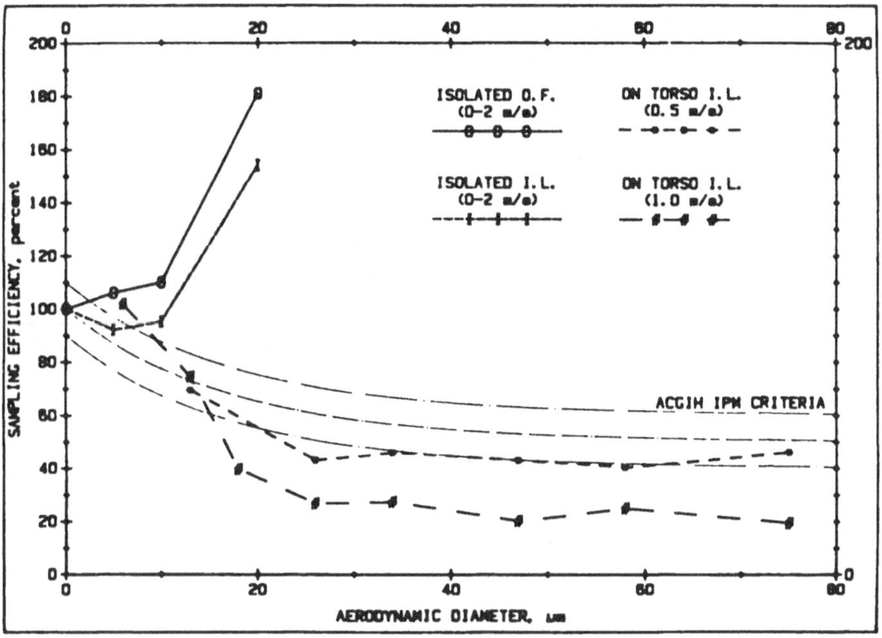

Figure 2. Sampling performance of 37-mm and Casella model T13037 filter cassettes.

U.K., have recently developed an *area sampler*, the IOM/STD1, that comes close to matching the American Conference of Governmental Industrial Hygienists (ACGIH) IPM criteria over the range of 0 to 100 µm aerodynamic diameter.[18] The sampling head for their device is a vertical axis cylinder about 5 cm in diameter and 6 cm high. A horizontal axis, oval-shaped inlet slot (about 3 mm high × 16 mm wide) is located midway up the side of the cylinder. The device samples at 3 lpm through a 37-mm filter mounted in a weighable cassette inside the cylinder. The sampling head is mounted on a larger vertical axis cylinder about 15 cm in diameter and about 18 cm high which houses batteries, pump, and flow control. The sampling head rotates continuously at about 2 rpm. The whole unit weighs 2.5 kg. Results of preliminary tests are shown in Figure 3 for wind speeds of 1–3 m/s. Overall, the agreement with the ACGIH IPM criteria envelope is good. At a wind speed of 1 m/s, sampling efficiency is within the envelope for 0- to 50-µm particles and very close to it for 50- to 100-µm particles. At 3 m/s the device undersamples in the range of 10 to 50 µm, but it is within the envelope for other sizes. When data for 1 and 3 m/s wind speeds are averaged, the sampling

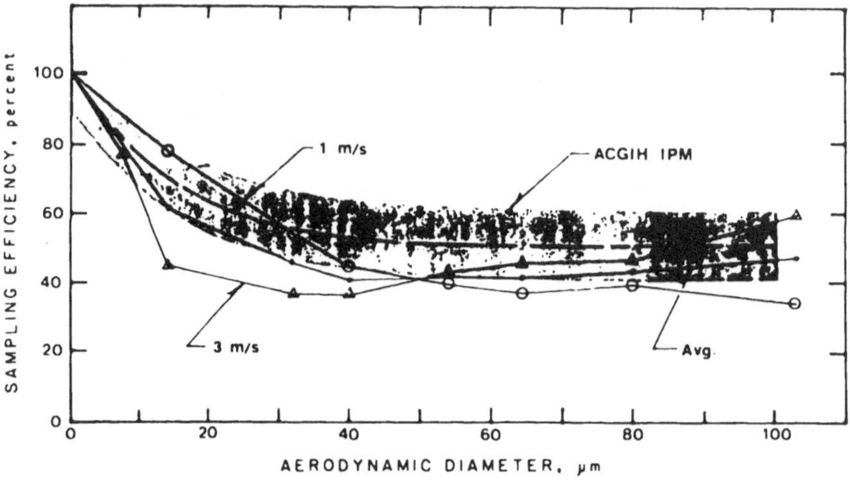

Figure 3. Sampling performance of IOM area inspirable particulate matter sampler.

efficiency shows only a slight departure from the envelope in the range of 30 to 55 μm.

Mark and Vincent[16] have developed and evaluated a new *personal sampler* designed to approximate the ACGIH IPM curve shown in Figure 4. The device has a cylindrical body, is 37 mm in diameter and 27 mm long, and has a 15-mm diameter inlet, which faces forward. It uses a 25-mm filter contained in a light metal cassette having a 15-mm diameter inlet tube that protrudes 1.5 mm beyond the body of the sampler. The filter and cassette are weighed together. Tests were conducted at a sampling flow rate of 2 lpm. Orientation-averaged performance of this device at wind velocities of 0.5 and 1.0 m/s is shown in Figure 5 where sampling efficiency is calculated from their data assuming mannequin inspiration follows the IPM criteria. Overall the agreement is very good with 13 out of 14 points falling within the IPM band.

RECOMMENDATIONS

Inspirable particulate mass (IPM) sampling, as defined by the ACGIH Air Sampling Procedures Committee,[18] is recommended for airborne wood dust. Because of the presence of local sources and the rapid change in concentration and size distribution with position, it is recommended that personal sampling be used. At present the best sampler for this purpose is the IOM personal sampler[16] or one with similar geometry and

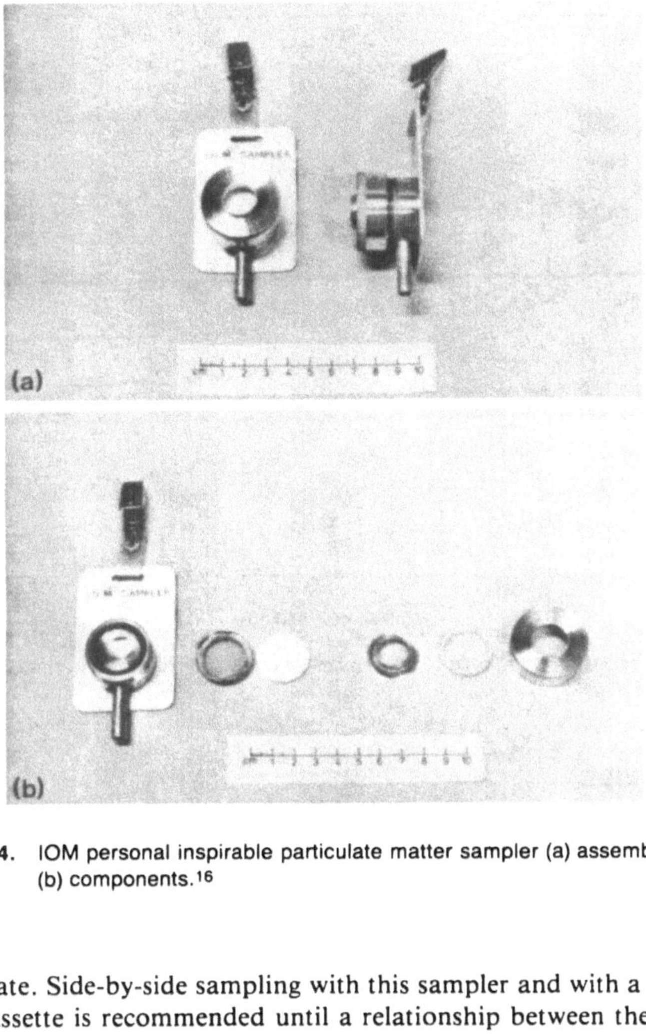

Figure 4. IOM personal inspirable particulate matter sampler (a) assembled; (b) components.[16]

flow rate. Side-by-side sampling with this sampler and with a 37-mm in-line cassette is recommended until a relationship between the two sampling methods is established.

In setting the TLV for wood dust, there is an important issue that the TLV committee must recognize. Because most of the mass is contributed by particles larger than 10 μm, the measured mass concentration will be strongly influenced by local air currents and by the equipment and method used for sampling. Consequently, historical data should be evaluated with great caution. As a first approximation, one might assume that earlier concentration measurements are within a factor of four (either way) of the equivalent IPM measurement.

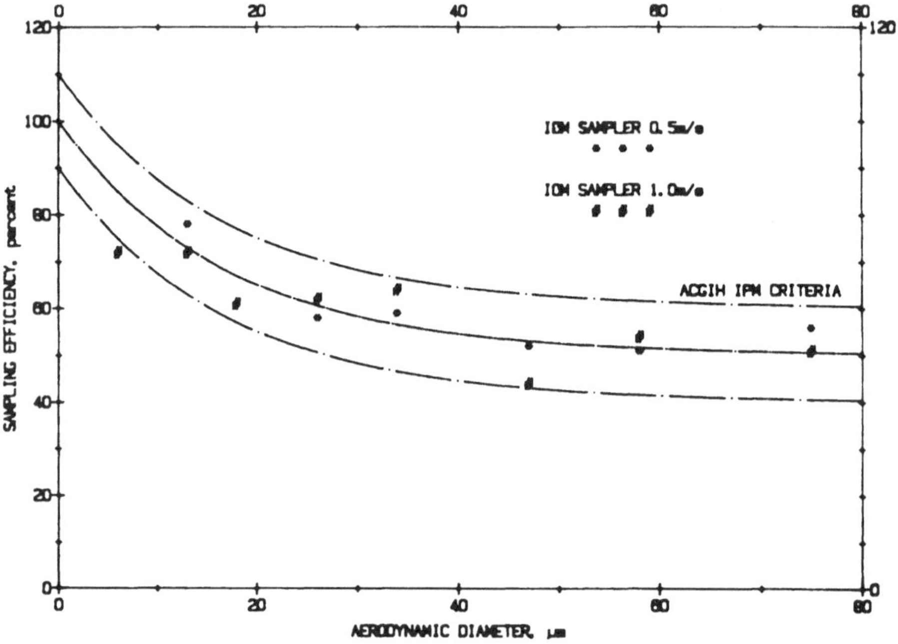

Figure 5. Sampling performance of IOM personal inspirable particulate matter sampler.

REFERENCES

1. Hausen, B.: *Woods Injurious to Human Health: A Manual.* Walter deGruyter, New York (1981).
2. Montgomery, R.R.: Polymers, *Patty's Industrial Hygiene and Toxicology*, 3rd ed., Vol. IIC, *Toxicology*. G.D. Clayton and F.E. Clayton, Eds. Wiley-Interscience, New York (1982).
3. NBOSH: Scientific Basis for Swedish Occupational Standards II: Consensus Report on Wood Dust. *Arbete och Halsa*, Vol. 9 (1982).
4. OML: *The Health Effects of Exposure to Wood Dusts.* Special Studies and Services Branch, Ontario Ministry of Labour, Toronto, Ontario, Canada (1986).
5. Holliday, M.G., P. Dranitsoris, P.W. Strahlendorf et al.: *Wood Dust in Ontario Industry: The Occupational Health Aspects.* Report to the Occupational Health and Safety Branch, Ontario Ministry of Labour, Toronto, Ontario, Canada (1985).
6. American Conference of Governmental Industrial Hygienists: *Documentation of Threshold Limit Values and Biological Exposure Indices*, 5th ed., pp. 635–636. Cincinnati, Ohio (1986).

7. Darcy, F.J.: *Physical Characteristics of Wood Dust*. Ph.D. Thesis, University of Minnesota, Minneapolis (1982).
8. International Agency for Research on Cancer: *IARC Monographs on the Evaluation of the Carcinogenic Risk of Chemicals to Humans, Vol. 25, Wood, Leather and Some Associated Industries*. Lyon, France (1981).
9. Whitehead, L.W., T. Ashikaga and P. Vacek: Pulmonary Function Status of Workers Exposed to Hard Wood or Pine Dust. *Am. Ind. Hyg. Assoc. J. 42*:178–186 (1981).
10. National Institute for Occupational Safety and Health: *Proposed National Strategies for the Prevention of Leading Work-related Diseases and Injuries*, Part 1. Assoc. Sch. of Public Health and NIOSH (1986).
11. McCammon, C.S., C. Robinson, R.J. Waxweiler and R. Roscoe: Industrial Hygiene Characterization of Automotive Wood Model Shops. *Am. Ind. Hyg. Assoc. J. 46*:343–349 (1985).
12. Darcy, F.J.: Wood Working Operations—Furniture Manufacturing. *Industrial Hygiene Aspects of Plant Operations*, Vol. 2, Chap. 25. L.J. Cralley and L.V. Cralley, Eds. Macmillan, New York (1984).
13. Whitehead, L.W., T. Freund and L.L. Hahn: Suspended Dust Concentration and Size Distributions, and Qualitative Analysis of Inorganic Particles, from Wood Working Operations. *Am. Ind. Hyg. Assoc. J. 42*:461–467 (1981).
14. Hinds, W.C.: Addendum to Chapter 5A, *Particle Size-Selective Sampling in the Workplace*. Technical Report of the Air Sampling Procedures Committee, ACGIH, Cincinnati, OH (1986).
15. Fairchild, C.I., M.I. Tillery, J.P. Smith and F.O. Valdez: Collection Efficiency of Field Sampling Cassettes. LA-8640-MS. Los Alamos Scientific Laboratory, Los Alamos, NM (1980).
16. Mark, D. and J.H. Vincent: A New Personal Sampler for Airborne Total Dust in Workplaces. *Ann. Occup. Hyg. 30*:89–120 (1986).
17. Buchan, R.M., S.C. Soderholm and C.I. Fairchild: Collection Efficiency of Personal Cassette Samplers for "Inspirable" Size Aerosols. Paper number 241. Presented at the American Industrial Hygiene Conference, Las Vegas, NV (May 23, 1985).
18. Hinds, W.C.: Sampler Efficiencies: Inspirable Mass Fraction. *Particle Size-Selective Sampling in the Workplace*, Chap. 5A. American Conference of Governmental Industrial Hygienists, Cincinnati, OH (1985).

CHAPTER 6

Basis for Particle Size-Selective Sampling for Sulfuric Acid Aerosol*

MORTON LIPPMANN, JEFFERY M. GEARHART and RICHARD B. SCHLESINGER

Institute of Environmental Medicine, New York University Medical Center, Tuxedo, New York 10987

BACKGROUND

Sulfuric acid (H_2SO_4) has had an eight-hour time-weighted average (TWA) exposure limit of 1 mg/m^3 since 1948, when it was raised from the original (1946) recommendation of 0.5 mg/m^3.[1] The Occupational Safety and Health Administration's (OSHA) Permissible Exposure Limit (PEL),[2] the National Institute for Occupational Safety and Health's (NIOSH) recommendation,[2] the West German Maximum Concentration Value (MAK),[3] and the USSR Maximum Allowable Concentration (MAC)[4] are all 1 mg/m^3 TWA.

The current American Conference of Governmental Industrial Hygienists (ACGIH) TLV Documentation cites literature published between 1950 and 1978, with only two of eleven citations after 1961.[5] The documentation concludes: "The TLV of 1 mg/m^3 is recommended to

* Based on research supported by Grants ES00881 and ES03213 from the National Institute of Environmental Health Science (NIEHS), Grant R-810101 and contracts 68-02-1726 and 504 014 MAEX from the U.S. Environmental Protection Agency, and by Contracts RP1157 and 2155 from the Electric Power Research Institute. It is part of a center program supported by Grant ES-00260 from NIEHS.

prevent pulmonary irritation and injury to the teeth. At particle sizes likely to occur in industrial situations it should be adequate to prevent harmful effects." The most relevant studies cited included: 1) Amdur et al.[6] who reported that the concentration producing a 50 percent increase in pulmonary flow resistance in anesthetized guinea pigs following a one-hour exposure was 0.3 mg/m^3 for 2.5 μm and 30 mg/m^3 for 7 μm; 2) Alarie et al.[7] who reported that monkeys exposed 23 hours/day for two years had histopathology regarded as slight (for 0.38 mg/m^3 at 2.15 μm MMAD [mass median aerodynamic diameter]), moderate (for 2.43 mg/m^3 at 3.6 μm) and moderate to severe at 4.79 mg/m^3; 3) Amdur et al.[8] who reported that 5- to 15-minute exposures of humans elicited reports of taste, odor, or irritation in two persons at 1 mg/m^3 and in all at 3 mg/m^3; and 4) Malcolm and Paul[9] who noted tooth erosion in workers exposed to 3 to 16 mg/m^3.

In June 1981, NIOSH published the *Review and Evaluation of Recent Literature – Occupational Exposure to Sulfuric Acid.*[10] This publication reviewed literature published between 1978 and 1981 on studies of functional responses in humans following one- to two-hour exposures, and it also reported a variety of acute and chronic animal inhalation studies published between 1973 and 1980. These studies and the more recent literature will be reviewed further in the next section. However, it may be of some interest at this point to cite some of the conclusions drawn in the 1981 NIOSH review.

> "The information now available from studies in experimental animals suggests possible alternative approaches to the current permissible limit for occupational exposure to sulfuric acid. For aerosols containing particle sizes that can penetrate to the lung, at least two mechanisms of action have been demonstrated. Aerosol particles that deposit in the upper lung appear to be more acutely harmful because reflexive bronchoconstriction occurs. Somewhat smaller aerosol particles, however, appear to cause greater alterations in pulmonary function and eventually in microscopic lesions. Exposure to very large particles would not lead to either of these effects. This all suggests that any future revision of the occupational exposure limit should consider aerosol particle-size."

REVIEW OF LITERATURE

Deposition, Growth, and Neutralization within the Respiratory Tract

The deposition pattern within the respiratory tract is dependent on the size distribution of the droplets. Acidic ambient aerosol typically has a mass median aerodynamic diameter (MMAD) of 0.3 to 0.6 μm, while industrial aerosols can have a MMAD as large as 14 μm.[11] With hygroscopic growth in the airways, submicrometer-sized droplets can increase in diameter by a factor of two to four and still remain within the fine particle range which deposits preferentially in the distal lung airways and airspaces. As droplet sizes increase above about 3 μm MMAD, deposition efficiency within the airways increases, with more of the deposition taking place within the upper respiratory tract, trachea, and larger bronchi.[12] For larger droplets, the residence time in the airways is too short for a large growth factor.

Some neutralization of inhaled acidic droplets can occur before deposition, due to the normal excretion of endogenous ammonia into the airways.[13] Once deposited, free H^+ reacts with components of the mucus of the respiratory tract, changing its viscosity.[14] Unreacted H^+ diffuses into surrounding tissues. The capacity of the mucus to react with H^+ is dependent on the H^+ absorption capacity, which is reduced in acidic saturated mucus as found in certain disease states, e.g., asthma.[14]

Effects on Experimental Animals

Acute Exposures

Respiratory mechanical function. Alterations of pulmonary function, particularly increases in pulmonary flow resistance, occur after acute exposure. Reports of the irritant potency of various sulfate species are variable, due in part to differences in animal species and strains and also to differences in particle sizes, pH, composition, and solubility.[15] H_2SO_4 is more irritating than any of the sulfate salts in terms of increasing airway resistance. For short-term (1-hour) exposures, the lowest concentration of H_2SO_4 shown to increase airway resistance was 100 $\mu g/m^3$ (in guinea pigs). The irritant potency of H_2SO_4 depends in part on droplet size, with smaller droplets having more effect.[6]

Some recent animal inhalation studies by Amdur et al. are of interest to this discussion because they demonstrate that effects produced by single exposures at very low acid concentrations can be persistent.[16] They exposed guinea pigs by inhalation for three hours to the diluted effluent from a furnace which simulates a model coal combuster.

Pulverized coal yields large particle mineral ash particles and an ultrafine (< 0.1 μm) condensation aerosol. The core of the ultrafine particles consists of oxides of Fe, Ca, and Mg, covered by a layer containing Na, As, Sb, and Zn. The Zn is important because it generally has the highest concentration on the surface of the solidified particle. As the particles cool further, there is surface formation and/or condensation of a layer of H_2SO_4.

In the initial experiments by Amdur et al.,[16] the model aerosol was a mixture containing SO_2, ZnO, and water vapor, and there was a single three-hour exposure to a mixture containing 1 ppm SO_2 and 5 mg/m^3 of ZnO passed through a humid furnace. The amount of H_2SO_4 on the surface of the ZnO particles was less than 40 $\mu g/m^3$. In control studies, neither 1 ppm of SO_2 nor 5 mg/m^3 of ZnO alone produced any significant responses. There were also no significant responses to the mixture in the absence of water vapor and passage through the furnace. However, the humid mixture, passed through the furnace, where it acquired a surface coating of H_2SO_4, produced significant decrements of lung diffusing capacity (DL_{CO}). At one hour after exposure, there was an increase in lung permeability. At 12 hours after exposure, there was distention of perivascular and peribronchial connective tissues and an increase in lung weight. The alveolar interstitium also appeared distended. At 72 hours after exposure, total lung capacity (TLC), vital capacity (VC), and functional residual capacity (FRC) had returned to baseline levels, but DL_{CO} was still significantly depressed. These changes are illustrated in Figure 1. Based upon prior experience with pure sulfur dioxide (SO_2) and pure H_2SO_4 exposures in the guinea pig model, Amdur et al. concluded that the humid furnace effluent effect was an acid aerosol effect because of its persistence.

The persistent changes in function and morphological changes following exposure to very low levels of acidic aerosol suggest that repetitive exposures could lead to chronic lung disease. However, the implications of these changes in guinea pigs to human disease remain highly speculative.

Particle clearance function. Donkeys exposed by inhalation for one hour to 0.3–0.6 μm H_2SO_4 at concentrations ranging from 100 to 1000 $\mu g/m^3$ exhibited slow bronchial mucociliary clearance function at concentrations $\geq$ 200 $\mu g/m^3$ while, as shown in Figure 2, rabbits undergoing similar exposures exhibited an acceleration of clearance at concentrations between 100 and 300 $\mu g/m^3$, and a progressive slowing of clearance at 500 $\mu g/m^3$.[17]

Subchronic Exposures

Particle clearance function. Donkeys exposed for one hour per day, five days per week for six months to an aerosol (0.3–0.6 μm) of H_2SO_4 at

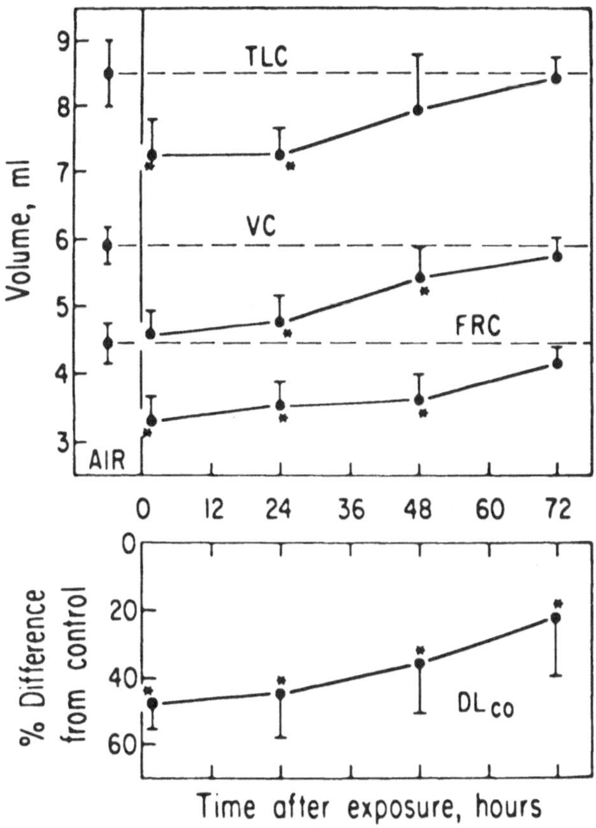

Figure 1. Changes in total lung capacity (TLC), vital capacity (VC), functional residual capacity (FRC), and CO-diffusing capacity (DL_{co}) in guinea pigs following a 3-hr exposure to 1 ppm SO_2 and 5 mg/m³ ZnO in a humid furnace. The asterisks (*) indicate $p < 0.05$. (Reproduced with permission from Amdur et al.[16])

a concentration of 100 μg/m³ developed highly variable clearance rates and a persistent shift from baseline rate of bronchial mucociliary clearance during the exposures and for three months after the last exposure. As shown in Figure 3, two animals had much slower clearance than their baseline during the three months of follow-up while two had faster than baseline rates.[18] Rabbits exposed for one hour per day, five days per week for four weeks to 0.3 μm H_2SO_4 at 250 μg/m³ developed variable mucociliary clearance rates during the exposure period, and their clearance

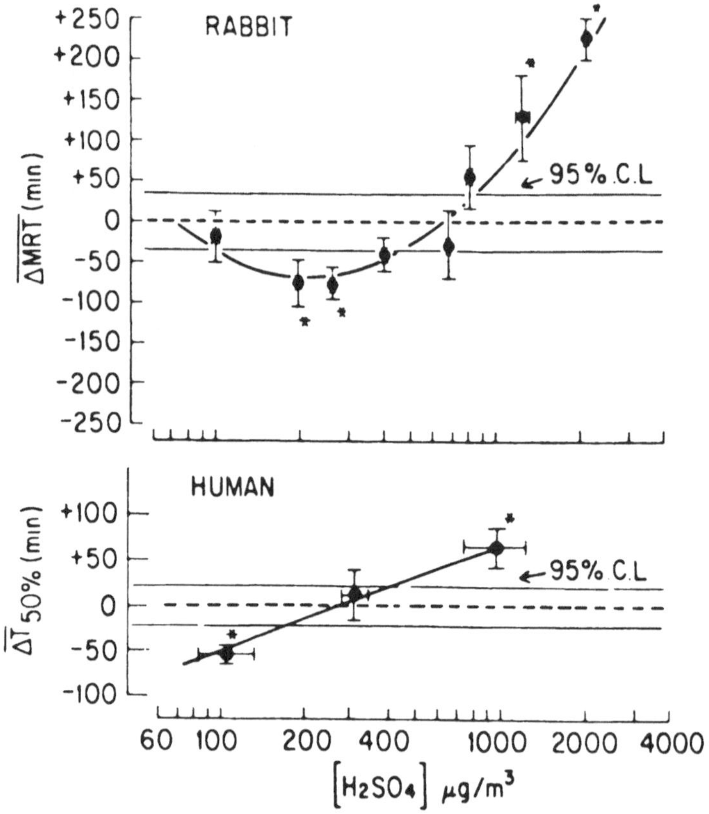

Figure 2. Exposure dependent changes in characteristic bronchial mucociliary clearance times in rabbits (mean residence time) and humans (clearance halftime) due to 1-hr exposures to submicrometer-sized H_2SO_4 aerosols. Each point represents the average for the group, with the vertical bars indicating ± 1 S.E. The horizontal bands represent the mean ± S.E. of the measurements for the sham exposure controls. Asterisks indicate a significant change ($p < 0.05$) at an individual concentration (paired t-test, 1-tailed). (Reproduced with permission from Lippmann.[72])

during a two-week period following the exposures was substantially faster than their baseline rates.[19] For a group of rabbits undergoing daily exposures via the nose at 250 μg/m³ for one year, Figure 4 shows that clearance was consistently slowed after the first few weeks, and it became even slower during a three-month period following the end of acid exposures.[20]

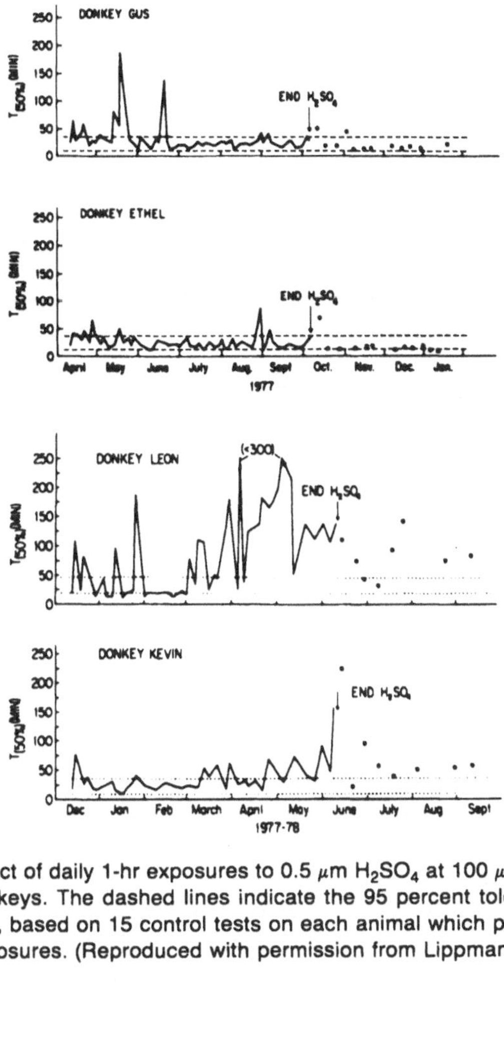

Figure 3. Effect of daily 1-hr exposures to 0.5 μm H_2SO_4 at 100 $\mu g/m^3$ on $T_{1/2}$ in four donkeys. The dashed lines indicate the 95 percent tolerance intervals for $T_{1/2}$, based on 15 control tests on each animal which preceded the H_2SO_4 exposures. (Reproduced with permission from Lippmann et al.[41])

During the course of a one-year series of one hour per day, five days per week nasal exposures to submicrometer H_2SO_4 at 250 $\mu g/m^3$, groups of rabbits were exposed on three occasions to ^{85}Sr-tagged latex aerosols for determination of the rates of clearance from the non-ciliated alveolar region.[21] The latex aerosols were administered on days 1, 57 and 240 following the start of the H_2SO_4 exposures, and particle retention was followed for 14 days after each latex administration. Compared to baseline rates of clearance in control animals, early alveolar clearance was

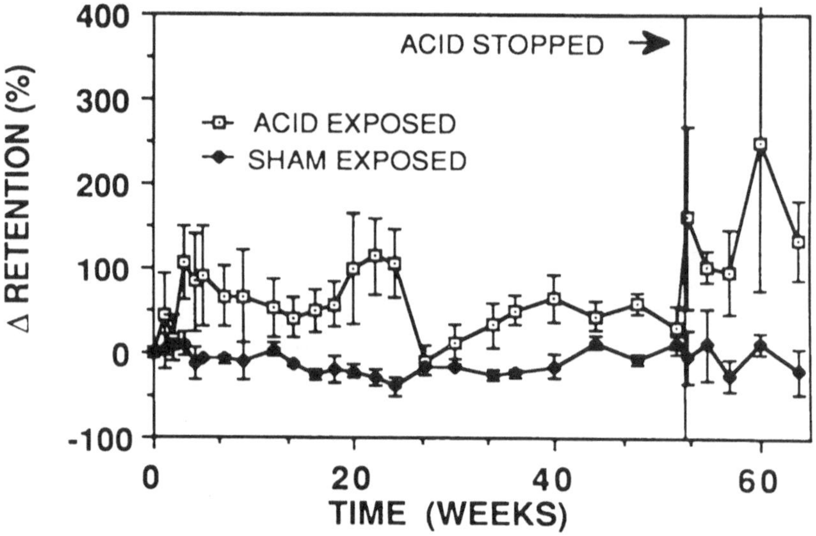

Figure 4. Mean change in percent retention of tracer particles (± S.D.) during intermittent exposure to H_2SO_4 in acid and sham exposed animals from that established in pre-exposure control tests.

accelerated to a similar degree in all three tests performed (Figure 5) during the chronic H_2SO_4 exposures.

Airway hyperresponsiveness. The effects of daily, one-hour exposures of rabbits to 250 $\mu g/m^3$ of H_2SO_4 on bronchial responsiveness were assessed at the end of 4, 8, and 12 months by "IV" administration of doubling doses of acetylcholine and measurement of pulmonary resistance (R_L), as shown in Figure 6.[22] Dynamic compliance (C_{dyn}) and respiratory rate (f) were also measured following agonist challenge. Those animals exposed for four months showed increased sensitivity to acetylcholine (i.e., the dose required to produce a 150% increase in R_L), and there was an increase in reactivity (i.e., the slope of dose vs. change in R_L) by eight months, with a leveling off of the response after this time. No changes in C_{dyn} or f were noted at any time. Thus, repeated exposures to H_2SO_4 resulted in the production of hyperresponsive airways in previously healthy animals. This has implications for the role of nonspecific irritants in the pathogenesis of airway disease.

Histology. In the study of Schlesinger et al.[19] in which rabbits were exposed to 250 $\mu g/m^3$ for four weeks and sacrificed two weeks later, histological examination showed increased numbers of secretory cells in

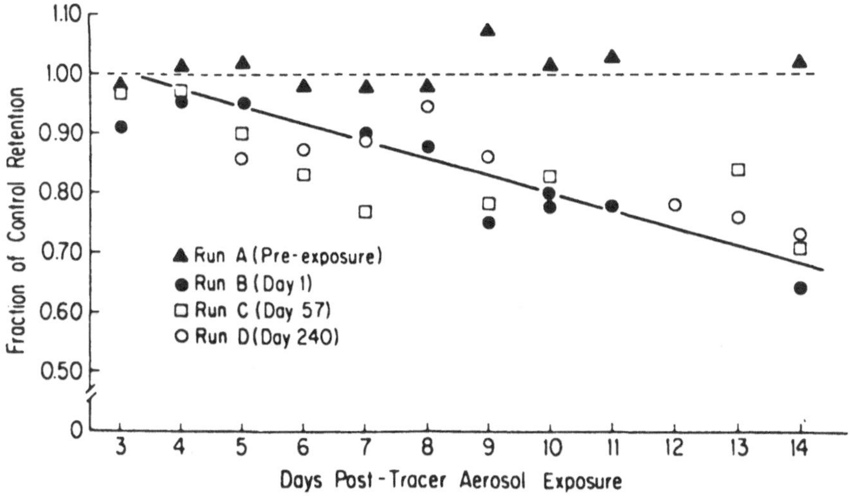

Figure 5. Retention of tracer particles during intermittent exposure to H_2SO_4, expressed as the ratio of retention in acid-exposed animals to that in sham control animals for each test run. The solid line represents the common regression of clearance runs B, C, and D. The dotted line represents the regression for run A. (Reproduced with permission from Schlesinger and Gearhart.[21])

distal airways and thickened epithelium in airways extending from mid-sized bronchi to terminal bronchioles. There were no corresponding changes in the trachea or other large airways. In the follow-up study,[23] in which rabbits received daily exposures for one year via the nose at 250 $\mu g/m^3$, the secretory cell density was elevated in some lung airways at four months and in all lung airways at eight months (Figure 7). At 12 months the increased density remained in small and midsized airways but not in large airways. Partial recovery was observed at three months after the last exposure.

In a study in which dogs were exposed daily for five years to 1100 $\mu g/m^3$ SO_2 plus 90 $\mu g/m^3$ H_2SO_4 and then allowed to remain in unpolluted air for two years, there were small changes in pulmonary functions during the exposures, which continued following the termination of exposure. Morphometric lung measurements made at the end of a two-year post-exposure period showed changes analogous to an incipient stage of human centrilobular emphysema.[24]

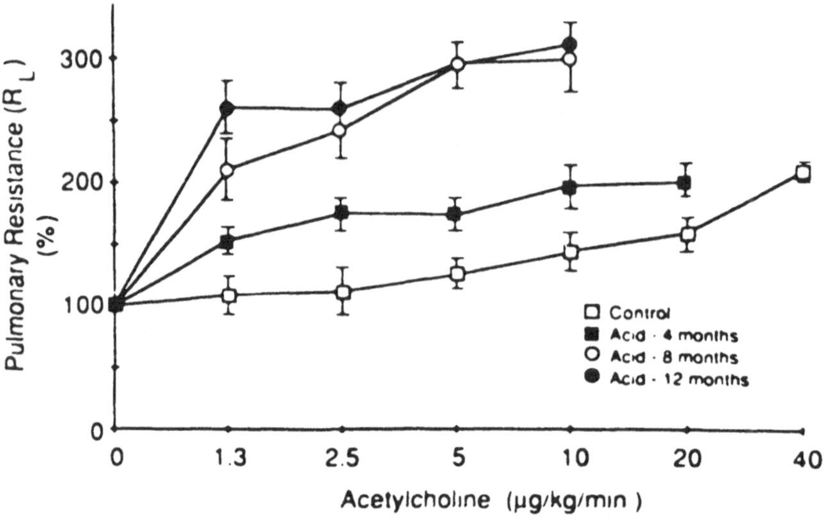

Figure 6. Effect of serial bronchoprovocation challenge on pulmonary resistance (R_L). The abscissa is expressed in terms of doubling doses of acetylcholine. Data are expressed as the group mean (± S.E.) percentage of baseline R_L at each dose (n = 12 for control; n = 4 for each acid group). (Reproduced with permission from Gearhart and Schlesinger.[22])

Effects on Humans

Acute Effects

Respiratory mechanical function. H_2SO_4 and other sulfates have been found to affect both sensory and respiratory function in humans. Respiratory effects from exposure to H_2SO_4 (350 to 500 μg/m³) have been reported to include increased respiratory rates and tidal volumes.[8,25] However, other studies of pulmonary function in nonsensitive healthy adult subjects indicated little effect on pulmonary mechanical function when subjects were exposed to submicrometer H_2SO_4 at 10 to 1000 μg/m³ for 10 to 120 minutes. In one study, the bronchoconstrictive action of carbachol was potentiated by 0.8-μm H_2SO_4 and other sulfate aerosols, more or less in relation to their acidity.[26]

Asthmatics are substantially more sensitive in terms of changes in pulmonary mechanics than healthy people, and vigorous exercise potentiates the effects at a given concentration. The lowest-demonstrated-effect level was 100 μg/m³ of 0.6-μm H_2SO_4 via mouthpiece inhalation in exercising adolescent asthmatics.[27] As shown in Figure 8, the effects were

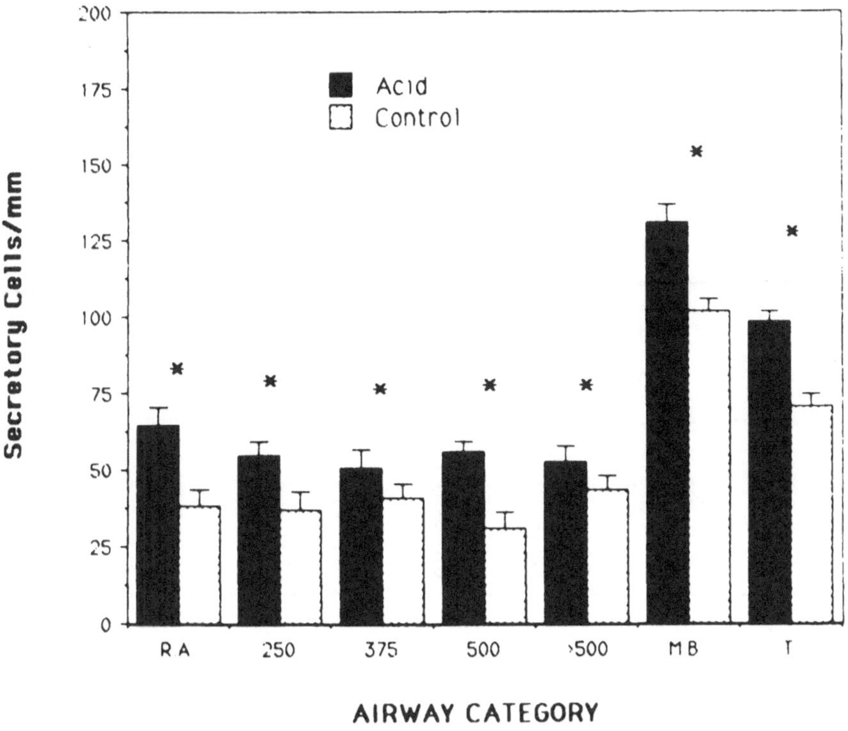

Figure 7. Epithelial secretory cell density in airways of various diameter ranges (in mm) after 8 months of daily 1-hr exposures (5 d/wk) to either 250 $\mu g/m^3$ H_2SO_4 or sham (control) exposures. R.A. = respiratory airways; M.B. = main bronchi; T = trachea.

relatively small and disappeared within about 15 minutes. Table I shows that in adult asthmatics undergoing similar protocols, the lowest-observed-effect level was 450 $\mu g/m^3$,[28] but in this case, the effects persisted over 24 hours (Figure 9).

The effects of acid fog droplets on respiratory function and symptoms have been studied by Linn et al.[29] They exposed both normal and mild asthmatic adult volunteers for 60 minutes to 8-μm MMAD fog droplets containing 0, 150 and 680 $\mu g/m^3$ of H_2SO_4, with alternating ten-minute periods of rest and heavy exercise. Both normals and asthmatics reported more symptoms with increasing concentration, and the asthmatics showed an increase in airway resistance at the higher acid concentration. There were no significant differences in either forced expiratory function

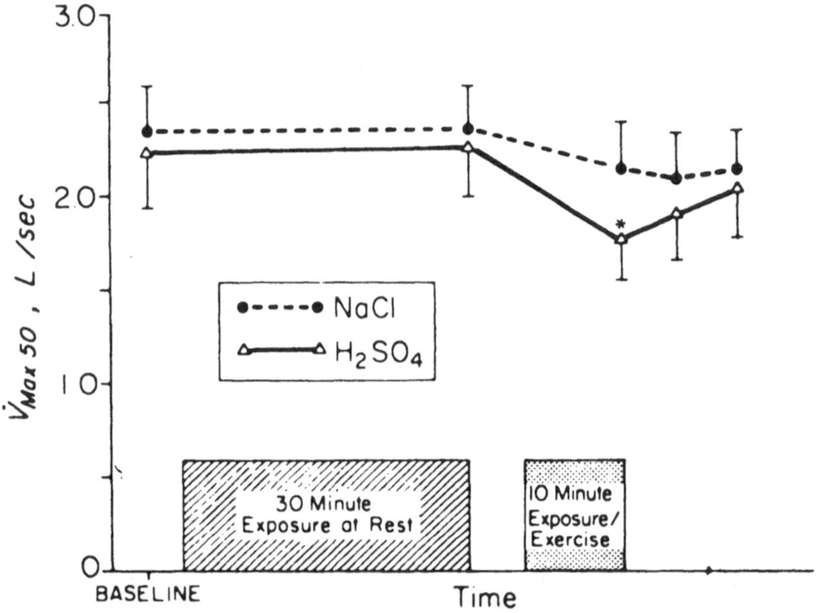

Figure 8. Changes in V_{max50} in asthmatic adolescent subjects after 30 min of exposure at rest followed by 10 min of exposure during moderate exercise; mean ± standard error; n = 10. *Statistically different from baseline. (Reproduced with permission from Koenig et al.[27])

or airway reactivity to methacholine between the sham and acid exposures.

Particle clearance function. In healthy nonsmoking adult volunteers exposed to 0.5 μm H_2SO_4 at rest at 100 $\mu g/m^3$ for one hour, there was an acceleration of bronchial mucociliary clearance of tracer particles which deposited primarily in the larger bronchial airways and a slowing of clearance when the exposure was raised to 1000 $\mu g/m^3$ (Figure 2).[30] For tracer particles which deposited primarily in midsized to small conducting airways, there was a small but significant slowing of clearance at 100 $\mu g/m^3$ H_2SO_4 and a greater slowing at 1000 $\mu g/m^3$.[31] These changes are consistent with the greater deposition of acid in midsized to smaller airways. Exposures to 100 $\mu g/m^3$ for two hours produced slower clearance than the same exposure for one hour, indicating a cumulative relationship to dose.[32]

TABLE I. Pulmonary Effects of Sulfuric Acid Aerosols in Asthmatics

Investigator	Design	Effects
Sackner et al.[68]	Six asthmatics inhaled 10, 100, 1000 $\mu g/m^3$ H_2SO_4 for 10 min orally at rest (MMAD = 0.1 μm)	No pulmonary function changes; no alteration in gas exchange
Avol et al.[69]	Six asthmatics inhaled 75 $\mu g/m^3$ H_2SO_4 for 2 hr in chamber with exercise (MMAD = 0.3 μm)	Two of 6 subjects showed "possibly meaningful changes in respiratory resistance"
Koenig et al.[27]	Ten adolescent asthmatics with exercise-induced bronchospasm inhaled 100 $\mu g/m^3$ H_2SO_4 for 40 min via mouthpiece with exercise (MMAD = 0.6 μm)	Significant reduction in V_{max}, FEV_1, and total respiratory resistance compared to NaCl exposure or pre-exposure
Utell et al.[70]	17 carbachol-sensitive asthmatics inhaled 100, 450, 1000 $\mu g/m^3$ H_2SO_4 for 16 min via mouthpiece at rest (MMAD = 1.1 μm)	100 $\mu g/m^3$ H_2SO_4 and NH_4HSO_4 caused significant decline in SG_{aw} and FEV_1; 450 $\mu g/m^3$ H_2SO_4 produced significant fall in airway conductance, no change after 100 $\mu g/m^3$ H_2SO_4
Utell et al.[71]	12 asthmatics inhaled 100, 450 $\mu g/m^3$ for 4 hr in chamber with exercise (MMAD = 0.8 μm)	Reduction in FEV_1 and SG_{aw} after 1 and 2 hr of 450 $\mu g/m^3$ inhalation; no change after 100 $\mu g/m^3$ H_2SO_4
Spektor et al.[34]	Ten asthmatics inhaled 100, 300, 1000 $\mu g/m^3$ for 1 hr via nasal mask at rest (MMAD = 0.5 μm)	100 $\mu g/m^3$ H_2SO_4 caused significant decline in flow rates and airway conductance in 6 of 10 asymptomatic asthmatics which persisted 3 hr after exposure. Four asthmatics showed large variability in response; no effect after 300 or 100 $\mu g/m^3$

The results of these studies have been used by Yu et al. to construct a model for the effects of surface deposition of acidic droplets on mucus transport velocity along the tracheobronchial airways.[33] Based on this model, mucus velocities are increased when less than about 10^{-7} g/cm^2 of H_2SO_4 is deposited, while clearance is retarded when the acid deposition exceeds this limit.

The effects of a one-hour inhalation of submicrometer H_2SO_4 aerosols

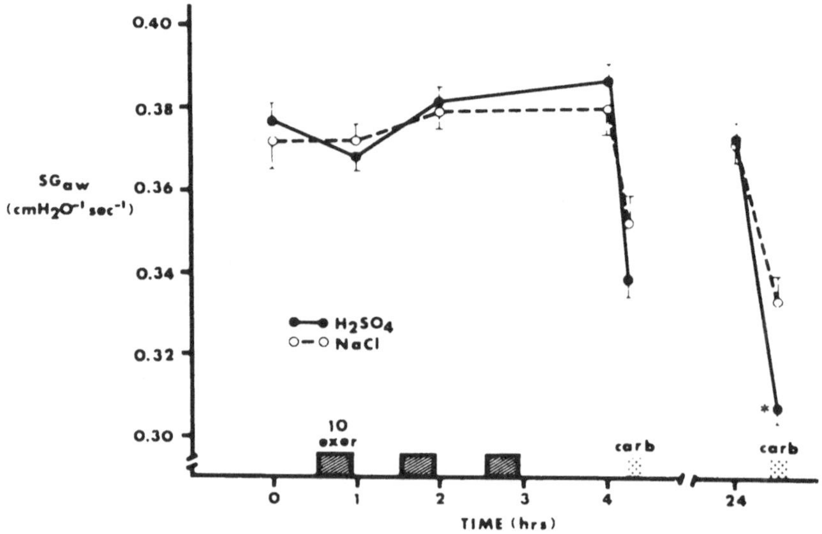

Figure 9. Mean values of specific airway conductance (SG_{aw}) before, during, immediately after and 24 hr after the 4-hr exposure to 450 $\mu g/m^3$ H_2SO_4 and NaCl aerosols. The prior inhalation of H_2SO_4 significantly potentiates ($p < 0.002$) the effect of barbachol on SG_{aw} 24 hr later compared to NaCl plus carbachol. Bars represent standard error of mean. (Reproduced, with permission, from Utell and Morrow.[28])

via nasal mask on tracheobronchial mucociliary particle clearance and respiratory mechanics were studied by Spektor et al. in subjects with asthmatic histories.[34] A brief inhalation of tagged aerosol preceded the one-hour H_2SO_4 or a sham exposure. Respiratory function was measured before, 15 minutes after, and three hours after the H_2SO_4 or sham exposure. After exposure to 1000 $\mu g/m^3$ of H_2SO_4, the six subjects not on routine medication exhibited a transient slowing of mucociliary clearance and also decrements in SG_{aw}, FEV_1, MMEF, and V_{25} ($p < 0.05$) in both sets of measurements. The four asthmatics on daily medication exhibited stepwise mucociliary clearance that was too variable to allow detection of any H_2SO_4 effect on clearance. Mucociliary clearance rates in both groups in the sham exposure tests were significantly slower than those of healthy nonsmokers studied previously by Leikauf et al. using the same protocols.[31] The extent of mucociliary clearance slowing following the 1000 $\mu g/m^3$ exposure in the nonmedicated subjects was similar to that in the healthy nonsmokers. This similar change, from a reduced baseline rate of clearance, together with the significant change in respiratory function, indicates that asymptomatic asthmatics may respond to H_2SO_4 exposures with functional changes of greater potential health significance than do healthy nonsmokers.

Effects of Longer-Term Exposure

Kitagawa identified sulfuric acid as the probable causal agent for approximately 600 cases of acute respiratory disease in the Yokkaichi area in central Japan between 1960 and 1969.[35] The patients' residences were concentrated within 5 km of a titanium dioxide plant with a 14-m stack, which emitted from 100,000 to 300,000 kg/month of H_2SO_4 in the period 1961–1967. The average concentration of SO_3 in February 1965 in Isozu, a village 1-2 km from the plant, was 130 $\mu g/m^3$, equivalent to 159 $\mu g/m^3$ of H_2SO_4. Kitagawa estimated that the peak concentrations might be up to 100 times as high with a north wind. Electrostatic precipitators were installed to control aerosol emissions in 1967, and after 1968 the number of newly found patients with "allergic asthmatic bronchitis" or "Yokkaichi asthma" gradually decreased. Although Kitagawa's quantitative estimates of exposure to H_2SO_4 and the criteria used to describe cases of respiratory disease may differ from current methods, the unique aspect of this report is the identification of H_2SO_4 as the likely causal agent for an excess in morbidity.

In an independent analysis of mortality from asthma and chronic bronchitis associated with changes in sulfur oxide air pollution in Yokkaichi from 1963 to 1983, Imai et al. correlated mortality with sulfation index (lead peroxide candle measurements) and focused on reductions in SO_x emissions from a petroleum refinery in the harbor area in 1972.[36] Thus, it is not clear from their analysis what the SO_2 or H_2SO_4 exposures to the population from these emissions were. In any case, mortality rates for bronchial asthma were significantly elevated in Yokkaichi in the period 1967–1970, and the mortality rates due to chronic bronchitis were significantly elevated for the periods 1967–1970 and 1971–1974. There was a greater lag between the reduction in SO_x pollution and reduction in mortality rate for chronic bronchitis than for bronchial asthma.

Other evidence of links between high concentrations of ambient H_2SO_4 and human health effects is more circumstantial. For example, H_2SO_4 concentrations in the ambient air were certainly much higher than current levels during the classic episodes in London, Meuse Valley, and Donora, but so were those of many other pollutants. Similarly, the decline in the prevalence of chronic bronchitis in the U.K. over the past three decades could have been due to the decline in any of several pollutants. However, on mechanistic grounds and known exposure-response relationships, H_2SO_4 is a more plausible candidate than SO_2, carbonaceous particles, or other known constituents.[37]

IMPLICATIONS OF THE EFFECTS OF ACIDIC AEROSOLS ON RESPIRATORY FUNCTION IN THE EXACERBATION OF ASTHMA AND CHRONIC BRONCHITIS

The studies of Utell et al. demonstrate that brief exposures to acidic aerosols reduce airway conductance in healthy humans and that asthmatic subjects are more sensitive than healthy individuals.[26,38] The lowest concentration that produced a significant response in the group as a whole was 450 $\mu g/m^3$. Koenig et al. reported a 40 percent increase in total airway resistance in a group of exercising asthmatic adolescents when they were exposed to 100 $\mu g/m^3$ of H_2SO_4.[27] The responses were similar to those reported by Koenig et al. for the same protocols and kinds of subjects for exposure to 0.5 ppm of SO_2 (1300 $\mu g/m^3$).[39,40]

Moderate exercise appears to enhance the response by increasing the dose of irritant delivered to epithelial surfaces. With increasing exercise, more pollutant is inhaled. The greater inspiratory flowrates also act to increase the percentage of the highly soluble SO_2 vapor which can penetrate through the upper airways into those bronchial airways where reflex responses are most likely to be initiated. The greater flowrate also produces a thinner boundary layer around the airway bifurcations, enhancing "hot spots" of deposition of particles and vapors from the airstream. Thus, exercise results in increased deposition in this region for the submicrometer sized H_2SO_4 droplets which would have minimal deposition in such airways at lower flowrates.

The irritant dose delivered to the larger bronchial airways is greater in asthmatics and bronchitics than in healthy individuals because the former groups have airways with smaller diameters. This may account for some, or perhaps all, of the greater responsiveness of asthmatics to inhaled irritants. They may also have a greater responsiveness at the sites of deposition to the delivered dose, but this has not been clearly established in *in vivo* tests. In any case, an irritant-induced narrowing of the conducting airways of the lung can increase the surface deposition of subsequently inhaled irritant, resulting in further airway constriction.

The subjective responses to inhaled acid aerosols may include a feeling of chest tightness, and the work of breathing is increased. For individuals with chronic respiratory disease, any increment of work in breathing may be considered an adverse effect. For asthmatic individuals, the major concern is the induction of bronchospasm. The few clinical laboratory studies on carefully selected asthmatics cannot be expected to generate data on the exact conditions that provoke bronchospasm and acute respiratory insufficiency. It would be highly desirable to have an animal model for bronchial asthma so that this important issue could be systematically studied.

IMPLICATIONS OF THE EFFECTS OF ACIDIC AEROSOLS ON MUCOCILIARY CLEARANCE IN THE PATHOGENESIS OF CHRONIC BRONCHITIS

In three series of repetitive daily one-hour rabbit exposures to H_2SO_4 for four weeks, Schlesinger et al. found an acceleration of group mean clearance from the tracheobronchial tree on days in which the animals were exposed.[19] It occurred to some extent during the period of acid aerosol exposures and, to a greater extent, during the post-exposure period. The oral inhalation series at 250 $\mu g/m^3$ also showed an increased variability in clearance times. In the follow-up studies extending the daily rabbit exposures out to one year, mucociliary clearance was measured at one day after the last exposure. During these studies, there was greater variability than in the pre-exposure control or sham exposure tests in almost all animals; some individual animals showed accelerated clearance, while the group as a whole showed persistently slowed clearance.[20] In the group followed for three months after the last exposure, the mean clearance was slower during the exposures and was further slowed after the end of the exposure (Figure 4). These results are consistent with the study by Schlesinger et al. in which four donkeys were exposed for one hour per day, five days per week for six months to submicrometer H_2SO_4 at approximately 100 $\mu g/m^3$.[18] In all animals, clearance rates become very variable within the first few weeks of exposure. In two donkeys, clearance times fell consistently within the normal range during the last few months of exposure. However, after the end of the exposures, these two animals exhibited rates which were significantly faster than those occurring before the exposure series began (Figure 3), similar to the observations in the rabbits after 20 days of exposure.[41] The altered clearance rates during and after the exposure period may be an adaptive response of the mucociliary system to acid exposures. On the other hand, they may be early stages in the progression toward more serious dysfunctions, e.g., those found in chronic bronchitis, which may result from continued irritant exposures. The implications of chronic acid aerosol exposure to the pathogenesis of chronic bronchitis have been discussed by Schlesinger et al.[19] The following is a summary which reflects the knowledge gained from recent studies.

Chronic bronchitis is a disease of the conducting airways, characterized by a persistent production of excess mucus.[42,43] In addition, human bronchitics and experimental animals having spontaneous or induced chronic bronchitis show altered mucociliary clearance function.[44-47] Thus, chronic bronchitis involves dysfunction of the mucociliary system, and altered clearance may be an initial stage in disease progression. Retarded mucociliary clearance has been demonstrated in bronchitics who showed no sign of airway obstruction, while young smokers having various degrees of impairment of tracheal mucus transport rates had no overt bronchitic symptoms and had normal pulmonary function, including tests of airway obstruction.[48,49]

Unfortunately, there are few data concerning the response of the human mucociliary clearance system under prolonged insult by potentially harmful pollutants such as H_2SO_4. The most direct evidence for an association between chronic bronchitis and exposure to H_2SO_4 comes from occupational exposures, but these were at high levels. Williams observed an excess incidence of chronic bronchitis in workers occupationally exposed to H_2SO_4 levels above 1 mg/m^3 (probable diameter = 14 μm); however, the excess was actually in increased incidence of episodes in affected workers, rather than an increase in the number of workers affected.[11]

While available evidence suggests that exposure to H_2SO_4 may exacerbate disease, it has not been clearly established whether it can initiate it. Some limited evidence indicates that it can. For example, in two previously healthy human subjects, Sim and Pattle found the development of what appeared to be long lasting symptoms of bronchitis due to repeated exposure to H_2SO_4 lasting one hour and given no more than twice a week, with at least 24 hours between exposures;[50] however, concentrations were high, ranging from 3 to 39 mg/m^3.

The suggestion for a role of H_2SO_4 in the development of chronic bronchitis is given added strength when results of studies of submicrometer H_2SO_4 or whole fresh cigarette smoke exposures, both conducted in the New York University laboratory with animals and humans, are compared.[41] Cigarette smoke is an agent known to be involved in the etiology of human chronic bronchitis. As shown in Figure 10, the effects of both agents on the mucociliary clearance of tracer particles are essentially the same in terms of: 1) transient acceleration of clearance in single low dose exposures, and 2) transient slowing following single high dose exposures. Furthermore, there are alterations in clearance rates persisting for several

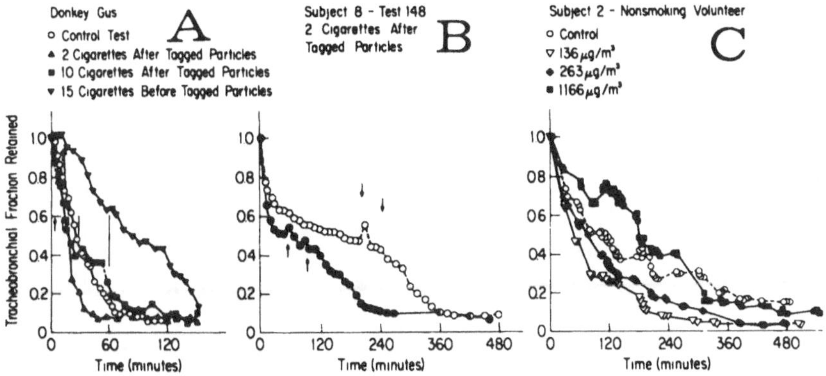

Figure 10. A: Tracheobronchial particle retention vs. time for donkey Gus in a control test and in tests involving exposure to whole fresh cigarette smoke from the indicated number of cigarettes. The dashed lines indicate the period of smoke exposure. B: Tracheobronchial particle retention vs. time for a 38-year-old nonsmoking man for two tagged aerosols inhaled 2.5 hr apart on the same day. Smoke from two cigarettes, which was inhaled beginning 1 hr after the inhalation of the second tagged aerosol, accelerated the clearance of both. C: Tracheobronchial particle retention vs. time for the same man as in Figure 10B, but 7 yr later. In the tests, he was exposed during the first hour to 0.5 μm H_2SO_4 via nasal mask at four different concentrations, i.e., 0 (sham or control), 136, 263, and 1166 $\mu g/m^3$. (Reproduced with permission from Lippmann et al.[41])

months following multiple exposures to both agents (Figures 3, 4, and 11). Thus, although direct evidence for an association between intermittent low level exposures to H_2SO_4 and chronic bronchitis is lacking, the similarity in response between H_2SO_4 and cigarette smoke exposures suggests that such an association is possible.

Human chronic bronchitis is a clinically diagnosed disease, but one which is characterized by certain morphological changes associated with these clinical symptoms.[45,51-54] One of the basic stigmata is an increase in the number and/or size of epithelial mucus secretory cells in both proximal bronchi as well as in peripheral airways where such cells are normally absent or few in number; this change is accompanied by an increase in the volume of secretion.[51] In the Schlesinger et al. subchronic rabbit studies and the Gearhart and Schlesinger chronic studies, an increase in epithelial secretory cell proportions in smaller airways was noted in all series.[19,23]

An increase in the relative amount of epithelial secretory cells may be

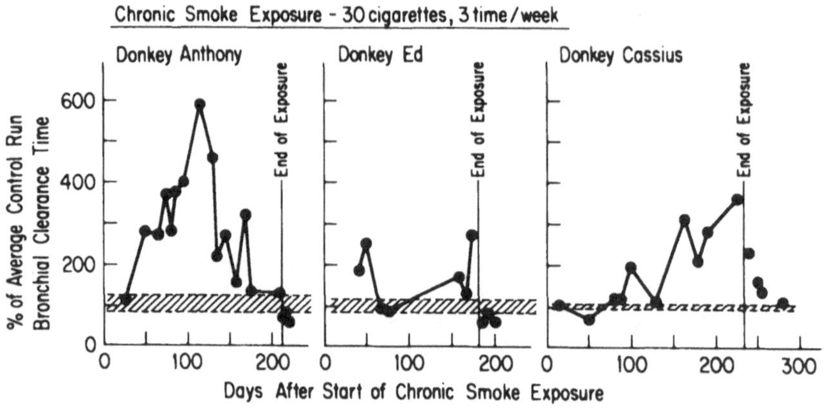

Figure 11. Effect of exposures to the whole fresh smoke from 30 cigarettes, three times/week, on the mean residence time for tagged particles on the tracheobronchial airways in three donkeys. The dashed lines indicate the range of the three control tests for each animal which preceded the smoke exposures. (Reproduced with permission from Lippmann et al.[41])

the initial response of the mucociliary system to acid exposure. Repeated or prolonged exposures of various experimental animals to other common irritants, e.g., SO_2 and cigarette smoke, have been shown to result in proliferation of secretory cells and increased production of mucus, often as the first response.[51,55-59] In rabbits, exposure to formaldehyde for three hours per day 50 days resulted in an increase in the population of mucus cells in both bronchi and bronchioles.[60] In another study using rabbits, the first change noted after exposure to synthetic smog (1.5 ppm total oxidant) was an increase in the number and secretory activity of goblet cells, which progressed until there was large-scale replacement of the epithelium by mucus-secreting cells.[61] In experimental bronchitis induced in rabbits by immunologic stimuli, epithelial secretory cell hyperplasia was observed extending into bronchioles.[53]

Hamsters exposed to H_2SO_4 at 1.1 mg/m^3 for three hours by Schiff et al. showed a marked increase in PAS positive material (an indication of mucus glycoproteins) in tracheal explants prepared 70 hours after exposure.[62] Consistent results were found in the Gearhart and Schlesinger study, where the rabbits exposed to H_2SO_4 had increased alcian blue staining of the mucus secretory cells.[23] Bronchiolar epithelial hyperplasia was observed in cynomolgus monkeys exposed to 0.38 mg/m^3 H_2SO_4

(2.15 μm) for 78 weeks, but secretory cells were not specifically examined in this study.[7]

In the Schlesinger et al. subchronic study, histological changes in the airways were still present two weeks after the last exposure.[19] In the Gearhart and Schlesinger chronic study, such changes were reduced in extent, but they were still present three months after the last exposure.[23] The persistence of histological effects during this time is consistent with studies using other irritants. For example, Reid exposed rats to 300–400 ppm SO_2 for five hours per day, five days per week for up to six weeks.[51] After three to four weeks of exposure, goblet cells in exposed animals were found further in the periphery than in controls. Three months after exposures ended, these goblet cells were found to persist, although there was some degree of reversal.

The appearance of persistently increased secretory cell numbers in peripheral airways due to H_2SO_4 is a finding of major importance since excessive mucus production in small airways, which is consistent with an increase in the propagation of secretory cells, may be an early feature in the pathogenesis of bronchitis.[63] Furthermore, it demonstrates an underlying histological change consistent with the observed physiological effects of the H_2SO_4, i.e., altered mucociliary clearance.

In addition to a change in the relative number of secretory cells in different airway levels of acid-exposed animals in the study by Schlesinger et al., two other changes were noted after H_2SO_4 exposures.[19] There was an increase in epithelial thickness and a decrease in airway diameter. A significant increase in epithelial thickness of small bronchi and bronchioles occurred in rabbits exposed orally at approximately 250 $\mu g/m^3$ and nasally at approximately 500 $\mu g/m^3$. In addition, in the oral exposure series, the lumen diameter of the smallest airways was significantly less than in the sham controls.

In human chronic bronchitis and in experimental bronchitis in laboratory animals, an initial change in secretory cell number or size is followed by intrabronchial narrowing, especially in small bronchi and bronchioles; in part, this is due to a thickening of the bronchial wall.[64,65] For example, in rabbits exposed to formaldehyde, thickening of airway walls occurred after the initial mucus cell hyperplasia.[53] In rats exposed to cigarette smoke, thickening of epithelium and increased goblet cell numbers were the only histological changes observed in the bronchial tree.[56]

In summary, the first stage of effect of acid exposure may be a change in secretory cell proportions in the airways; thickening of the epithelium may occur later. Thus, the studies of Schlesinger et al. and Gearhart and Schlesinger provide further support for the role of H_2SO_4 in the pathogenesis of chronic bronchitis, via effects on the mucociliary clearance system.[19,23] However, the progression of clearance dysfunction in the pathogenesis of chronic bronchitis is not known.

The Albert et al. schema describing the pathogenesis of chronic bronchitis in man due to cigarette smoking is illustrated in Figure 12.[66] It may also apply to repeated exposures to other irritants, such as H_2SO_4. According to this schema, irritant inhalation initially results in a tendency toward some acceleration of clearance as excess mucus is produced but mucosal damage has not occurred. The H_2SO_4 dose delivered to the rabbit in nasal breathing at 250 $\mu g/m^3$ may have been enough to initiate this first stage. An increase in the number of airways containing epithelial secretory cells is consistent with increased mucus production. However, the degree of clearance rate change could vary with the individual rabbit and with the time after exposure at which the clearance was measured. This may account for the fact that a significant acceleration was often observed when clearance was measured immediately after H_2SO_4 exposure, while retardation was more commonly observed when clearance was measured one day after the last H_2SO_4 exposure.[20]

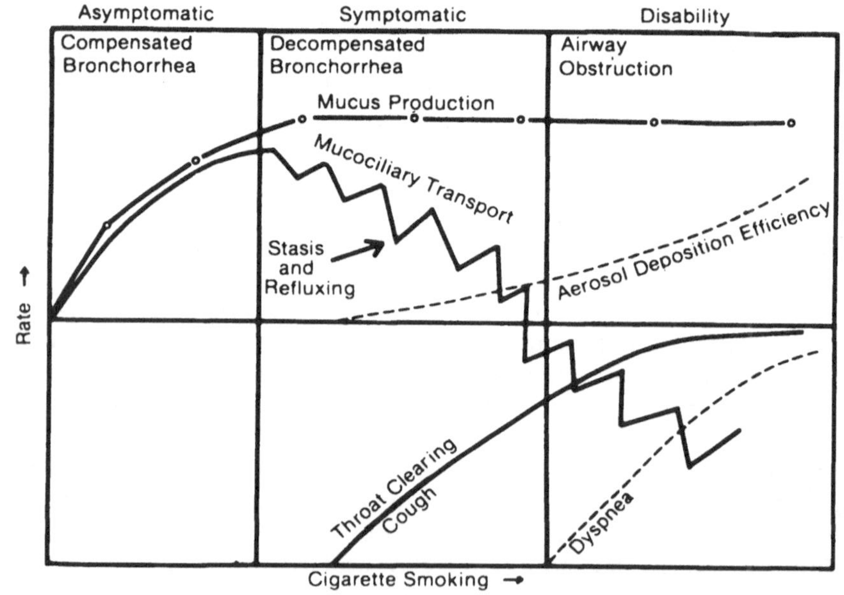

Figure 12. Tentative schema for the pathogenesis of obstructive lung diseases resulting from exposure to inhaled airborne irritants. (Reproduced with permission from Albert et al.[66])

In the next stage of the Albert et al. schema, a further increase in the level of secretion, coupled with some mucosal damage, results in an overloading of transport mechanisms; the result is a retardation of clearance. Such a retardation was observed in the study by Schlesinger et al., involving six months of daily one-hour H_2SO_4 exposures at 100 $\mu g/m^3$ in donkeys; it was also noted in the study of Gearhart and Schlesinger involving daily one-hour H_2SO_4 exposures in rabbits for one year.[18,20]

Since H_2SO_4 produces essentially the same sequence of effects on mucociliary bronchial clearance as cigarette smoke, both following short-term and chronic exposures, it may be capable of contributing to the development of bronchitis. But the question still remains whether variable clearance rates and persistent clearance rate changes merely predispose to chronic bronchitis or are the actual initiating events in a pathogenic sequence leading to its development. Furthermore, the response of the mucociliary clearance system observed in the rabbits may be adaptive, rather than pathological. Many irritants may stimulate clearance at low doses or after exposure for a short time, and then retard it at higher doses or with prolonged exposures.[67] An increase in secretory cell proportions is consistent with hypersecretion. Thus, low level exposures may initially increase secretion, which can be coped with, and may even be protective. However, pathological changes appear when adaptive capacity is overloaded. Thus, with increasing exposure time or dose, the degree of enhanced secretion may be too great, resulting in overwhelming of clearance, leading to retardation and, eventually, bronchitis.[19]

RECOMMENDATION FOR REVISED TLV

Criteria for Size Selection

The human health effects of major concern with respect to the inhalation of H_2SO_4 are bronchospasm in asthmatics and chronic bronchitis in all exposed persons. The former relates to acute exposure, while the latter can be related more closely to chronic or cumulative exposures. In either case, the effects are produced by droplets depositing on the surface of the conductive airways of the lungs. In view of this, Thoracic Particulate Matter (TPM) is the appropriate aerosol fraction for a particle size-selective TLV.

Concentration Limits

The data related to the provocation of bronchospasm are difficult to apply to the selection of a TLV. Evidence for increased airway resistance in exercising mild to moderate asthmatics has been reported at 100 mg/m^3 in adolescents (0.6 μm), at 450 $\mu g/m^3$ in adults (0.8 μm), and at 680 $\mu g/m^3$ in adults (8.0 μm). While the increases in airway resistance were small on average, the populations studied were small, and some subjects had much greater responses than the average. Also, the populations were carefully selected and did not include the more unstable and potentially more reactive asthmatics in the population. On the other hand, the more highly reactive asthmatics are unlikely to be employed in industrial jobs involving potential exposure to H_2SO_4.

The data related to the induction of chronic bronchitis are limited by the paucity of relevant data on humans. The reports of Williams that large droplet H_2SO_4 above 1 mg/m^3 increased the frequency of episodes of bronchitis, and that of Sim and Pattle in which bronchitic symptoms were produced after a few scattered one-hour exposures above 3 mg/m^3, are useful in indicating likely causality, but they are of limited value for quantitative risk estimation.[11,50] A firmer mechanistic basis for linkage between chronic exposure to H_2SO_4 and the pathogenesis of chronic bronchitis lies in the series of studies involving chronic exposures of animals and persistent histological alterations in lung structure. These structural changes in the rabbit model have correlates in terms of clearance function changes. These, in turn, are indicative of changes in mucus secretion leading to mucus stasis, a hallmark of bronchitic disease.

The animal studies can be related to human responses in two ways. One is the concordance in functional and morphometric responses of animals to H_2SO_4 and cigarette smoke, a known causal factor for human chronic bronchitis. The other is that humans, rabbits, and donkeys all have essentially the same transient mucociliary clearance function responses to single one-hour exposures to H_2SO_4. The fact that daily one-hour H_2SO_4 exposures in rabbits and donkeys produce persistent changes in clearance function makes it highly likely that they would also produce such changes in similarly exposed humans. Furthermore, a comparison of the human and rabbit responses to single exposures indicates that humans respond at lower concentrations than rabbits.[32]

The effects of H_2SO_4 on the airways are very likely to be cumulative, during each exposure day, at least in part. Thus, the daily one-hour exposures at 250 $\mu g/m^3$ in the rabbits may be equivalent to < 50 $\mu g/m^3$ for a seven- to eight-hour day and to a still lower concentration for equivalent effects in humans. On the other hand, the effects produced by the one-year series of exposures in the rabbits were less severe than the condition corresponding to a clinical diagnosis of chronic bronchitis in humans.

While further experimental inhalation and epidemiological studies are needed to provide a more definitive basis for a TLV, it appears almost certain that the current TLV of 1 mg/m³ does not provide an adequate margin of protection against either bronchospasm in asthmatics or chronic bronchitis in the general working population. A reduction in the TLV, expressed as a TPM-TLV, to 100 μg/m³ is recommended at this time.

REFERENCES

1. *Ann. Am. Conf. Govt. Ind. Hyg.*, Vol. 9, *Threshold Limit Values—Discussion and Thirty-five Year Index with Recommendations.* M.E. LaNier, Ed. American Conference of Governmental Industrial Hygienists, Cincinnati, OH (1984).
2. National Institute for Occupational Safety and Health: *Pocket Guide to Chemical Hazards.* DHHS (NIOSH) Pub. No. 85-114, U.S. Government Printing Office, Washington, DC (1985).
3. Deutsche Forschungsgemeinschaft: *Maximum Concentrations at the Workplace and Biological Tolerance Values for Working Materials—1985.* Report No. XXI. VCH Verlagsgesellschaft, Weinheim, FRG (1985).
4. United Nations Environment Programme: *Maximum Allowable Concentrations and Tentative Safe Exposure Levels of Harmful Substances in the Environmental Media, Moscow.* Centre of International Projects. Ghent, Belgium (1984).
5. American Conference of Governmental Industrial Hygienists: *Documentation of the Threshold Limit Values and Biological Exposure Indices*, 5th ed. Cincinnati, OH (1986).
6. Amdur, M.O., M. Dubriel and D. Creasia: Respiratory Response of Guinea Pigs to Low Levels of Sulfuric Acid. *Environ. Res. 15*:418–423 (1978).
7. Alarie, Y., W.M. Busey, A.A. Krumm and C.E. Ulrich: Long-term Continuous Exposure to Sulfuric Acid Mist in Cynomolgus Monkeys and Guinea Pigs. *Arch. Environ. Health 27*:16–24 (1973).
8. Amdur, M.O., L. Silverman, and P. Drinker: Inhalation of H_2SO_4 Mist by Human Subjects. *Arch. Ind. Hyg. Occup. Med. 6*:305–313 (1952).
9. Malcolm, D. and E. Paul: Erosion of the Teeth Due to Sulphuric Acid in the Battery Industry. *Br. J. Ind. Med. 18*:63–69 (1961).
10. National Institute for Occupational Safety and Health: *Review and Evaluation of Recent Literature—Occupational Exposure to Sulfuric Acid.* DHHS (NIOSH) Pub. No. 82-104. U.S. Government Printing Office, Washington, DC (1981).

11. Williams, M.K.: Sickness, Absence and Ventilatory Capacity of Workers Exposed to Sulphuric Acid Mist. *Br. J. Ind. Med. 27*:61–66 (1970).
12. Lippmann, M., D.B. Yeates and R.E. Albert: Deposition, Retention, and Clearance of Inhaled Particles. *Br. J. Ind. Med. 37*:337–362 (1980).
13. Larson, T.V., D.S. Covert, R. Frank and R.J. Charlson: Ammonia in the Human Airways: Neutralization of Inspired Acid Sulfate Aerosols. *Science 197*:161–163 (1977).
14. Holma, B.: Influence of Buffer Capacity and pH-dependent Rheological Properties of Respiratory Mucus on Health Effects Due to Acidic Pollution. *Sci. Toxic Environ. 41*:101–123 (1985).
15. Environmental Protection Agency: *Air Quality Criteria for Particulate Matter and Sulfur Oxides*, Volume III. EPA-600/8-82-029c. Research Triangle Park, NC (1982).
16. Amdur, M.O., A.F. Sarofim, M. Neville et al.: Coal Combustion Aerosols and SO_2: An Interdisciplinary Analysis. *Environ. Sci. Technol. 20*:138–145 (1986).
17. Schlesinger, R.B.: Effects of Inhaled Acids on Respiratory Tract Defense Mechanisms. *Environ. Health Perspect. 63*:25–38 (1985).
18. Schlesinger, R.B., M. Halpern, R.E. Albert and M. Lippmann: Effect of Chronic Inhalation of Sulfuric Acid Mist Upon Mucociliary Clearance from the Lungs of Donkeys. *J. Environ. Pathol. Toxicol. 2*:1351–1367 (1979).
19. Schlesinger, R.B., B.D. Naumann and L.C. Chen: Physiological and Histological Alterations in the Bronchial Mucociliary Clearance System of Rabbits Following Intermittent Oral or Nasal Inhalation of Sulfuric Acid Mist. *J. Toxicol. Environ. Health 12*:441–465 (1983).
20. Gearhart, J.M. and R.B. Schlesinger: Effects of Intermittent Exposure to Sulfuric Acid Aerosol on Mucociliary Clearance in Rabbits. *Toxicologist* (in press).
21. Schlesinger, R.B. and J.M. Gearhart: Early Alveolar Clearance in Rabbits Intermittently Exposed to Sulfuric Acid Mist. *J. Toxicol. Environ. Health 17*:213–220 (1986).
22. Gearhart, J.M. and R.B. Schlesinger: Sulfuric Acid-induced Airway Hyperresponsiveness. *Fund. Appl. Toxicol. 7*:681–689 (1986).
23. Gearhart, J.M. and R.B. Schlesinger: A Morphometric Analysis of Rabbit Airways After Repeated Exposures to Sulfuric Acid Aerosol. *Toxicologist 7* (in press).
24. Stara, J.F., D.L. Dungworth, J.G. Orthoefer and W.S. Tyler: *Long-term Effects of Air Pollutants in Canine Species*. EPA-600/8-80-014. U.S. Environmental Protection Agency, Research Triangle Park, NC (1980).
25. Ericsson, G. and P. Camner: Health Effects of Sulfur Oxides and Particulate Matter in Ambient Air. *Scand. J. Work Environ. Health 9*:52 (1983).
26. Utell, M.J., P.E. Morrow and R.W. Hyde: Airway Reactivity to Sulfate and Sulfuric Acid Aerosols in Normal and Asthmatic Subjects. *J. Air Poll. Contr. Assoc. 34*:931–935 (1984).
27. Koenig, J.Q., W.E. Pierson and M. Horike: The Effects of Inhaled Sulfuric

Acid on Pulmonary Function in Adolescent Asthmatics. *Am. Rev. Respir. Dis. 128*:221-225 (1983).
28. Utell, M.J. and P.E. Morrow: Effects of Inhaled Acid Aerosols on Human Lung Function. *Aerosols*, pp. 671-681. S.D. Lee, T. Schneider, L.D. Grant and P.J. Verkerk, Eds. Lewis Publishers, Chelsea, MI (1986).
29. Linn, W.S., E.L. Avol, D.A. Shamoo et al.: Respiratory Effects of Acid Fog Exposure in Normal and Asthmatic Volunteers. *Am. Rev. Resp. Dis. 133*:A214 (1986).
30. Leikauf, G., D.B. Yeates, K.A. Wales et al.: Effects of Sulfuric Acid Aerosol on Respiratory Mechanics and Mucociliary Particle Clearance in Healthy Nonsmoking Adults. *Am. Ind. Hyg. Assoc. J. 42*:273-282 (1981).
31. Leikauf, G.D., D.M. Spektor, R.E. Albert and M. Lippmann: Dose Dependent Effects of Submicrometer Sulfuric Acid Aerosol on Particle Clearance from Ciliated Human Lung Airways. *Am. Ind. Hyg. Assoc. J. 45*:285-292 (1984).
32. Schlesinger, R.B.: The Effects of Inhaled Acids on Lung Defenses. *Aerosols*, pp. 617-635. S.D. Lee, T. Schneider, L.D. Grant and P.J. Verkerk, Eds. Lewis Publishers, Chelsea, MI (1986).
33. Yu, C.P., J.P. Hu, B.M. Yen et al.: Models for Mucociliary Particle Clearance in Lung Airways. *Aerosols*, pp. 569-578. S.D. Lee, T. Schneider, L.D. Grant and P.J. Verkerk, Eds. Lewis Publishers, Chelsea, MI (1986).
34. Spektor, D.M., G.D. Leikauf, R.E. Albert and M. Lippmann: Effects of Submicrometer Sulfuric Acid Aerosols on Mucociliary Transport and Respiratory Mechanics in Asymptomatic Asthmatics. *Environ. Res. 37*:174-191 (1985).
35. Kitagawa, T.: Cause Analysis of the Yokkaichi Asthma Episode in Japan. *J. Air Poll. Contr. Assoc. 34*:743-746 (1984).
36. Imai, M., K. Yoshida and M. Kitabatake: Mortality from Asthma and Chronic Bronchitis Associated with Changes in Sulfur Oxides Air Pollution. *Arch. Environ. Health 41*:29-35 (1986).
37. Lippmann, M.: Airborne Acidity: Estimates of Exposure and Health Effects. *Environ. Health Persp. 63*:62-70 (1985).
38. Utell, M.J., P.E. Morrow and R.W. Hyde: Comparison of Normal and Asthmatic Subjects' Response to Sulfate Pollutant Aerosols. *Ann. Occup. Hyg. 26*:691-697 (1982).
39. Koenig, J.Q., W.E. Pierson, M. Horike and R. Frank: Effects of Inhaled SO_2 Plus NaCl Aerosol Combined with Moderate Exercise on Pulmonary Function in Asthmatic Adolescents. *Environ. Res. 25*:340-348 (1981).
40. Koenig, J.Q., W.E. Pierson, M. Horike and R. Frank: A Comparison of the Pulmonary Effects of 0.5 ppm versus 1.0 ppm Sulfur Dioxide Plus Sodium Chloride Droplet in Asthmatic Adolescents. *J. Toxicol. Environ. Health 11*:129-139 (1983).
41. Lippmann, M., R.B. Schlesinger, G. Leikauf et al.: Effects of Sulphuric Acid Aerosols on Respiratory Tract Airways. *Ann. Occup. Hyg. 26*:677-690 (1982).

42. Thurlbeck, W.M.: *Chronic Airflow Obstruction in Lung Disease.* W.B. Saunders, Philadelphia (1976).
43. Snider, G.L.: Pathogenesis of Emphysema and Chronic Bronchitis. *Med. Clin. MA 65*:647–665 (1981).
44. Holma, B.: Lung Clearance of Mono- and Di-disperse Aerosols Determined by Profile Scanning and Whole-body Counting. A Study on Normal and SO_2 Exposed Rabbits. *Acta. Med. Scand. 473*: (Suppl.) (1967).
45. Lourenco, R.V.: Distribution and Clearance of Aerosols. *Am. Rev. Respir. Dis. 101*:460–461 (1969).
46. Iravani, J. and A. Van As: Mucus Transport in the Tracheobronchial Tree of Normal and Bronchitic Rats. *J. Pathol. 106*:81–93 (1972).
47. Melville, G.M., S. Ismail and C. Sealy: Tracheobronchial Functions in Health and Disease. *Respiration 40*:329–336 (1980).
48. Mossberg, B. and P. Camner: Impaired Mucociliary Transport as a Pathogenetic Factor in Obstructive Pulmonary Diseases. *Chest 77*: (Suppl.) 265 (1980).
49. Goodman, R.M., B.M. Yergin, J.F. Landa et al.: Tracheal Mucous Velocity (TMV) in Non-smokers, Smokers and Patients with Obstructive Lung Disease. *Fed. Proc. 36*: (Abstr.) 607 (1977).
50. Sim, V.M. and R.E. Pattle: Effect of Possible Smog Irritants on Human Subjects. *J. Am. Med. Assoc. 165*:1908–1913 (1957).
51. Reid, L.: An Experimental Study of the Hypersecretion of Mucus in the Bronchial Tree. *Br. J. Exp. Pathol. 44*:437–445 (1963).
52. Mitchell, R.S.: Clinical and Morphologic Correlations in Chronic Airway Obstruction. *Am. Rev. Respir. Dis. 97*:54–62 (1967).
53. Suhs, R.H., J.L. Lumeng and M.H. Leppe: An Experimental Immunologic Approach to the Induction and Perpetuation of Chronic Bronchitis. *Arch. Environ. Health 18*:564–573 (1969).
54. Jefferey, P.K.: The Effects of Irritation on the Structure of Bronchial Epithelium. *The Lung in the Environment*, pp. 303–313. G. Bonsignore and G. Cummings, Eds. Plenum Press, New York (1982).
55. Lamb, D. and L. Reid: Mitotic Rates, Goblet Cell Increase and Histochemical Changes in Mucus Rat Bronchial Epithelium During Exposures to Sulfur Dioxide. *J. Pathol. Bacteriol. 96*:97–111 (1968).
56. Lamb, D. and L. Reid: Goblet Cell Increase in Rat Bronchial Epithelium After Exposure to Cigarette and Cigar Tobacco Smoke. *Br. Med. J. i*:33–35 (1969).
57. Jones, R., P. Boldue and L. Reid: Protection of Rat Bronchial Epithelium Against Tobacco Smoke. *Br. Med. J. ii*:142–144 (1972).
58. Mawdesley-Thomas, L.E., P. Healey and D.H. Barry: Experimental Bronchitis in Animals Due to Sulfur Dioxide and Cigarette Smoke. An Automated Quantitative Study. *Inhaled Particles III*, Vol. 1, pp. 509–525. W.H. Walton, Ed. Unwin Bros., London (1971).
59. Spicer, S.S., L.W. Chakrin and J.B. Wardell: Effect of Chronic Sulfur

Dioxide Inhalation on the Carbohydrate Histochemistry and Histology of the Canine Respiratory Tract. *Am. Rev. Respir. Dis.* *110*:13-24 (1974).

60. Ionescu, J., D. Marinescu, V. Tapu and A. Eskenasy: Experimental Chronic Obstructive Lung Disease. I. Bronchopulmonary Changes in Rabbits by Prolonged Exposure to Formaldehyde. *Morphol. Embryol.* *24*:233-242 (1978).
61. Falk, H.L., P. Koton and W. Rowlette: The Response of Mucus Secretory Epithelium Mucus to Irritants. *Ann. N.Y. Acad. Sci.* *130*:583-608 (1966).
62. Schiff, L.J., M.M. Byrne, J.D. Fenters et al.: Cytotoxic Effects of Sulfuric Acid Mist, Carbon Particulates and Their Mixtures on Hamster Tracheal Epithelium. *Environ. Res.* *19*:359-354 (1979).
63. Hogg, J.C., P.T. Macklem and W.M. Thurlbeck: Site and Nature Airway Obstruction in Obstructive Lung Disease. *New Eng. J. Med.* *278*:1355-1360 (1968).
64. Matsuba, K. and W.M. Thurlbeck: Disease of the Small Airways in Chronic Bronchitis. *Am. Rev. Respir. Dis.* *107*:552-558 (1973).
65. McKenzie, H.I., M. Glick and K.G. Outhred: Chronic Bronchitis in Coal Miners: Antemortem/post-mortem Correlations. *Thorax* *24*:527-535 (1969).
66. Albert, R.E., M. Lippmann, H.T. Peterson, Jr. et al.: Deposition and Clearance of Aerosols. *Arch. Intern. Med.* *131*:115-127 (1973).
67. Wolff, R.K., J.L. Mauderly and J.A. Pickrell: Chronic Bronchitis and Asthma: Biochemistry, Rheology and Mucociliary Clearance. *Lung Connective Tissue: Location, Metabolism and Response to Injury*, pp. 169-183. J.A. Pickrell, Ed. CRC Press, Boca Raton, Florida (1981).
68. Sackner, M.A., D. Ford, R. Fernandez et al.: Effects of Sulfuric Acid Aerosol on Cardiopulmonary Function in Dogs, Sheep, and Humans. *Am. Rev. Respir. Dis.* *118*:497-510 (1978).
69. Avol, E.L., M.P. Jones, R.M. Bailey et al.: Controlled Exposures of Human Volunteers to Sulfate Aerosols. *Am. Rev. Respir. Dis.* *120*:319-326 (1979).
70. Utell, M.J., P.E. Morrow, D.M. Speers et al.: Airway Responses to Sulfate and Sulfuric Acid Aerosols in Asthmatics: An Exposure-response Relationship. *Am. Rev. Respir. Dis.* *128*:444-450 (1983).
71. Utell, M.J., P.E. Morrow, J.A. Mariglio et al.: Exercise, Age, and Route of Inhalation Influence Airway Response to Sulfuric Acid Aerosols in Exercising Asthmatics (abstract). *Am. Rev. Respir. Dis.* *129*:A175 (1984).
72. Lippmann, M.: Respiratory Tract Deposition and Clearance of Aerosols. *Aerosols*, pp. 43-57. S.D. Lee, T. Schneider, L.D. Grant and P.J. Verkerk, Eds. Lewis Publishers, Chelsea, MI (1986).

SECTION II

Sampling Gases and Vapors for Analysis

CHAPTER 7

Basic Factors in Gas and Vapor Sampling and Analysis

BERNARD E. SALTZMAN

Department of Environmental Health, University of Cincinnati, Cincinnati, Ohio 45267–0056

INTRODUCTION

Accurate sampling and analysis of gases and vapors requires proper regard for a number of basic factors. The final result can be no better than the weakest link in the series of steps in the procedures used. Among the important factors are the sampling strategy in relation to the objectives of the study, the measurement of the air flows, the convenience and efficiency of the sampling device, proper selection and evaluation of analytical methods, and quality control. The various techniques and instruments available are comprehensively described in the ACGIH manual, *Air Sampling Instruments.*[1-7] This chapter will not attempt to summarize these details, but it will deal with some significant aspects which merit more consideration.

SAMPLING STRATEGY

Concentrations in air may vary widely in three dimensions of space and in time; thus, when and where to collect a sample depends upon the objectives of the study. One could seek the worst conditions or representative average conditions for the exposed population. Without some knowledge of the variability of the concentration, analysis of a single brief sample may be totally misleading. Although such a result may be 50 percent or 90 percent of the time-weighted average (TWA), and thus may be regarded as acceptable, it is not unlikely that if one sampled on a number of other occasions the TWA might be exceeded. Various sampling strategies may be used to allow for sample variability and analytical errors and to decide whether the standard was met.[8] If necessary, analytical errors can be reduced by collecting a series of consecutive samples during the sampling interval and averaging the result. Alternatively, an assessment of the actual health risk, rather than of the likelihood of meeting the standard, may be made if the geometric mean and geometric standard deviation of the concentrations are determined and a dose-effect relationship is known.[9]

Another decision that must be made is the sampling time, which is related to the sensitivity achieved. Commonly the 8-hour TWA is used, but there are some 15-minute short-term exposure limits (STEL) used in industrial hygiene. There are air pollution standards requiring 1-hour, 3-hour, 24-hour, 3-month, and 1-year average values. A factor not often considered in the choice of sample averaging time is the biological half-life of the contaminant and the pharmacokinetic relationships between the external concentrations and those in the body. For some solvents with long biological half-lives, concentrations in the body may build up in a weekly cycle to a peak on Fridays; the evaluation of the exposure must include a cumulative calculation for the entire working week.[10] For carbon monoxide, a series of two-hour average values have been recommended as useful for evaluating its health hazard.[11] When the highest sensitivity is desired, long sampling times are employed, but one must remember to stay within the limits of time and flow rates that each sampling method has. If the averaging time is much longer than the biological half-life of the contaminant, significant short peaks may be averaged out and the health hazard may be underestimated.

MEASUREMENT OF PULSATING AIR FLOWS

Accurate measurement of the sampling flow rate is essential because the analysis is the ratio of contaminant to air, and it is equally affected by the errors in each. The pulsating flows produced by the commonly used diaphragm pumps may cause substantial errors in the readings of flowmeters. The lifting force on a rotameter ball, as well as the pressure difference reading of an orifice meter or pitot tube, are proportional to the mean of the pulsating air velocity raised to an exponent close to two, whereas the mean flow is proportional to the mean of the velocity to the first power.

Figure 1 illustrates the pulsating flow of an MSA model G pump without a pulsation damper as determined with a hot wire anemometer and oscilloscope.[12] The instantaneous flows for each millisecond in the 59 Hz cycle have been measured, and the mean is calculated as 2.12 lpm. However, the root mean square value is 2.58 lpm, and the flowmeters thus would be expected to read higher than the correct value by the factor $(2.58/2.12)^2$, or 1.48. Most pumps now have pulsation dampers. The actual wave form of the flow would be modified by the resistance and capacitance of the flow system as well as by the damper; in many cases, the error caused by pulsations would be smaller. However, the analyst should be aware of this possible error when using flowmeters calibrated under steady flow conditions. Calibration with a soap film burette under operating conditions should minimize this error.

SAMPLE COLLECTION EFFICIENCY

The collection efficiency of the sampling method is very important for a good analysis. A variety of methods have evolved for collecting gases and vapors. Early freeze-out traps were abandoned as inconvenient and replaced with liquid absorbents. At first the large Greenburg-Smith impingers were used, but they soon were replaced by more convenient midget impingers or fritted bubblers when higher absorption efficiencies were needed. The stability of the liquid absorbent both before and after sampling must be carefully evaluated at relevant temperatures.

Evacuated sampling flasks or bottles were replaced by plastic bags, made of Teflon®, Saran® or Mylar®. Surface adsorption and/or desorption from a previous sampler are serious problems which must be controlled by proper selection of bag materials and preconditioning with a similar mixture.

More recently solid absorbents have become popular, and adequate sensitivity using a gas chromatograph can be obtained with small devices such as the common tubes containing 100 mg of absorbent carbon followed by a backup layer of 50 mg of carbon. The second layer may be analyzed separately to check on whether the capacity of the first layer has been exceeded. However, it should be noted that even if breakthrough in the first layer has not occurred, upon long storage before analysis some volatile analytes (such as vinyl chloride) will diffuse into the second layer until the concentrations equalize (at which point it will contain one-third of the total quantity). The kinetic breakthrough capacity is less than the

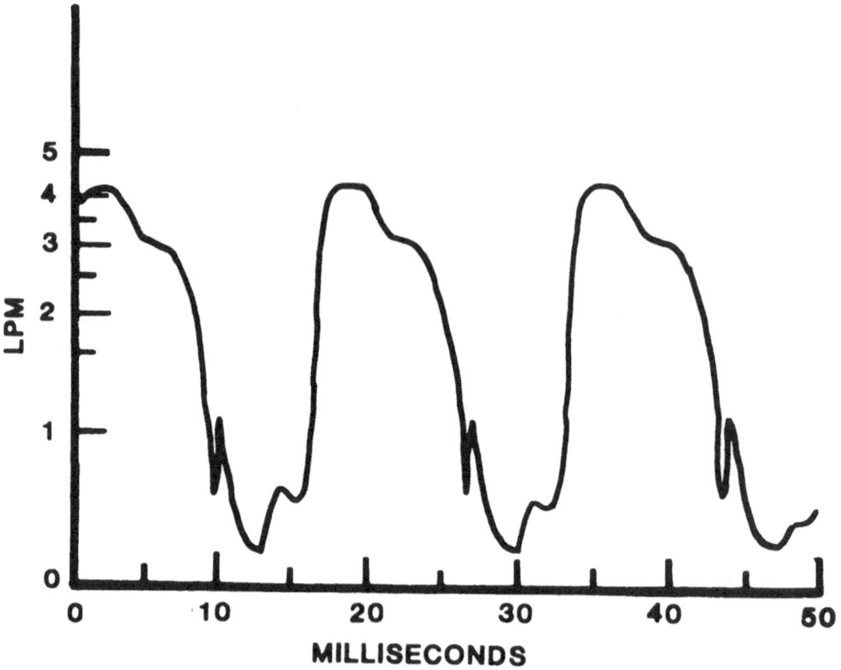

Figure 1. Pulsating flow of an MSA Model G pump without a pulsation damper.[12]

equilibrium (saturation) capacity and is decreased at higher sampling flow rate and higher humidities. Incomplete desorption from the absorbent for the final analysis may be another source of error which must be carefully checked for each batch of absorbent.

Passive sampling devices have become increasingly popular because of their convenience and compactness. Several new problems have arisen with them. The seals are certainly not as good as those in a sealed glass tube containing carbon or other absorbents, both before and after use. Thus in some models contamination or loss of sample may occur. Also unless there is sufficient air motion, depletion around the entrance of the device (the "starvation effect") may result in low recoveries. More recently direct reading passive monitoring tubes have been marketed for a variety of gases. These devices produce a length of stain which is related to the quantity of gas absorbed and which can be read immediately in the field. The accuracy for some types of passive samplers clearly could be improved, as will be shown below.

INTEGRATION OF FLUCTUATING CONCENTRATIONS

Analytical methods should be tested to see if they properly integrate fluctuating concentrations to give correct TWAs. Figure 2 illustrates the calibration curves for a method that will yield incorrect TWA values. The abscissa represents the quantities of analyte in air samples at constant concentrations and the ordinate represents the quantity of spike in the final step of the determination yielding the same response. The curvature at the low end may be due to losses of small quantities due to incomplete absorption, reverse equilibrium, trace impurities in the reagents, or interferences; that on the high end may be due to partial saturation of the reagents. Other losses may occur upon storage of the collected samples. The disturbing aspect is that the calibration for one-hour samples differs from that for eight-hour samples. For the same total quantity, the concentrations sampled for eight hours would be one-eighth of those sampled for one hour. Thus there are greater proportional losses at the low concentrations, even if the linear portion of the curve, AB, is used. The recovery percentage at point A is the slope of the line OA, which differs

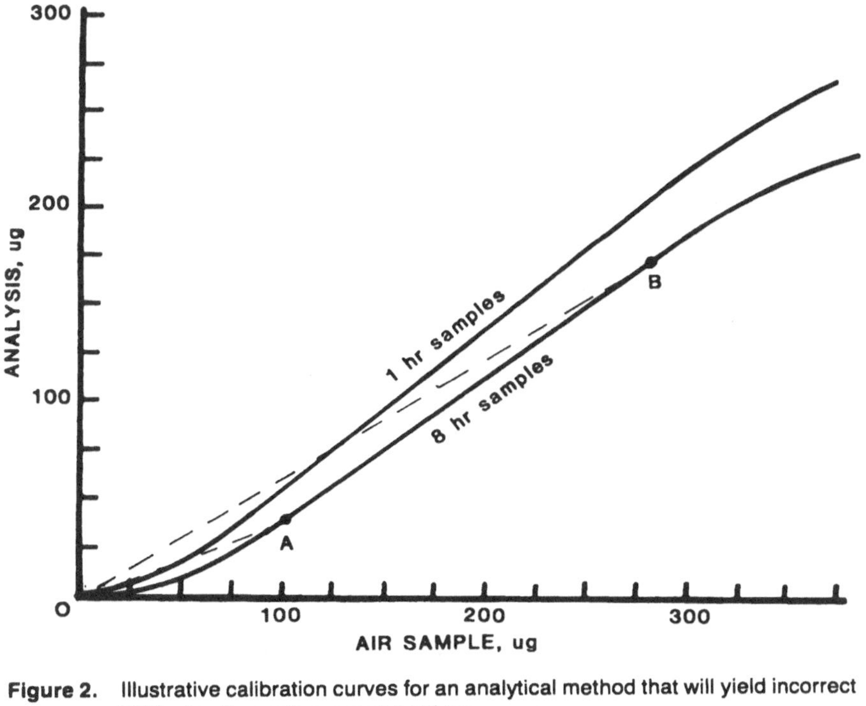

Figure 2. Illustrative calibration curves for an analytical method that will yield incorrect TWAs for fluctuating concentrations.

from that of line OB. If the concentration fluctuates, the recovery will be at an unknown level within the corresponding range of slopes, depending upon the fluctuation pattern, and an accurate result cannot be calculated. Additional errors may occur at the curved portion of the calibrations in the blank corrections and in the higher values. On the other hand, if the one-hour and eight-hour curves coincide, it might be possible that using them adequately integrates the fluctuating concentrations, provided that the variable recovery does not depend upon the concentration but only upon the total quantity of analyte. The major point of this discussion is that a critical examination of the validity of a method should not be overlooked.

EVALUATION OF METHODS AND QUALITY CONTROL

It is interesting to review the status of the different types of industrial hygiene analytical methods in use that could be applied to gases and vapors. A comprehensive listing is in the second edition of the National Institute for Occupational Safety and Health (NIOSH) *Manual of Analytical Methods*, which describes over 500 methods. Of these, 51 percent are by gas chromatography (GC), 11 percent by UV-Visible, 4 percent by infrared spectrophotometry, and 8 percent by high performance liquid chromatography. Of the solid absorbents listed, 61 percent were activated charcoal, 17 percent silica gel, and 6 percent Tenax GC. Of all the methods listed, about two-thirds were "validated" by a contract laboratory.[13] Validation consisted of minimal tests in a single laboratory of pure samples in a specified range about the Threshold Limit Value (TLV), usually without checking problems in field use, such as effects of interferences and humidity. Only a handful of methods have been collaboratively tested.

We have tested the validity of some recently marketed, direct reading passive dosimeter tubes for carbon monoxide.[14] These devices do not require pumps but operate on a diffusion principle to produce stain length proportional to TWA concentrations. Since they can be read immediately on the site, they offer great advantages of convenience, simplicity, and compactness. A flow dilution system was set up using 2.6 percent carbon monoxide from a cylinder, diluted to 25 to 400 ppm at relative humidities of 20 percent and 82–83 percent and a temperature of 26°C. Tubes in triplicate were exposed for periods up to six hours and were read by two observers independently at intervals. Effects of humidity were to slightly decrease the response. Initially Dräger and Gastec tubes were exposed simultaneously. Some results at 83 percent relative humidity are shown in Figure 3. The slope of the fitted line shows that the Dräger readings were in good agreement with the metered quantities. However, the slope of the Gastec calibration line was found to be only 0.22, and these results must be regarded as unsatisfactory. In a second series of tests a different batch of Gastec tubes were exposed, simultaneously with MSA tubes. Some results at 82 percent relative humidity are shown in Figure 4. This time the Gastec slope was 0.16, and the slope of the calibration line for MSA tubes was 2.16. The Gastec distributor has informed us that they will recall these batches of tubes and take corrective action. These tube types are difficult to read because the stain length boundaries are diffuse. However, it is evident that there are serious problems with the accuracy of some types of these tubes, and that improvements in calibration, quality control, and perhaps design are in order.

As the economic and legal importance of correct analyses grows, the major task of a more thorough examination of methods will have to be

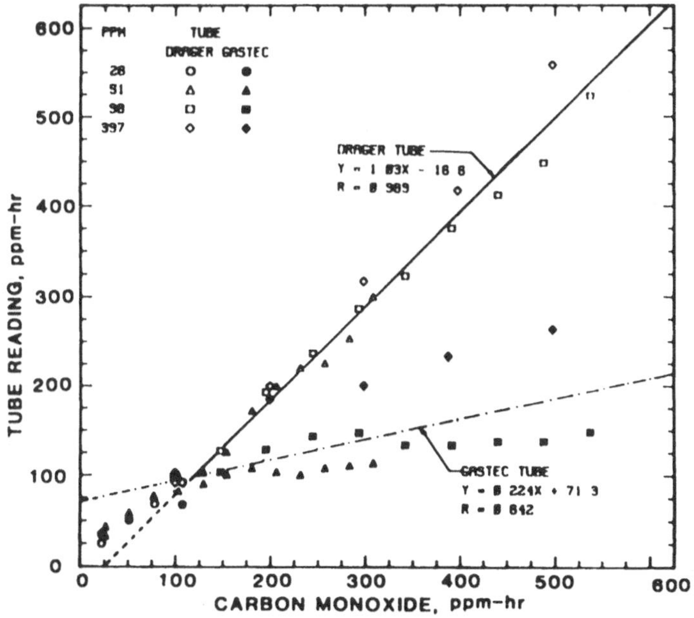

Figure 3. Performance tests of Dräger and Gastec passive colorimetric dosimeter tubes for carbon monoxide at 83 percent relative humidity.[14]

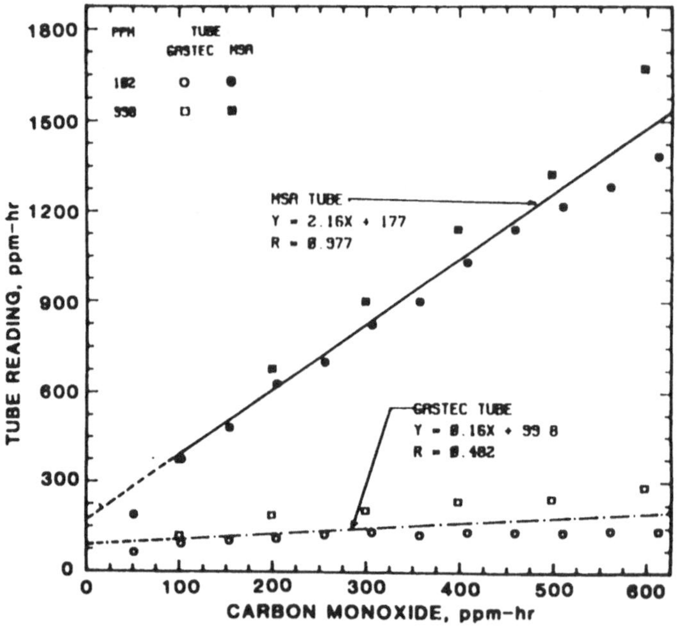

Figure 4. Performance tests of Gastec and MSA passive colorimetric dosimeter tubes for carbon monoxide at 82 percent relative humidity.[14]

undertaken. It is not enough for a researcher to demonstrate good results. Collaborative tests should involve people with the usual qualifications of industrial hygiene analysts to see if the instructions are clear and if acceptable results can be obtained in their hands. After early defeats in court, the Food and Drug Administration recognized the importance of methods evaluation and collaborative testing. It long has strongly supported the Association of Official Analytical Chemists, whose methods testing procedures can serve as an excellent model for conducting such activities. The Environmental Protection Agency (EPA) had a debacle in June 1973, when it withdrew the previously published Federal Reference Method for nitrogen dioxide as unreliable and delayed or rescinded control actions based upon its values.[15] EPA now has a formal methods testing and quality control program.[16] Data have been analyzed from interlaboratory programs conducted by four groups for 12 common types of industrial hygiene analyses.[17] The majority failed to meet the NIOSH criteria of ± 25 percent with 95 percent confidence (a coefficient of variation of less than 12.8 percent). In the introductory sections of the 3rd edition of the NIOSH *Manual of Analytical Methods*,[18-20] procedures for development and evaluation of methods and of quality control are described. However, their implementation for the hundreds of methods in use will proceed at snail's pace unless major new resources are allocated to this task.

REFERENCES

1. First, M.W.: Air Sampling and Analysis for Contaminants in Workplaces. *Air Sampling Instruments*, 6th ed., pp. A1-A13. P.J. Lioy and M.J.Y Lioy, Eds. American Conference of Governmental Industrial Hygienists, Cincinnati (1983).
2. Billings, C.E.: Gas Stream Sampling. *Ibid.*, pp. B1-B37.
3. Hinton, D.O.: Community Air Sampling. *Ibid.*, pp. C1-C6.
4. Thompson, R.J.: Air Monitoring for Organic Constituents. *Ibid.*, pp. D1-D7.
5. Pagnotto, L.D.: Gas and Vapor Sample Collectors. *Ibid.*, pp. S1-S23.
6. Saltzman, B.E.: Direct Reading Colorimetric Indicators. *Ibid.*, pp. T1-T29.
7. Nader, J.S., J.F. Lauderdale and C.S. McCammon: Direct Reading Instruments for Analyzing Airborne Gases and Vapors. *Ibid.*, pp. V1-V118.
8. Leidel, N.A., K.A. Busch and J.R. Lynch: *Occupational Exposure Sampling Strategy Manual*. DHEW (NIOSH) Pub. No. 77-173. Cincinnati (1977).
9. Saltzman, B.E.: Lognormal Model for Health Risk Assessment of Fluctuating Concentrations. *Am. Ind. Hyg. Assoc. J.* (in press, 1987).
10. Paustenback, D.J.: Occupational Exposure Limits, Pharmacokinetics, and Unusual Work Schedules. *Patty's Industrial Hygiene and Toxicology*, 2nd ed., Vol. 3A, *Rationale of Industrial Hygiene Practice: The Work Environment*, pp. 111-277. L.J. Cralley and L.V. Cralley, Eds. John Wiley, New York (1985).
11. Saltzman, B.E.: Biological Significance of Fluctuating Concentrations of Carbon Monoxide. *Env. Sci. Tech. 20*:916 (1986).
12. LaViolette, P.A. and P.C. Reist: Improved Pulsation Dampener for Respirable Dust Mass Sampling Devices. *Am. Ind. Hyg. Assoc. J. 33*:279 (1972).
13. Taylor, D.G., R.E. Kupel and J.M. Pryant: *Documentation of NIOSH Validation Tests*. DHEW (NIOSH) Pub. No. 77-185. Cincinnati (1977).
14. Hossain, M.A. and B.E. Saltzman: Laboratory Evaluation of Passive Colorimetric Dosimeter Tubes for Carbon Monoxide. Paper for presentation at AIHC, Montreal (1987).
15. Environmental Protection Agency: Reference Methods for Determination of Nitrogen Dioxide. *Fed. Reg. 38*:15174 (June 8, 1973).
16. Environmental Monitoring and Support Laboratory: *Quality Assurance Handbook for Air Pollution Measuring Systems*, Vol. I, *Principles*, EPA-600/9-76-005 (1976); Vol. II, *Ambient Air Specific Methods*, EPA 600/

4-77-027a (1977); Vol. III, *Stationary Source Specific Methods*, EPA 600/4-77-027b (1977). [Revisions to Vols. I-III are issued at intervals.] Office of Research and Development, EPA, Research Triangle Park, NC.

17. Saltzman, B.E.: Variability and Bias in the Analysis of Industrial Hygiene Samples. *Am. Ind. Hyg. Assoc. J. 46*:134 (1985).
18. Eller, P.M., Ed.: *NIOSH Manual of Analytical Methods,* 3rd ed., pp. 5-14. DHHS (NIOSH) Pub. No. 84-100. Cincinnati (1984).
19. Smith, D.L. and M.L. Bolyard: Quality Assurance. *Ibid.*, pp. 5-14.
20. Hull, R.D.: Development and Evaluation of Methods. *Ibid.*, pp. 29-35.

CHAPTER 8

Long-Term Passive Sampling of Environmental Airborne Contaminants

GHAZI F. HOURANI and DWIGHT W. UNDERHILL

Center for Environmental Epidemiology, Graduate School of Public Health, University of Pittsburgh, 130 DeSoto Street, Pittsburgh, Pennsylvania 15261

DIFFICULTIES IN "VALIDATING" PASSIVE SAMPLERS FOR ENVIRONMENTAL MONITORING

Passive samplers in the workplace are commonly exposed to relatively high concentrations of contaminants for time periods ranging up to eight hours. Some commercial passive samplers have been "validated" for such use, which according to the applicable National Institute for Occupational Safety and Health (NIOSH) criteria[1] requires that in the laboratory they show a standard error (S.E.) of 25 percent or less at one, two, and five times some standard concentration (usually the Threshold Limit Value, TLV) and a S.E. of 35 percent or less when tested at one-half the standard concentration. The difficulty in applying the NIOSH validation to the long-term sampling of environmental contaminants is that the concentrations we wish to measure may be several factors of ten below the concentrations found in industrial exposure, and the sampling periods may be one week or longer, i.e., more than twenty times the sampling period in the original validation. Generally the NIOSH validation cannot be used, and the passive sampler must be revalidated for environmental monitoring.

Difficulties that any investigator faces in attempting to validate a passive sampler for environmental monitoring include:

1. In validating a passive sampler for industrial monitoring, we generally have a primary exposure standard, the TLV, which we can use as our "Gold Standard" about which to test the performance of the sampler. For environmental monitoring, no such "Gold Standard" exists. So how can we, in advance of taking field samples, focus our validation studies over a concentration range that will be of importance in decision making?
2. Obtaining the necessary validation can be very time-consuming. Conceivably, it could take years to validate the performance of a passive sampler for several compounds, at different concentrations, for exposures ranging from days to months. In addition, while making these tests, the concentration of test agent, the temperature, and the relative humidity must be carefully controlled.
3. Laboratory validation by exposure to constant concentrations of contaminant may not be realistic if the sampler is to be used to measure fluctuating concentrations of environmental contaminants. Data obtained from constant concentrations of adsorbate do not represent the "worst case" sampling scenario which is exposure to a peak concentration at the start of the sampling period, with essentially all the adsorbed contaminant at risk for desorption over the entire sampling period. Thus special consideration must be given to the retention of peak concentrations of contaminant at the initial part of sampling period.
4. The current validation procedure gives no clear-cut way to determine whether or not a sampler is performing adequately. Suppose, for example, we attempt to validate a passive sampler at two concentrations of contaminant and are able to prove, after many time-consuming trials, a significant difference in the sampling constants (in cc/sec) at these two concentrations. Is one value "right" and the other "wrong" and if so, why? The basic problem is that we are trying to validate the performance of a passive sampler by comparing results obtained *from the same sampler* under different sampling conditions. Thus the sampler, which may be faulty, is serving as its own standard. Such a procedure is not objective.

DEVELOPMENT OF A NEW VALIDATION PROCEDURE

Before taking the plunge of committing ourselves to long-term experiments, the dynamics of passive sampling were examined to see what information is really needed and how best to acquire it. In order to develop a validation procedure appropriate for environmental monitor-

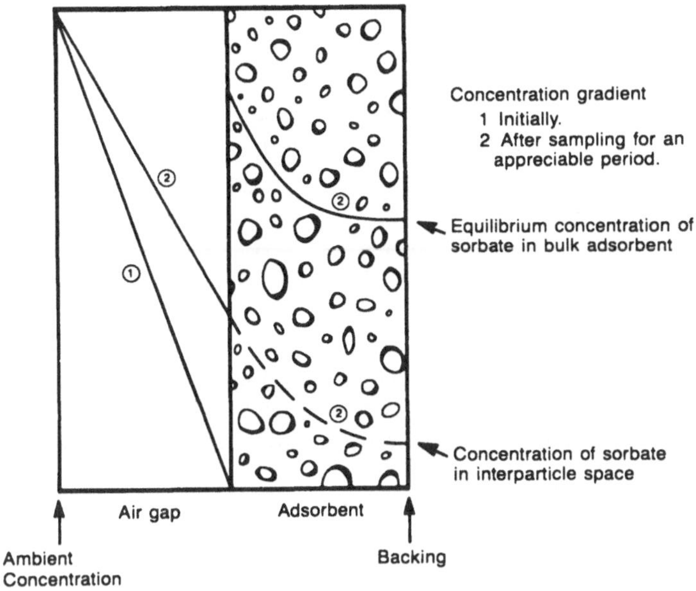

Figure 1. Flux in and out of a passive sampler.

ing, we need to consider the fundamental question of what controls the performance of a passive monitor. Such a sampler is shown schematically in Figure 1. In this passive sampler there is an adsorbent layer separated by an air gap from the ambient atmosphere. At the time sampling is started, the forward flux of adsorbate into the sampler is equal to:

$$J_+ = DC_o/L \tag{1}$$

where: J_+ = flux of adsorbate into the sampler (g/sec-cm^2)
D = diffusion coefficient of adsorbate in air (cm^2/sec)
C_o = ambient concentration of adsorbate (g/cm^3)
L = width of air gap (cm)

As the adsorbate accumulates, the vapor pressure of adsorbate builds up at the exposed surface of the adsorbent, producing a reverse flux equal to:

$$J_{-} = DC_{*}/L \tag{2}$$

where: J_{-} = flux of effluent adsorbate from the adsorbent (g/sec-cm[2])
C_{*} = airborne concentration of adsorbate at the exposed surface of the adsorbent (g/cm[3])

At this point the net flux is:

$$J = D(C_o - C_{*})/L \tag{3}$$

If, while the passive sampler is exposed to a concentration of contaminant, the forward flux is > > than the reverse flux, then Equation 1 permits a valid calculation of the time-weighted average (TWA) exposure. Line "2" in Figure 1 shows the concentration of adsorbate across the air gap after an appreciable buildup of adsorbate has occurred at the exposed surface of the adsorbent. The diminished slope of this line, compared with that of line "1," is proportional to the reduction in the sampling rate which occurs with the buildup of adsorbate vapor pressure over the adsorbent.

As the forward flux depends only on the external concentration of adsorbate, the diffusion coefficient of the adsorbate in air, and the geometry of the sampler, the biases that occur in extending the sampling period do not affect the forward flux. It is the reverse flux, i.e., the loss of collected adsorbate, that is variable and the source of error in long-term passive sampling of a constant ambient concentration of contaminant. Thus accurate measurement of back diffusion is the key to determining precisely the feasibility of the long-term passive sampling of contaminants.

An important point is that it is very easy to measure this back diffusion, thus confirming directly the usefulness of a passive sampler for use over this same period of time. The difference between measuring the loss directly and indirectly (indirect measurement = the sampler is challenged until the uptake is found to be no longer linear with time and/or concentration) can be compared to the difference between weighing a flea and weighing a flea on top of an elephant. Furthermore, if the loss by back diffusion begins to become significant in comparison with the total uptake, we then have an objective criterion to reject the use of that sampler for field use where similar loadings might be expected over the same period of time.

EXPERIMENTAL PROCEDURE

The experimental technique was to expose a set of passive samplers for one day, then place a carbon backing over each sampler and place each one in a separate closed container. Afterwards, by analyzing the uptake on the carbon strips by gas chromatography, both the total uptake and the fractional loss by back diffusion can be measured directly. As the exposure period in the chamber lasts only for one day, we can expose a new set of passive samplers each working day. Thus instead of gathering one set of data on a per week or perhaps on a per month basis, we can, on the average, take one set of data each working day, even for long-term experiments which individually require a number of days or weeks. These data represent the worst case scenario of a very high exposure at the initial part of the sampling period.

In the exposure system, shown schematically in Figure 2, laboratory-supplied compressed air was cleansed of particulate and organic matter by passing through, in sequence, an oil trap, a desiccant, and a charcoal

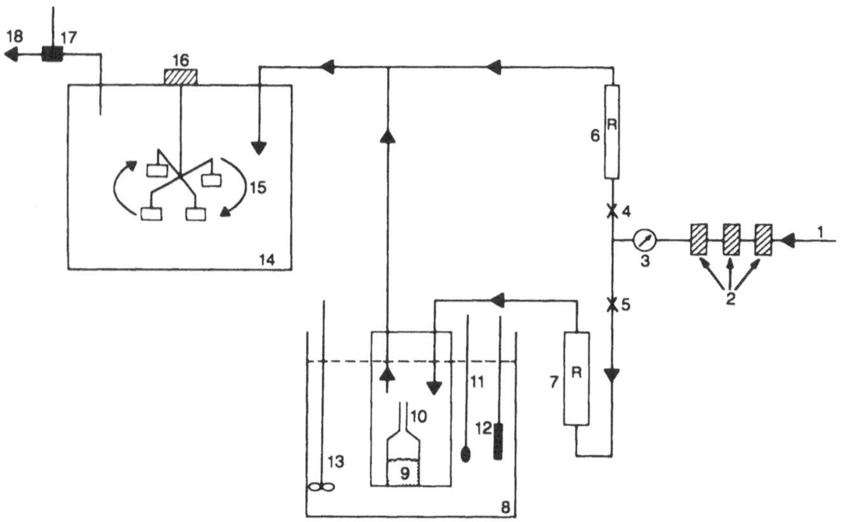

Figure 2. The experimental test system. 1) Compressed air from laboratory utilities; 2) air filters; 3) pressure gage; 4,5) flow control valves; 6) rotameter (0–25 lpm); 7) rotameter (0–2.5 lpm); 8) hot water bath; 9) TCE in diffusion cell; 10) capillary; 11) thermometer; 12) electric heater; 13) stirrer; 14) exposure chamber; 15) rotating chandelier (holding samplers); 16) motor; 17) wet bulb thermometer; and 18) vent to laboratory hood.

adsorbent bed. Following a pressure regulator, the flow was split into two unequal streams. One stream, at about a flow of 1.0 lpm, passed through a small (about 50 cc in volume) glass chamber immersed in a stirred hot water bath maintained at 50°C. Inside this glass chamber was a diffusion cell which released a small constant stream of test agent. For concentrations of trichloroethylene (TCE) above 50 ppm, the largest diffusion cell that we had (Part #303–4013, Thermoelectron, Waltham, MA) proved to be inadequate and an infusion pump was used to supply a fixed input of TCE. The bypass air stream, with a flow of up to 15 lpm, was mixed with the first air stream at the T junction and passed into the exposure chamber. A constant air velocity over the exposed faces of the passive samplers in the exposure chamber was maintained by attaching the badges to the 5-cm diameter stirrer blade driven by a motor at a rate of approximately one revolution per second. Levine et al.[2] have described a similar system.

TCE was selected as the challenge contaminant as it seems typical of many man-made environmental organic contaminants. TCE was once widely used as a solvent and often disposed of carelessly. But recently TCE has been listed as a "cancer suspect" chemical, and now comparatively small environmental concentrations of TCE are not acceptable.

The passive sampler used in these experiments, the Mine Safety Appliances Co. "Organic Vapor Detector" (MSA-OVD) contains two layers of adsorbent in series (Figure 3). The presence of contaminant on the second layer demonstrates movement of adsorbate from the exposed layer and serves to warn of saturation of the primary layer. Earlier experiments by Tidwell[3] at higher concentrations of various contaminants, for time periods up to eight hours, make it appear that if the uptake of contaminant on the secondary strip was 25 percent or more of that on the exposed strip, then significant loss of contaminant from the passive sampler may have occurred. It was important to determine in our more recent experiments, which involved sampling at much lower concentrations of contaminant, if the uptake on the secondary layer can also serve as a warning of loss by back diffusion.

DATA AND RESULTS

Figure 4 shows the results in terms of the distribution of TCE on 1) the exposed sampling surface, 2) the secondary sampling layer, 3) the carbon impregnated strip placed over the passive sampler after the initial 24-hour exposure, and 4) on a carbon strip placed alongside the passive sampler in the closed container. The period of back diffusion was six

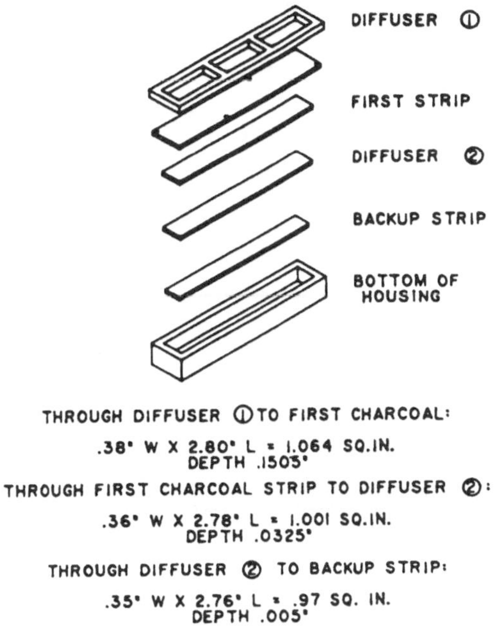

Figure 3. The Mine Safety Appliances Co. "Organic Vapor Detector."

days. This latter strip of adsorbent serves as a scavenger to pick up any TCE which escaped through the carbon strip placed over the sampler. The x-axis in Figure 4 gives the total weight of TCE as measured on the four adsorbent strips. These results are from the first series of tests (completed in early February 1987) and should be considered to be preliminary. We are in the process of repeating these experiments to determine the reproducibility and reliability of the test system.

The results shown in Figure 4 indicate strongly that at low uptakes (i.e., 50 mg or less) the TCE is strongly bound to the primary sampling layer. As the uptake increased beyond 50 mg TCE, the TCE increasingly redistributed itself between the primary strip, the secondary strip, and the carbon strip placed over the sampler to monitor back diffusion. The ability of the primary adsorption strip to retain adsorbate decreased sharply as the uptake rose from 50 mg to 100 mg TCE. At the highest loading (108 mg), there was nearly as much TCE on the secondary strip as on the primary strip. From the viewpoint of field sampling with these particular passive samplers, these results indicate that, should the total

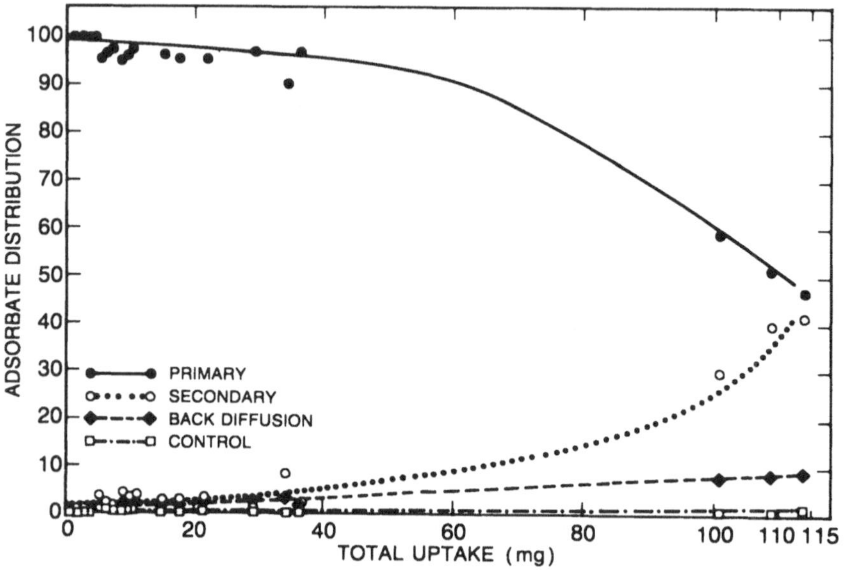

Figure 4. Distribution of adsorbate after one week as a function of initial uptake.

uptake of adsorbate (assumed to be TCE) be less than 50 mg, then one could have confidence that the sampler would retain the input for the one-week period. Should the input be higher, then one should consider either using a different passive sampler or using the same sampler for shorter periods of time.

DISCUSSION

The fact that the distribution of adsorbate is a strong function of uptake is clear proof that the adsorption isotherm is nonlinear—with a linear isotherm there would be absolutely no change in the fractional distribution of adsorbate as a function of loading. Because of the nonlinearity of the adsorption isotherm for most adsorbates on activated carbon, any estimates of sampling efficiency based on a linear isotherm, especially if they cover a wide range of concentrations, are dubious if they are applied to passive sampling by activated carbon adsorbates.

In field measurements, we cannot measure back diffusion, but we can look for the presence of adsorbate on the secondary sampling strip. If, however, we were to use the presence of TCE on the secondary as a means of determining loss of TCE from the primary sampling layer, we would be able to determine the presence of such a loss before the sampler had lost an appreciable quantity of sorbate. In these tests, as the quantity of adsorbate increased and the primary strip began to lose adsorbate by back diffusion, it also began to lose adsorbate to the secondary strip. This shows the usefulness of having two layers of adsorbent and allowing the fraction of the adsorbate on secondary layer to serve as a warning of saturation of the primary layer.

Possibly the most useful property of the secondary strip has yet to be investigated. We have shown from measurements such as those presented here that as the primary strip approaches saturation, there is transfer of TCE to the secondary strip. In environmental sampling we are seldom faced with "pure" compounds. Suppose in the environmental setting, the primary layer of adsorbent were to become saturated by an unknown compound "A," and that this weakened the retention of compound "B" and permitted compound "B" to migrate to the secondary layer. Can a passive sampler be designed so that if the adsorptive capacity of the primary layer were to approach being exhausted, a sufficient fraction of compound "B" would migrate to the secondary strip to serve as a warning of possible loss of adsorbate from the primary strip by back diffusion? It would seem that what is necessary and sufficient for the secondary strip to give such a warning is that the diffusion path from the exposed (outer) surface of the primary strip to the secondary strip be shorter than the diffusion path of adsorbate out of the sampler. Then were interference to occur, leading to back diffusion, such interference could be determined (and possibly corrected for) by the presence of adsorbate on the secondary strip. In other words, the secondary strip can serve to give proof of the ability of the primary strip to serve as a sink for the contaminants of interest—and thus we can determine whether or not interference in passive sampling has occurred even if we do not know the composition or concentration of the interfering compounds. In the environmental setting, then each sampler can take its own validation data. Additional laboratory tests, and possibly numerical modeling, are needed to determine the sampler geometries permitting such "fail safe" passive sampling.

SUMMARY

In this chapter the means for "validating" passive samplers for the long-term measurement of environmental contaminants were examined. The focus was on directly measuring the back diffusion from a passive sampler as a means to determine the performance of a passive sampler under the worst possible condition of all the input being essentially at the initial portion of the sampling period. This procedure may decrease enormously the amount of laboratory work needed to validate the accuracy of passive samplers for specific conditions of time, concentration, and composition. In the field, the presence of contaminant on a second layer of adsorbate placed in the series with the primary adsorbent strip could serve as a secondary means of validating the continual activity of the primary strip over the entire sampling period.

DISCLAIMER

Although the information in this document has been funded wholly or in part by the United States Environmental Protection Agency under assistance agreement CR 812761–01 to the Center for Environmental Epidemiology, University of Pittsburgh, it does not necessarily reflect the views of the Agency and no official endorsement should be inferred.

ACKNOWLEDGMENT

The authors wish to thank the Mine Safety Appliances Co. for their donation of the passive samplers used in this study.

REFERENCES

1. *Fed. Reg. 38*:11458 (given as a standard for detector tubes, but commonly accepted as a basis of performance for passive samplers) (May 8, 1973).
2. Levine, S.P., J.A. Gonzalez and E.V. Kring: A Dynamic Exposure System for Evaluating Passive Exposure Dosimeters. *Am. Ind. Hyg. Assoc. J. 47*:347 (1986).
3. Tidwell, C.J.: *Evaluation of the Mine Safety Appliances Company Organic Vapor Dosimeter.* Master's Essay submitted to the faculty of the Graduate School of Public Health, University of Pittsburgh, Pittsburgh, PA (1985).

CHAPTER 9

Recent Developments in Passive Sampling Devices

JAMES D. MULIK and ROBERT G. LEWIS

Environmental Monitoring Systems Laboratory, U.S. Environmental Protection Agency, Research Triangle Park, North Carolina 27711

INTRODUCTION

Passive sampling devices (PSDs) have been used extensively over the past decade by industrial hygienists to assess the effects of respiratory exposures to hazardous pollutants on workers. Only recently, however, has the United States Environmental Protection Agency (EPA) become interested in personal exposure monitoring to support the ambient air quality standards set under the Clean Air Act. Since ambient air levels of most pollutants are several orders of magnitude lower than those normally found in the workplace, more sensitive monitoring systems are required. These systems are generally not portable and are usually located at fixed, outdoor sites. Such fixed-site monitoring may not accurately reflect the average daily exposure of the general populace to air pollutants. This is particularly true in the United States where the average person spends an estimated 90 percent of the time indoors.[1]

To obtain an accurate estimate of individual exposure to air pollution, the person must carry or wear the monitor. Therefore, the device must be unobtrusive and lightweight. It should operate quietly and place little or no burden on the individual. It must also be inexpensive because many more units may be required for personal monitoring than for fixed-station monitoring. Active (pump-based) sampling systems used for occupational exposure assessment can and have been used successfully outside of the workplace.[2] However, the pumps required weigh from 0.5 to 1.5 kg and may not be comfortable to wear, especially for small persons. In addition, they must be battery-powered and most cannot be continuously operated for more than 6 to 12 hours. Passive devices, which require no pump, are much lighter in weight and are not power-limited. They have the additional advantages of small size and relatively low cost which make them ideally suited as personal exposure monitors for toxic chemicals in air and for unattended area monitoring, especially when electrical power sources are not readily accessible, e.g., at remote hazardous waste sites.

RESULTS AND DISCUSSION

Passive air monitors may be either permeation- or diffusion-controlled. In each case, a collector or sorbent material is separated from the external environment by a physical barrier that determines the sampling characteristics of the device. Permeation-limited devices employ a membrane in which the test compounds are soluble. Because of this solubility requirement, it is possible to achieve some selectivity with permeation devices by choice of the membrane material. However, because of solubility variation even within a congeneric series of compounds, permeation devices must be calibrated for each individual chemical that is sampled.

With diffusion-limited devices, the collector is isolated from the environment by a porous barrier containing a well-defined series of channels or pores. The purpose of these channels is to provide a geometrically well-defined zone of essentially quiescent space through which mass transport is achieved solely by diffusion. As a general criterion for this condition, the length/diameter ratio (L/d) of the pores should be at least three. However, some investigators feel that the ratio should be at least eight.[3] Under such conditions, the mass flow rate to the collector is given by Fick's first law

$$\dot{m} = (DA/L)\,(C_{\infty} - C_0)$$

where:
- D = the diffusion coefficient for the compound of interest
- A = the total area of the diffusion channel openings
- L = the diffusion path length
- C_{∞} = the external concentration of the species
- C_0 = the concentration at the surface of the collector

If the collector has any appreciable efficiency for sorption of the compound of interest, C_0 is generally taken to be zero.

The quantity DA/L has units of volume/time. This quantity is often referred to as the sampling rate, analogous to that of active samplers. Preferably, the sampling rate of a given device should be determined independently before it is used for monitoring purposes. Usually this is not practical, however, and one must rely on the manufacturer's specifications of A/L and/or DA/L. For most commercial diffusion-controlled devices, the effective sampling rate varies from 1 to 150 cm^3/min depending on the molecular species. Pump-based personal monitors may sample at rates up to 8000 cm^3/min. Consequently, longer exposure times are often required for passive monitors in order to achieve equivalent sensitivities.

The effects of ambient temperature and pressure fluctuations on the sampling rates of passive devices may be estimated from the equation:

$$D \propto \frac{T^{3/2}}{P}$$

where: T = absolute temperature
P = atmospheric pressure

Therefore, the theoretical temperature coefficient would be approximately 0.6 percent per °C and the pressure coefficient approximately 0.1 percent per mm Hg.[3] Humidity effects are less predictable, but they may be pronounced for hydrophilic collectors or sorbents. Sampling rates are also dependent on the velocity of air movement over the face of the device, particularly if there are protrusions around the channel openings or if one side of a two-sided badge is obstructed. Protrusions can contribute to the formation of secondary layers of stagnant air which reduce uptake rates.[3] For chemicals that are weakly sorbed, significant equilibrium vapor pressures may exist at the face of the sorbent, which effectively reduce sampling rates according to Fick's law, i.e., $C_0 > 0$. Theoretical predictions suggest that the magnitude of this decrease will depend on ambient air concentrations.[3] Because most passive samplers have relatively large time constants (e.g., 11 sec for benzene, where L = 1 cm) and the rates of migration into the sorbent bed are slow compared to the time constant, diffusional samplers may not respond accurately to rapidly fluctuating air concentrations. However, such fluctuations are not usually characteristic of pollutant levels in ambient air.

Despite the potential limitations of applying passive monitors to quantitatively measure air pollutants, these devices have in recent years been used quite successfully in occupational environments.[4] Their application to ambient atmospheres, which requires detection limits from 0.1 to 50 ppbv, presents a greater challenge. Most commercial devices use activated carbon as the collector; therefore, for most organic chemicals sorption is thermally irreversible in a practical sense and C_0 is essentially zero. Solvents such as carbon disulfide (CS_2) or a mixture of CS_2 in methanol must be used to desorb the chemicals for analysis. Concentration by evaporation of the solvent extract is impractical for the analysis of volatile organic compounds. Consequently, carbon-based commercial dosimeters generally do not have adequate sensitivity for ambient air monitoring.

PSDs FOR VOLATILE ORGANIC CHEMICALS

Coutant and Scott[5] studied several commercial badges containing activated charcoal for measuring ambient air exposures to volatile organic chemicals (VOCs). The study indicated that contamination problems (probably arising in the manufacturing process) would have to be eliminated before they could be used for nonoccupational applications. In another study,[6] a popular carbon-based PSD (DuPont Pro-Tek® G-AA badge) was shown to be very adversely affected by atmospheric humidity. The response of these devices was reduced by ca. 50 percent at 80–90 percent relative humidities.

Lewis et al.[7] reported on the development and evaluation of a high-efficiency PSD designed for thermal desorption. While the device is capable of using any granular sorbent, Tenax®-GC (Enka NV, The Netherlands) was used because it can be desorbed at relatively low temperatures (150°C). Use of Tenax-GC also permitted direct comparison with popular pump-based sampling systems.

The PSD has recently become commercially available from Scientific Instrumentation Specialists, Inc. (SIS) of Moscow, Idaho. An exploded view of the device is shown in Figure 1. The PSD is 3.8 cm in diameter and 1.2 cm in depth, and weighs 36 g. Pairs of 200-mesh wire screens and perforated plates placed on each side of the sorbent bed serve as diffusion barriers. It will hold 0.4 g of Tenax-GC. The SIS passive sampler may be easily assembled and disassembled with snap-ring pliers and reused without changing sorbent. After use, the PSD was thermally desorbed in a specially constructed oven coupled with a cryogenic trap (77°K). Desorbed chemicals are subsequently released into a gas chromatograph for analysis.

Direct comparisons between the passive sampler and pumped Tenax

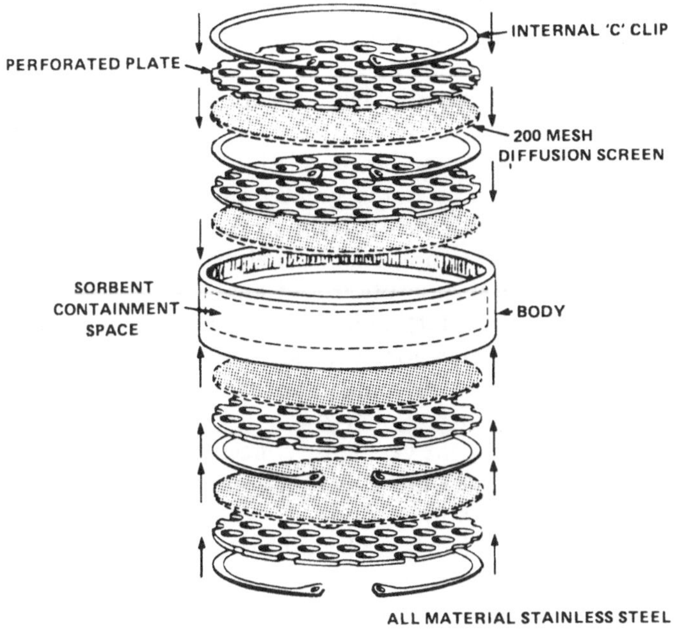

Figure 1. SIS passive sampling device.

tubes (28 cm^3/min) for one to five hours showed very good agreement for several chlorinated aliphatic and aromatic hydrocarbons.[7] The effects of air velocity on performance were studied under carefully controlled laboratory and field conditions. These studies showed that sampling rates were linear with wind velocity at face velocities above 15 cm/sec (0.3 mph). However, under very quiescent conditions, boundary layer formation across the face of the PSD drastically reduces the effective sampling rate.[7] Almost exactly the same velocity dependence was observed for the DuPont Pro-Tek badges.[6] This velocity dependence may be predicted accurately using standard boundary layer theory[8] and is directly proportional to the effective sampling rate of the PSD. For low-rate PSDs, such as the Palmes tube, air velocity effects are negligible. High-rate devices, such as the SIS passive sampler and most other commercial organic vapor dosimeters, should not be used when the air velocity is less than 15 cm/sec. Such conditions should not exist out-of-doors, but they may be encountered indoors, especially in poorly ventilated enclosures. Normal body movements usually provide adequate face velocity when the PSD is worn. For indoor area monitoring purposes, a small electric fan may be placed near the PSD to prevent stagnation. When hydrophobic sorbents such as Tenax-GC are used, atmospheric humidity should not affect PSD sampling rates. The absence of humidity effects at 85–95 percent relative humidity has been confirmed for the SIS device.[7]

A major advantage of the Tenax-based passive sampler over the charcoal-based commercial devices is the greatly enhanced sensitivity achieved by thermal desorption. When solvent desorption is required, only 0.5 percent of the collected sample can be analyzed (e.g., 5 μL of 1 ml extraction solvent). Thermal desorption and analysis of the entire sample affords a 200-fold (or greater) increase in sensitivity, permitting the measurement of ppbv concentrations with less than one-hour exposure times. However, by definition thermally-desorbable sorbents exhibit reversible sorption, while sorption on charcoal is irreversible for most organic chemicals. Consequently, the gas-phase concentration at the surface of the sorbent (C_o in the Fick's law equation presented earlier) may become significant for more volatile compounds with increasing exposure time. In studies reported by Lewis et al.,[7] no reversible sorption was observed for loadings below approximately 1.0 μg of the tested chemicals. However, at higher loadings, reversible sorption probably accounts for the nonlinear behavior shown in Figure 2 for all but the least volatile compounds examined. The effect of increasing sampling time on the time-weighted average sampling rate can also be predicted theoretically when velocity effects are taken into account if retention volumes for the chemicals of concern are assumed to be equivalent to those determined for actively-pumped sorbent systems. The relative sampling rate R_t at any given time after exposure of the passive device is initiated can be approximated by the equation:

$$\overline{R}_t = (1 - e^{-kt})/kt$$

The parameter k is equal to R_o/WV_b, where R_o is the intrinsic diffusion-limited rate under irreversible conditions, or DA/L; W is the weight of the sorbent, and V_b is the retention volume. Using published[9] values of V_b for Tenax-GC, the SIS device (for which W = 0.4 g) would be expected to behave as shown in Figure 3. It can be estimated from this plot that in one hour the apparent sampling rate for 1,1,1-trichloroethane would decrease by about 20 percent, while that for chlorobenzene would be essentially unchanged. Exposure data[10] have shown that this equation can be used to predict the sampling rate within experimental error. For the more volatile organic compounds (i.e., those with $V_b <$ 30–40 L/g), multiple-hour exposures with the Tenax-based passive

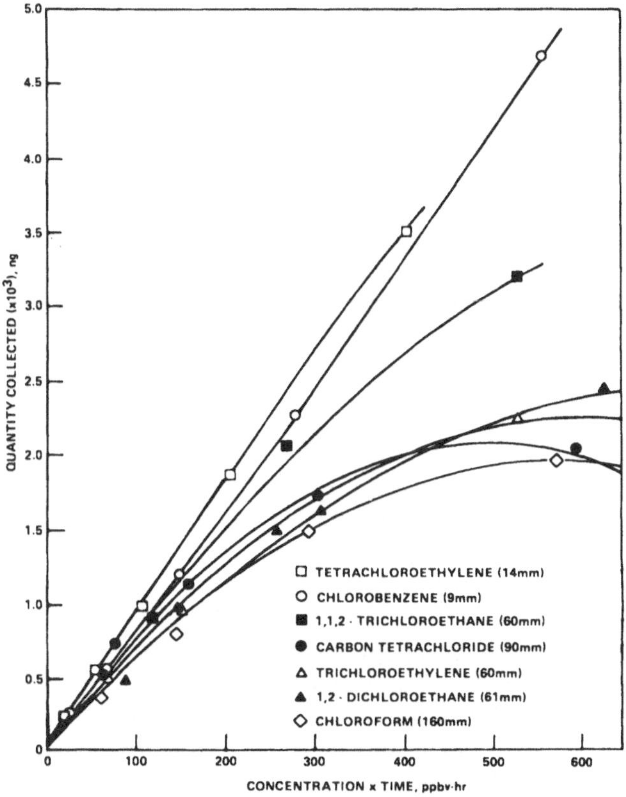

Figure 2. Response of PSD at long exposure times or high sorbent loadings.

sampler would require mathematical correction to an extent that accurate data may be very difficult to obtain. However, because of the much greater sensitivity of this device, exposures of one hour or less provide lower detection limits than do 24-hour exposures with carbon-based commercial devices.

In an effort to develop a sensitive PSD for long-term (24-hour) exposure monitoring of VOCs, two approaches were explored. The first involved the use of thermally-desorbable sorbents with higher retention volumes for the more volatile compounds. A carbonized molecular sieve, Spherocarb® (Analabs, North Haven, CT), was substituted for Tenax-GC. Evaluation of this sorbent resulted in its rejection due to difficulty of purification and the requirement for high desorption temperatures

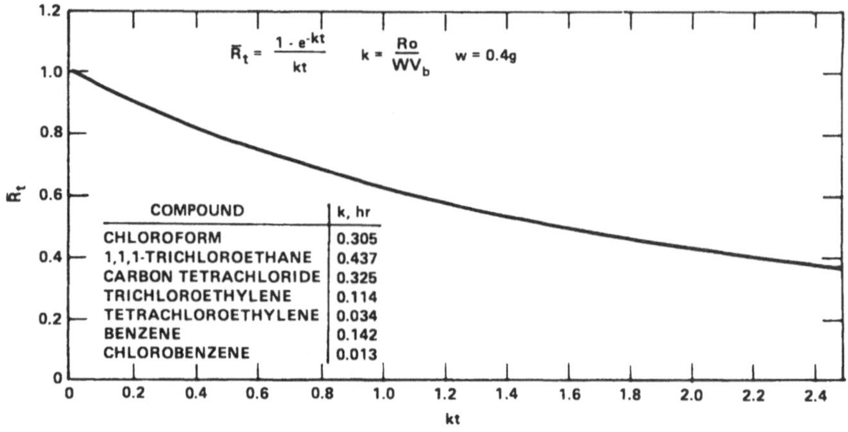

Figure 3. Effect of exposure time on sampling rate.

(> 350°C), which led to thermal cracking of some of the more labile chloroparaffins. Other sorbents were not tested as similar results could be expected.

The other, more successful approach involved alteration of the diffusion barrier to reduce the effective sampling rate. This was accomplished by replacing the outer screens and perforated plates with stainless-steel disks provided with a single hole 0.76 mm in diameter. The PSD sampling rate was thereby reduced to 3 to 5 percent of the original rate (ca. 2–3 cm^3/min). Exposure chamber testing of the low-rate PSD showed that 24-hour sampling could be performed for compounds with V_b as low as 15 L/g and 8-hour sampling for those with V_b of ca. 5 L/g.[10]

Both the high- and low-rate PSDs were intercompared with pumped distributed-volume Tenax-tube samplers[11] and integrated whole air canisters[12] in a furnished, but unoccupied, residence.[13] The HVAC system of the house was spiked with chloroform, 1,1,1-trichloroethane (1,1,1-TCE), tetrachloroethane (TeCE), bromodichloromethane (BDCM), trichloroethylene (TCE), benzene, toluene, styrene, p-dichlorobenzene (p-DCB), and hexachlorobutadiene (HCBD) so as to result in concentrations of these chemicals inside the air of the living space which were in the 3 to 30 $\mu g/m^3$ range. All sampling was carried out continuously over 12 hours, and all samples were analyzed by cryo-focusing and measurement on the same gas chromatograph–mass spectrometer to minimize analytical bias. The canister collection was used as the reference method. The comparison study was carried out over ten sampling days and at three indoor air concentrations, nominally 3, 10, and 30 $\mu g/m^3$. Agreement between all three methods was generally good, with measurement differences largely predictable from breakthrough volume considerations. The data from one day of sampling (spike level 30 $\mu g/m^3$) are summarized in Table I. The results of all sampling data (ten daily comparisons) showed better precision for the low-rate PSD than for the high-rate devices. Slope (s) estimates for the low-rate PSDs were < 1 and all intercept estimates were within 2 σ. Precision for the high-rate PSD was predictably poorer. For three of the compounds (benzene, chloroform and BDCM) s was < 1, while s was > 1 for the remaining seven. HCDB was the only compound with an intercept estimate greater than two standard errors from one. Agreements between the PSDs and canisters (6-L, SUMMA®-polished stainless steel) were good but not entirely predictable. The mean slope of the low-rate PSD versus the canister, however, was about 35 percent low, while that of the high-rate device was about 14 percent high. Better agreement would have been expected for the low-rate PSDs. The results for benzene and trichloroethylene are shown in Figures 4 and 5 for each of the ten sampling days.

The low-rate PSD design proved to be problematic. The single-orifice plates had to be precisely machined and press-fitted in order to prevent leakage at the edges. This made disassembly difficult and encumbered

TABLE I. Comparison of 12-hour Exposures of Low- and High-rate Passive Sampling Devices with Canister and Tenax Tube Samplers in Indoor Air[A]

Measured Air Concentration $\mu g/m^3$	Sampler			
	Canister	Low-rate PSD	High-rate PSD	Pumped Tenax®
Chloroform	23.4 ± 0.3[B]	15.05 ± 0.3[C]	15.8 ± 0.1[C]	17.8 ± 0.9[D]
1,1,1-TCE	20.9 ± 0.2	15.5 ± 0.0	57.1 ± 0.5	17.6 ± 0.9
TeCE	29.9 ± 0.4	15.3 ± 0.5	41.1 ± 2.0	24.9 ± 2.8
BDCE	15.8 ± 0.1	11.5 ± 0.2	13.3 ± 0.1	13.0 ± 0.1
TCE	15.8 ± 0.2	10.3 ± 0.4	19.6 ± 0.6	14.4 ± 0.4
Benzene	27.1 ± 0.7	18.4 ± 0.4	19.9 ± 1.0	25.2 ± 1.1
Toluene	33.0 ± 0.7	15.3 ± 0.5	48.9 ± 1.6	33.5 ± 1.4
Styrene	11.4 ± 0.2	6.8 ± 0.2	17.2 ± 0.3	12.3 ± 0.3
p-DCB	11.5 ± 0.2	6.9 ± 0.3	18.9 ± 0.4	12.0 ± 0.2
HCBD	17.8 ± 0.8	6.2 ± 5.8	28.0 ± 3.4	17.2 ± 0.2

[A]Summary of data from one day of sampling; t = 27°C, RH = 37%, exchange rate 0.045 hr^{-1}.
[B]Avg. of two determinations on the same canister sample (6-L stainless-steel).
[C]Avg. of two collocated PSDs.
[D]Avg. of two collocated distributed-volume samplers (5, 10, 20 and 40 L).

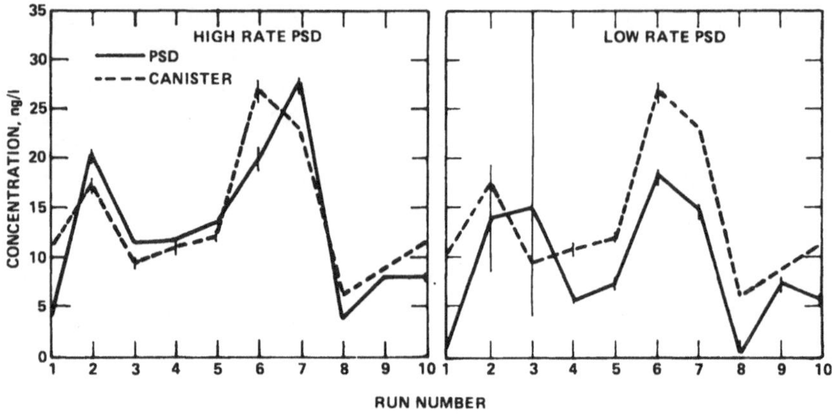

Figure 4. Comparison of canister and PSD data for benzene.

the desorption process. The PSD is currently undergoing redesign in conjunction with SIS, Inc.

PSDs FOR NITROGEN DIOXIDE

Commercially-available PSDs for NO_2 also lack sufficient sensitivity to obtain 8- to 24-hour time-weighted average (TWA) measurements in non-occupational indoor air. The sampling rate for the Palmes tube is

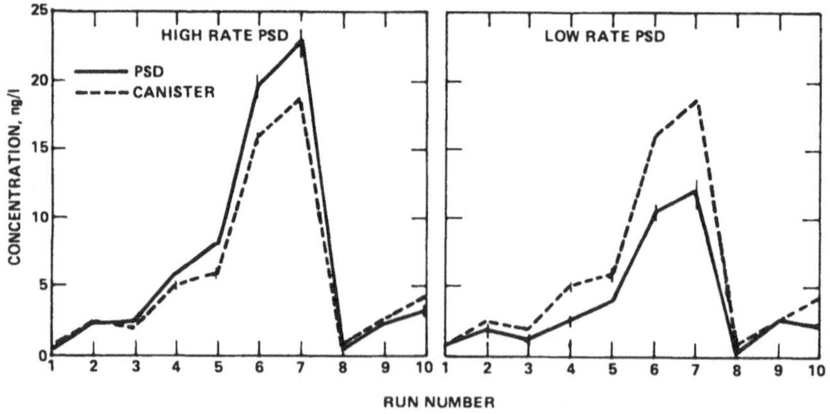

Figure 5. Comparison of canister and PSD data for trichloroethylene.

approximately 1.0 cm^3/min, which affords a sensitivity of 300 ppbv-hour when analyzed spectrophotometrically. Consequently, 5- to 7-day exposures are required to determine much lower levels of NO_2 found in ambient and indoor air.

In an attempt to lower the minimum detection limit of the Palmes tube sufficiently to achieve 8- to 24-hour utility, an ion chromatographic (IC) analytical finish was explored. The application of ion chromatography, which is inherently more sensitive than UV spectrophotometry, combined with a specially-designed concentrator column, improved the overall sensitivity to 30 ppbv-hr $\pm$ 20 percent.[14] While this analytical improvement theoretically permitted one day or less exposure times, the concentration step caused chloride ion interference with resolution of the nitrite ion. It became necessary to build a special extraction apparatus to extract the nitrite ion from the triethanolamine(TEA)-coated screens in order to prevent handling contamination. The acrylic tube of the Palmes NO_2 device was found to act as a sink or a source of NO_2 depending on how well it was cleaned. Finally, stainless steel tubes were recommended to minimize contamination in order to obtain the necessary minimum detectable quantity of NO_2 for ambient air measurements. These added steps made this method too cumbersome to recommend for routine analysis of NO_2 in ambient air environments.

A second method that appeared to have the sensitivity required was based on a Thermosorb® cartridge originally designed for nitrosamine collection and analysis by GC-chemiluminescence analysis.[15] The feasibility of converting the pumped Thermosorb cartridge to the passive mode was investigated. However, problems with the chemistry of reaction of NO_2 with the morpholine-based sorbent resulted in lack of repeatability of the method.

A third approach was directed at modification of the SIS passive sampler to make it responsive to NO_2.[16] This was accomplished simply by replacing the Tenax sorbent bed with TEA-coated glass fiber filters. Ion chromatography was employed for analysis. The effective sampling rate of the SIS passive sampler for NO_2 is 154 cm^3/min, making it potentially 150 times more sensitive than the standard Palmes tube.

The SIS devices were evaluated in a NO_2 exposure chamber in sets of five at concentrations ranging from 20 $\mu g/m^3$ to 460 $\mu g/m^3$. The NO_2 concentration in the exposure chamber was continuously monitored with a Bendix NO_x chemiluminescent monitor. Figure 6 shows a linear regression plot of NO_2 generated (as measured by a Bendix NO_x monitor) versus NO_2 found by IC analysis. Each point is an average of five values obtained at the indicated concentration for a 24-hour exposure. The correlation coefficient was 0.9955 over the range studied. Nitric oxide (91

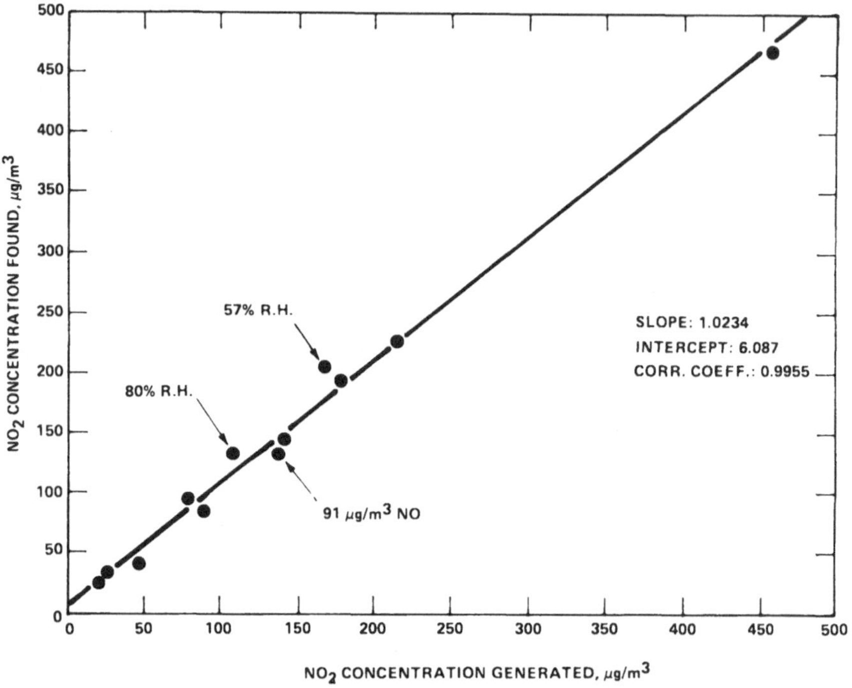

Figure 6. Linear regression plot of PSD response to NO_2.

$\mu g/m^3$) and relative humidities as high as 80 percent caused no deleterious effects on the efficiency of the PSD. The response was also shown to be linear over exposure times from 2 to 24 hours at constant air concentrations (109 $\mu g/m^3$ or 58 ppbv). The minimum detection limit for the SIS passive sampler was shown to be 30 ppbv hour when IC analysis was performed without the aid of a concentrator column.

The modified SIS device was tested in ambient air by direct comparison with a tunable diode laser system.[17] Collocated outdoor air exposures of 22 hours each showed excellent comparability between the two methods over 13 consecutive days (Table II). In addition to the direct outdoor exposures, another set of PSDs was simultaneously exposed inside a chamber through which the outside air was passed at a known constant velocity (100 cm/sec). While these measurements were generally slightly higher than those made outside the chamber, they agreed within experimental error.

The modified SIS passive sampler is currently undergoing further eval-

TABLE II. Comparison of Modified PSD with a Tunable Diode Laser for NO_2 in Ambient Air

	Modified PSD		
	Outside	Inside Chamber*	Tunable Diode Laser
Day	Conc., PPB Uncertainty	Conc., PPB Uncertainty	Conc., PPB Uncertainty
1	11.1 ± 1.3	12.4 ± 3.8	10.2 ± 1.5
2	9.0 ± 1.1	9.3 ± 2.9	6.0 ± 0.9
3	10.9 ± 1.3	11.2 ± 3.5	7.8 ± 1.2
4	9.6 ± 1.2	11.2 ± 3.5	11.3 ± 1.7
5	9.3 ± 1.1	8.3 ± 2.6	6.2 ± 0.9
6	6.2 ± 0.7	5.9 ± 1.8	4.5 ± 0.7
7	9.1 ± 1.1	8.4 ± 2.6	8.4 ± 1.3
8	10.3 ± 1.2	10.3 ± 3.2	9.2 ± 1.4
9	12.1 ± 1.4	14.5 ± 4.5	13.4 ± 2.0
10	7.3 ± 0.9	11.1 ± 3.4	12.0 ± 1.8
11	7.1 ± 0.9	10.6 ± 3.3	12.2 ± 1.8
12	9.4 ± 1.1	9.9 ± 3.1	7.8 ± 1.2
13	3.5 ± 0.4	5.1 ± 1.6	4.7 ± 0.7
Avg.	8.8 ± 1.1	9.9 ± 2.8	8.8 ± 1.3

*Ambient outside air passed through chamber at 100 cm/sec.

uation for potential interferences from peroxyacetyl nitrate and nitrous acid. It is also being tested for temperature effects.

SUMMARY AND CONCLUSIONS

Carbon-based PSDs for VOCs lack the sensitivities required for nonoccupational exposure monitoring and are subject to problems associated with high atmospheric humidity. Likewise, commercial NO_2 monitors are not sensitive enough to meet the needs of personal exposure monitoring or indoor air sampling in most nonindustrial settings. The thermally-desorbable PSD developed by the EPA and marketed by Scientific Instruments Specialists, Inc., has been shown to be adequate for short-term (one to several hours) and long-term (24 hours) exposure monitoring of many VOCs at parts-per-billion concentrations. Exposure times are limited, however, for VOCs with breakthrough volumes (on Tenax-GC) of less than 30 L/g. Work is continuing toward development of a suitable low-rate PSD for those more volatile chemicals. A simple modification of the SIS passive sampler has also resulted in a very sensitive and convenient PSD for NO_2. Application to other gaseous chemicals is a distinct possibility in the future.

DISCLAIMER

The research described in this article has not been subjected to EPA review. Therefore, it does not necessarily reflect the views of the Agency and no official endorsement should be inferred. Mention of trade names or commercial products does not constitute endorsement or recommendation for use.

REFERENCES

1. Budiansky, S.: Indoor Air Pollution. *Environ. Sci. Tech. 14*:1023 (1980).
2. Wallace, L.A. and W. R. Ott: Personal Monitors: A State-of-the-Art Survey. *J. Air Poll. Control Assoc. 32*:601 (1982).
3. Coulson, D.M., S.J. Selover, E.C. Gunderson and B.A. Kingsley: Diffusional Sampling for Toxic Substances, Paper No. ENVR 32. *184th National Meeting of the American Chemical Society*. Kansas City, MO (September 1982).
4. *Detection and Measurement of Hazardous Gases*. C.F. Callis and J.G. Firth, Eds. Heinemann, London (1981).
5. Coutant, R.W. and D.R. Scott: Applicability of Passive Dosimeters for Ambient Air Monitoring of Toxic Organic Compounds. *Environ. Sci. Tech. 16*:410 (1982).
6. Lewis, R.G., R.W. Coutant, G.W. Wooten et al.: Applicability of Passive Monitoring Device to Measurement of Volatile Organic Chemicals in Ambient Air. *1983 Spring National Meeting*, American Institute of Chemical Engineers (March 1983).
7. Lewis, R.G., J.D. Mulik, R.W. Coutant et al.: Thermally Desorbable Passive Sampling Device for Volatile Organic Chemicals in Air. *Anal. Chem. 57*:214 (1985).
8. Bird, R.B., W.E. Steward and E.N. Lightfoot: *Transport Phenomena*, Chaps. 4 and 18. John Wiley, New York (1960).
9. Krost, K.J., E.D. Pellizari, S.G. Walburn and S.A. Hubbard: Collection and Analysis of Hazardous Organic Emissions. *Anal. Chem. 54*:810 (1982).
10. Coutant, R.W., R.G. Lewis and J.D. Mulik: Passive Sampling Devices with Reversible Adsorption. *Anal. Chem. 57*:219 (1985).
11. Walling, J.F.: The Utility of Distributed Air Volume Sets When Sampling Ambient Air Using Solid Adsorbents. *Atmos. Environ. 18*:855 (1984).
12. McClenny, W.A., T.A. Lumpkin, J.D. Pleil et al.: Canister Based VOC Samplers. *Proceedings of the 1986 EPA/APCA Symposium on Measurement of Toxic Air Pollutants*, pp. 402–407. Air Pollution Control Associa-

tion Publication VIP-7, U.S. Environmental Protection Agency Report 600/9-86-013. USEPA, Research Triangle Park, NC (1986).

13. Spicer, C.W., M.W. Holdren, L.E. Slivon et al.: Intercomparison of Sampling Techniques for Volatile Organic Compounds in Indoor Air. *Ibid.*, pp. 45–60.
14. Miller, D.P.: *Analysis of Nitrite in NO_2 Diffusion Tubes Using Ion Chromatography.* U.S. Environmental Protection Agency Report No. 600/54-87-013. USEPA, Research Triangle Park, NC (1987).
15. Fine, D.H. and D.P. Boundbehler: Trace Analysis of Volatile *N*-Nitro Compounds by Combined Gas Chromatography and Thermal Energy Analysis. *J. Chrom. 109*:271 (1975).
16. Mulik, J.D. and D.E. Williams: Passive Sampling Devices for NO_2. *Proceedings of the 1986 EPA/APCA Symposium on Measurement of Toxic Air Pollutants,* pp. 61–70. Air Pollution Control Association Publication VIP-7, U.S. Environmental Protection Agency Report 600/9-86-013. USEPA, Research Triangle Park, NC (1986).
17. Schiff, H.I., D.C. Hastie, G.I. Mackay et al.: Tunable Diode Laser Systems for Measuring Trace Gases in Tropospheric Air. *Environ. Sci. Technol. 17*:352 (1980).

CHAPTER 10

Tenax® Sampling of Volatile Organic Compounds in Ambient Air

SYDNEY M. GORDON

IIT Research Institute, 10 West 35th Street, Chicago, Illinois 60616

INTRODUCTION

The use of solid sorbents to selectively concentrate trace levels (i.e., ppt-ppb) of volatile organic compounds from ambient air for analysis is a well-established technique.[1-3] Of the various collection sorbents in use, high sensitivities and comprehensive analyses can be achieved using the organic polymer adsorbent Tenax®, along with thermal desorption and high-resolution gas chromatography/mass spectrometry/computer (GC/MS/COMP) analysis.[1,2]

Conceptually simple and quite convenient to use for active field sampling, the collection procedure relies upon a pump to draw air through a glass sampling cartridge containing Tenax. Thereafter, the volatile organics are recovered for analysis by thermally desorbing them with helium into a liquid nitrogen-cooled metal trap before introducing them into the GC/MS/COMP system. For trace-level work, however, several factors complicate the effective and reliable use of this technique. Often, relatively little is known about the composition of the atmosphere being sampled, making it necessary to identify the volatile organics present and, where possible, to obtain quantitative information about compounds of interest. Furthermore, in Tenax sampling, the entire sample is used in the analysis, so each sample can be analyzed only once. Coupled with the high cost of GC/MS/COMP analysis, every Tenax sample must be regarded as important. As a result, great care must be taken to ensure that the sampling is carried out effectively and that the analytical techniques used are acceptable.

This chapter presents a review of some of the more important problems associated with the use of Tenax for sampling ambient air and summarizes an approach to interpreting data obtained using the technique.

SAMPLING AMBIENT AIR USING TENAX

Several factors have been identified that complicate the quantitative interpretation of ambient air data obtained for volatile organics using Tenax:[4,5]

1. The Tenax may be contaminated and blank corrections may not apply equally to all cartridges in a set.
2. High moisture levels in the air being sampled may affect sample retention.
3. Compounds of interest may exceed the retention capacity of the Tenax bed.
4. Artifact formation may occur due to chemical reactions during sampling and/or thermal desorption.

Although all of these factors are undoubtedly important, 1 and 2 can be identified and dealt with in a straightforward manner. Our discussion will be confined to 3 and 4, since they are much more complex.

Sampling Efficiency: Quantitative Retention of Compounds by Tenax

The ability of Tenax to quantitatively retain trace organic compounds from air is a very important property which is expressed in terms of the breakthrough volume. It is usually defined[1,3] for a particular compound as the point in the volume-concentration profile at which a measurable amount (usually 50%) of the compound appears in the cartridge effluent (Figure 1). This quantity varies greatly from compound to compound. In principle, if the breakthrough volume for each compound in a sample is known, then limiting the volume of air drawn through the Tenax tube to a value less than that for the compound with the smallest breakthrough volume would result in quantitative retention of all compounds present.

Typical ambient air samples, however, generally give rise to well over

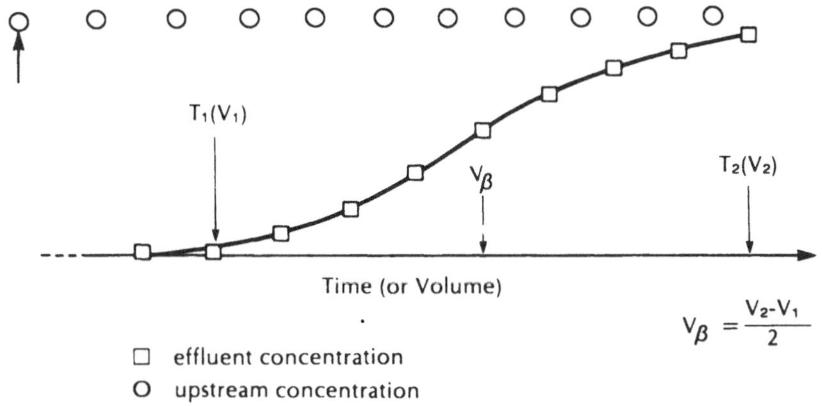

Figure 1. Schematic diagram of volume-concentration profile used to define the breakthrough volume for a compound.

100 identifiable compounds for which the breakthrough volumes are mostly unknown. Moreover, breakthrough volumes are complex functions of temperature and gas composition,[3,4,6] and a comprehensive data base representing such compositional complexity over typical temperature and concentration ranges simply does not exist.

In previous attempts to define a suitable sampling strategy with Tenax, one approach has been to group compounds into classes based on their known chromatographic retention behavior at a particular ambient temperature. Breakthrough volumes for a number of compounds, some of which are listed in Table I, have been measured by several workers.[1,3,6,7] Large differences are evident in the measured breakthrough volumes for certain substances. Nevertheless, a clear distinction can be made between those compounds with small values (e.g., chloroform, acetone, 1-propanol) and those with fairly large retentions (e.g., toluene, benzaldehyde). With these differences in mind, most conventional field sampling work using Tenax has been based on the following general strategy:[6]

1. Use a fixed cartridge size containing a given amount of Tenax.
2. Select the substance of interest that has the smallest breakthrough volume at a given temperature. If necessary, divide the list of substances into ranges of breakthrough volumes, and sample each range independently.
3. Select a suitable flow rate for sampling, in the range 50–500 ml/min.

TABLE I. Breakthrough Volumes for Selected Ambient Pollutants (liters/gram Tenax) at 20°C

	Breakthrough Volume			
Compound	Pellizzari[A]	Brown/Purnell[B]	Walling[C]	IITRI[D]
Benzene	25	31	50	
Toluene	111	190	268	105
Ethylbenzene	315			
Chloroform	11	9	17	
Carbon tetrachloride	10	31		
1,2-Dichloroethane	14	27	13	
1,1,1-Trichloroethane	7	19		
Trichloroethylene	23	28	44	
Tetrachloroethylene	89	240	203	
Chlorobenzene	215	130		
o-Dichlorobenzene	394			
Acetone	5	3		1
Acetophenone	808,1258			
Benzaldehyde	1594			
Ethoxyethylacetate				1800
Methanol	0.4			0.1
1-Propanol	6	4		1

[A]Krost et al.[1]
[B]Brown and Purnell.[3]
[C]Walling et al.[6]
[D]Dravnieks.[7]

In following this strategy, the traditional approach to check for loss of material from a Tenax cartridge has been to simply place two tubes in tandem.[6] If the substance of interest is found on the front collector with little or nothing on the back collector, this indicates quantitative retention of that substance by the first collector. If, however, the substance is not detected on either collector, this suggests that it was not present in the sample to begin with. When significant amounts of the compound are found on the second collector, this is usually because the breakthrough volume of the first collector has been exceeded, and it is assumed that everything that eluted from the first collector was retained by the second collector. The total amount of material in the sample is then taken to be the sum of the amounts measured on the two collectors. Aside from the fact that two analyses are required to obtain one datum point, the major problem with this approach is that it assumes that no significant loss occurs from the second collector.[6] This assumption is unverifiable in practice since complete retention of the material on the second collector cannot be conclusively demonstrated.

A more direct approach has been to simply "correct" the data when the total volume of air sampled is thought to exceed the breakthrough volume. Data are usually "corrected" when only one Tenax cartridge has been used and the results are needed for a variety of compounds with widely differing breakthrough volumes. The "correction" is carried out in two steps. For a given temperature, the chromatographically determined breakthrough volume is first found from tables or is estimated. The "corrected" concentration is then calculated as the amount of material found on the collector per unit breakthrough volume. The calculated concentration is therefore directly affected by the uncertainty in the value of the breakthrough volume, and the actual volume sampled plays no role at all.

Distributed Air Volume Sampling

These observations suggest that the standard approaches to the interpretation of Tenax data are insufficient to reveal special complications that may arise. The question of whether a compound with moderate breakthrough volume is being quantitatively retained cannot be answered in a straightforward way.

Walling[4] has proposed an empirical approach to Tenax sampling that is designed to reveal the presence of complications without, however, being able to indicate their origin. This approach, called distributed air volume (DAV) sampling, is based on the very simple idea that the amount of a substance adsorbed on Tenax should be a linear function of the volume of air sampled. If this is the case, then the concentration of the substance will be independent of the volume. Certain complicating factors, however, do not scale linearly with volume. When they occur, the consistency of the concentration for the data set is disturbed and serves as an indication that problems may be associated with the data.

This concept may be illustrated[8] by considering the hypothetical data in Table II. Four samples were collected over the same period of time, but each at a very different flow rate. After analysis, the amounts of material present on each collector were calculated to give the apparent concentrations shown in the table.

To judge the significance of these values, we need to know the repeatability of the sampling and analysis system. Previous studies[4,8] have shown that, in the concentration range 1–10 $\mu g/m^3$, the maximum difference to be expected between any two results at the $\alpha = 0.05$ level should be less than 2 $\mu g/m^3$ for benzene and toluene, and half that for other

TABLE II. Hypothetical Data Set for a Compound*

Air Volume (L)	Amount (ng)	Apparent Conc., g/m^3
5	25	5.0
10	40	4.0
20	70	3.5
40	130	3.3

*Walling.[4]

compounds. For concentrations greater than 10 $\mu g/m^3$, differences of 20 percent and 10 percent, respectively, are applicable.

In terms of these criteria, if it is assumed that the maximum difference between any two results is 1 $\mu g/m^3$, it is clear that some sort of complication is associated with the hypothetical data set of Table II. Three of the more important possible explanations are that:[4,8] 1) the breakthrough volume for the compound was exceeded, 2) the GC/MS system was calibrated incorrectly by a systematic amount, and 3) the first sample (5 L) in the table may have been contaminated. In this example, the use of DAV sampling sets has served to help identify corrupted data, but it does not identify the nature of the complication(s).

The DAV sampling approach has been used in several field studies[4,5,8] and provided data sets that give clear indications of various complicating factors that can have a significant impact on data quality. Because some acceptable data were obtained in each sampling situation, no generalizations can be made.

Although the DAV sampling approach is scientifically sound, a major practical problem associated with it is that it results in a high failure rate. The acceptance criteria mentioned earlier have recently been applied to a large number of ambient air values: in general only 22 percent of the individual data sets were found to satisfy the criteria.[9] If, however, the criteria are relaxed to the $\alpha = 0.01$ level, then the concentration differences must be less than 4 $\mu g/m^3$ for benzene and toluene, and less than 1.3 $\mu g/m^3$ for other compounds. When these criteria are applied to the same set of data, the acceptance rate increased to about 30 percent.[9]

Chemical Transformations During Sampling and/or Thermal Desorption

If the inconsistencies observed in DAV data sets are not due to factors such as experimental errors, contamination, inadequate blank corrections or low breakthrough volumes, then we are forced to consider chemical reactions during sampling or thermal desorption as a possible explanation.[5,10,11]

Under laboratory simulation of atmospheric conditions, Pellizzari et al.[10] examined the sampling of gas-phase organics in the presence of reactive inorganic gases. Even though Tenax does not concentrate reactive inorganic gases such as ozone, nitrogen oxides, and molecular halogens, *in situ* reactions may occur with the Tenax itself or with certain adsorbed compounds present. The investigation used deuterated analogs of toluene, styrene, and cyclohexene with thermal desorption GC/MS analysis to gain a better understanding of these reactions.

Table III lists several examples of chemical transformations involving Tenax that have been reported in the literature. The data show the formation of several deuterated oxidation and halogenated reaction products in the presence of ozone and molecular chlorine. Tenax itself also reacts with strong oxidizing agents to yield benzaldehyde, acetophenone, and phenol, lending weight to the conclusion that the analysis for these

TABLE III. Examples of Chemical Transformations Involving Tenax During Ambient Air Sampling for Volatile Organics

Ambient Air and Reactive Gases*			
Reactants			**Reaction Products**
Styrene-d_8	+ O_3	→	benzaldehyde-d_6
		→	benzoic-d_6
Cyclohexene-d_{10}	+ O_3	→	cyclohexadiene-d_8
		→	benzene-d_6
		→	$C_6D_{10}O$ isomers (3)
Styrene-d_8	+ Cl_2	→	styrene-d_8-dichloride
		→	chlorostyrene-d_7 (2)
Cyclohexene-d_{10}	+ Cl_2	→	2-chlorocyclohexanol-d_{10}
		→	dichlorocyclohexane-d_{10} (2)

Tenax and Reactive Gases**		
Reactants		**Reaction Products**
Tenax + Cl_2	→	benzaldehyde
Tenax + O_3, NO_x	→	acetophenone
Tenax + SO_2/SO_3	→	phenol

*Pellizzari and Krost.[11]
**Pellizzari, Domain and Krost.[10]

compounds in ambient air is not possible unless special precautions are taken.[10]

In addition to the reactions that can occur during sample collection, reaction products desorbed from Tenax have also been reported.[5] In these experiments, clean Tenax tubes were spiked with mixtures of bromotrichloromethane and pentachloroethane. Following thermal desorption and GC/MS analysis, both trichloroethylene and tetrachloroethylene were detected in significant amounts.

INHIBITING, IN SITU CHEMICAL REACTIONS ON TENAX

Pellizzari et al.[10] have shown that oxidants that appear to undergo *in situ* reactions with Tenax can be effectively removed before capturing volatile organics by using mild reducing agents. Sodium thiosulfate-impregnated glass fiber and Teflon filters quantitatively decompose ozone. However, further studies are needed to determine the effect of the impregnated filters on the composition of organic vapors collected.

Chemical reactions that occur as a result of thermal desorption may be avoided by using supercritical fluids to extract and recover volatile organics from Tenax. Hawthorne and Miller[12] have demonstrated the utility of supercritical CO_2 for the desorption of polynuclear aromatic hydrocarbons (PAHs) from Tenax. The amounts of material used in this study ranged from 100 to 2000 ng per compound, and quantitative recoveries were obtained for PAHs as large as coronene.

The ability of supercritical CO_2 to desorb compounds of relatively large molecular weight from Tenax has been further explored by Raymer and Pellizzari.[13] Compounds representing chlorinated hydrocarbons, chlorinated aromatic compounds, PAHs, and organophosphate pesticides showed recoveries exceeding 90 percent and, in certain cases, significantly better recoveries than were obtainable by thermal desorption. These experiments suggest that the future development of supercritical fluid extraction techniques for the recovery of organic pollutants from Tenax may provide an important alternative to the more traditional thermal desorption method.

CONCLUSIONS

Despite the documented problems with using Tenax to sample organic pollutants in ambient air, Tenax remains an important and valuable sorbent for the concentration and analysis of a wide variety of organic compounds at the ppt-ppb level. Nevertheless, it is clear that special care is required in using the sorbent, since it does not work equally well for all volatile organics.

To ensure the quality of ambient air data obtained with Tenax, a proposed approach for routine sampling is as follows:

1. Initiate every field study with the DAV sampling protocol.
2. Establish the overall sampling and analytical repeatability at the 99 percent confidence level.
3. Compare the experimental values from the DAV data sets with the pass/fail criteria.
4. Seek explanations for any inconsistencies.
5. For reasons of cost, once reliable data are available for a given situation, the analytical burden may be reduced by reverting to single or dual samples instead of full DAV data sets.
6. If standard quality assurance procedures reveal inconsistencies, reinstate the DAV sampling approach.

REFERENCES

1. Krost, K.J., E.D. Pellizzari, S.G. Walburn and S.A. Hubbard: Collection and Analysis of Hazardous Organic Emissions. *Anal. Chem. 54*:810 (1982).
2. Pellizzari, E.D.: Analysis for Organic Vapor Emissions near Industrial and Chemical Waste Disposal Sites. *Environ. Sci. Tech. 16*:781 (1982).
3. Brown, R.H. and C.J. Purnell: Collection and Analysis of Trace Organic Vapor Pollutants in Ambient Atmospheres: The Performance of a Tenax-GC Adsorbent Tube. *J. Chromatogr. 17*:879 (1979).
4. Walling, J.F.: The Utility of Distributed Air Volume Sets When Sampling Ambient Air using Solid Adsorbents. *Atmos. Environ. 18*:855 (1984).
5. Walling, J.F., J.E. Bumgarner, D.J. Driscoll et al.: Apparent Reaction Products Desorbed from Tenax Used to Sample Ambient Air. *Atmos. Environ. 20*:51 (1986).
6. Walling, J.F., R.E. Berkley, D.H. Swanson and F.J. Toth: *Sampling Air for Gaseous Organic Chemicals Using Solid Adsorbents. Application to Tenax.* EPA 600/4-82-059; U.S. Environmental Protection Agency (August 1982).
7. Dravnieks, A.: Unpublished results, IIT Research Institute (1977).
8. Walling, J.F.: *Experience from the Use of Tenax in Distributed Ambient Air Volume Sets.* Workshop on Quality Assurance in Air Pollution Measurements. APCA TP-8 Data Analysis/Quality Assurance Committee, Boulder, CO (October 1984).
9. Walling, J.F.: Private communication (August 16, 1985).
10. Pellizzari, E.D., B. Demian and K.J. Krost: Sampling of Organic Compounds in the Presence of Reactive Inorganic Gases with Tenax GC. *Anal. Chem. 56*:793 (1984).

11. Pellizzari, E.D. and K.J. Krost: Chemical Transformations during Ambient Air Sampling for Organic Vapors. *Anal. Chem. 56*:1813 (1984).
12. Hawthorne, S.B. and D.J. Miller: Extraction and Recovery of Organic Pollutants from Environmental Solids and Tenax-GC Using Supercritical CO_2. *J. Chromatogr. Sci. 24*:258 (1986).
13. Raymer, J.H. and E.D. Pellizzari: Toxic Organic Compound Recoveries from 2,6-Diphenyl-p-phenylene Oxide Porous Polymer Using Supercritical Carbon Dioxide and Thermal Desorption Methods. *Anal. Chem. 59*:1043 (1987).

CHAPTER 11

A Multisorbent Sampler for Volatile Organic Compounds in Indoor Air

ALFRED T. HODGSON,[A] JEHUDA BINENBOYM[B] and JOHN R. GIRMAN[C]

[A]Indoor Environment Program, Applied Science Division, Lawrence Berkeley Laboratory, University of California, Berkeley, California; [B]Israel Atomic Energy Commission, Soreq Nuclear Research Center, Yavne, Israel; [C]Indoor Air Quality Program, Air and Industrial Hygiene Laboratory, California Department of Health Services, Berkeley, California 94704

INTRODUCTION

The limited data on indoor/outdoor concentration ratios of volatile organic compounds (VOC) in residences, offices, and schools show that for many compounds these ratios are greater, and in some cases much greater, than one.[1-4] Considerably more data are needed on indoor concentrations of VOC for anything more than rudimentary assessments of population exposures and chronic health risks. Acute health problems, principally mucous membrane irritation, in some buildings have also been postulated to result from elevated concentrations of VOC. The present paucity of data can be largely attributed to the difficulty and expense of analysis for a wide range of VOC. Qualitative and quantitative methods for VOC which are rapid, reliable, and relatively inexpensive are needed to survey a variety of buildings, to identify sources of potentially harmful compounds, and to evaluate the efficacy of mitigation strategies.

Because most VOC occur at relatively low concentrations, even indoors, an enrichment or concentration step is required prior to chromatographic analysis. Solid adsorption is the most widely used method of sample preconcentration, and many NIOSH procedures use charcoal adsorption followed by solvent desorption. However, to avoid dilution by a solvent, methods have also been developed which use solid sorbents for sample collection in combination with thermal desorption of sample components onto a chromatographic column.

A number of solid sorbent materials have been evaluated for the collection and thermal desorption analysis of VOC. Of these, Tenax® is the most studied and widely used. It has the advantages of inertness, high thermal stability, and low affinity for water vapor. However, retention, or breakthrough volumes, for a variety of highly volatile compounds on Tenax are low.[5,6] A number of carbonaceous sorbents, such as activated charcoals, Ambersorb XE-340 (a pyrocarbon modified charcoal), and similar products, have also been evaluated for the collection and thermal desorption analysis of VOC.[7-9] These materials are thermally stable and have very high specific surface areas which make them well suited for the collection of low-boiling compounds; however, they adsorb water, and thermal desorption of high-boiling compounds is not always quantitative.

An obvious solution for a generalized sampler for VOC is to use several complementary materials, such as Tenax and a carbonaceous sorbent, in series in order of increasing affinity for low-boiling compounds.[5,9] The major objection to using carbonaceous backup sorbents has been the trapping of excessive amounts of water. However, some water is tolerable as long as it does not significantly decrease breakthrough volumes for analytes,[5] react with the sorbents or analytes, or otherwise interfere with the analysis, e.g., ice blockage of cryogenic traps. To minimize the difficulty and expense of analysis, it is most desirable to combine the sorbents within a single sampler.[10] For thermal desorption, the sampler must be reversed in the gas stream so that high-boiling compounds which are trapped by sorbents with low affinity do not contact sorbents with higher affinity as they elute. One such sampler which contains a series of Tenax-TA, Ambersorb XE-340, and activated charcoal is commercially available.

Using this sampler, an analytical method was devised for the quantitation of part-per-billion concentrations of VOC in ambient and indoor air; the method is described in this chapter. The authors assembled the instrumental components, including fabrication of an on-column cryogenic focusing attachment, and developed procedures for the method. Using this method in laboratory experiments, breakthrough volumes for low-boiling compounds under simulated use conditions were determined.

In a large office building, the method was used to evaluate the effect of ventilation rate on air quality.

EXPERIMENTAL

All sorbent tube samplers are constructed of borosilicate glass tubing, 203 mm in length with 6-mm O.D. and 1-mm wall thickness. A glass frit holds the sorbents in place at the inlet of a sampler. Small plugs of silanized glass wool separate sorbent layers and hold the sorbents in place at the exit of a sampler.

Multisorbent samplers (Part No. ST032, Envirochem, Inc., Kemblesville, PA) are packed in series with glass beads at the inlet followed by Tenax-TA, Ambersorb XE-340, and activated charcoal in order of increasing affinity for low-boiling compounds. The average lengths and weights of the sorbent layers are: glass beads—14.2 mm, 294 mg; Tenax-TA—30.1 mm, 85.5 mg; Ambersorb XE-340—29.4 mm, 167 mg; and activated charcoal—8.8 mm, 48.0 mg. The relative standard deviation (RSD) in the lengths of individual sorbent layers is approximately 10 percent. Weights of individual sorbent layers are more variable. The charcoal layer was removed from some multisorbent samplers prior to laboratory experiments.

Tenax samplers (Part No. ST023, Envirochem, Inc.) are packed with glass beads at the inlet followed by Tenax-TA. The average lengths of the layers of glass beads and Tenax-TA are 12.4 mm and 61.0 mm, respectively.

Prior to use, samplers are conditioned by heating them to 300°C for ten minutes with a helium purge flowing at 100 $cm^3 min^{-1}$ in the reverse direction of gas flow during sample collection. Samplers are sealed at each end with nylon tube caps and 6-mm Teflon ferrules. Capped samplers are stored at -10°C in elongated Kimax culture tubes with screw caps with Teflon liners.

After exposure, a sample is thermally desorbed from a sampler and introduced into a capillary gas chromatograph (GC) with a UNACON Model 810A (Envirochem, Inc.) sample concentrating and inletting system. The programmed sequence of events and temperatures for the UNACON is shown in Table I. First, the exposed sampler is inserted into the tube desorption chamber of this instrument so that helium flows through the sampler in the reverse direction of sample gas flow. After air has been purged from the sampler, it is heated over a period of several minutes to approximately 275°C. Next, the instrument passes the sample through dual sequential traps of decreasing internal diameter to concen-

TABLE I. Typical Event Times and Temperatures for Sample Concentration and Inletting with UNACON

Elapsed Time (min:sec)	Event	Max. Temp. (°C)
0:00 – 1:00	Initial carrier gas flow	
1:00 – 5:00	Tube chamber heating	275
5:00 – 10:00	Secondary carrier gas flow	
10:00 – 10:21	Trap 1 heating	310
10:21 – 12:21	Trap to trap transfer	
12:21 – 12:42	Trap 2 heating	310
12:42 – End	Trap to column transfer	
10:00 – 14:42	On-column cryogenic focusing	

trate the sample. These traps are packed with a series of sorbents similar to the sorbents in the samplers but do not contain activated charcoal. The first trap has a 4-mm I.D. enabling it to retain compounds from the relatively large volume of gas used in the desorption of the sampler. The second trap has a 1-mm I.D. to permit efficient desorption of the sample at typical carrier gas flow rates for capillary columns. A heated transfer line connects the instrument with the GC.

The UNACON is equipped with a flame-ionization detector (FID). At several points in the concentrating sequence, a small portion of the sample (4–8%) is automatically diverted to the FID for monitoring of trap desorption and for an indication of total organic carbon concentration.

Sample components are resolved with a GC (5790A series, Hewlett-Packard Co., Palo Alto, CA) equipped with liquid nitrogen (LN) subambient cooling and a crossed-linked 5 percent phenylmethylsilicone, fused-silica capillary column (Model 19091J Opt. 202, Hewlett-Packard Co.).

Chromatographic peak shape, peak resolution, and peak area response for low-boiling compounds are enhanced with an on-column cryogenic focusing attachment we developed for the GC. This device sprays LN directly on a 5-mm section of the capillary column, 30 mm from the within-oven juncture of the column and the UNACON transfer line. The LN at 60 kPa is obtained from the same line supplying the GC cryogenic valve assembly. Flow is controlled with a solenoid on-off valve (Part No. 8262A203, ASCO, Automatic Switch Co., Florham Park, NJ) positioned in the unoccupied detector compartment of the GC. The transfer line from the valve to the spray nozzle is 3.2-mm O.D. stainless-steel tubing insulated with ceramic wool. The nozzle is a 3.2- to 1.6-mm tube-fitting reducer bored through to 1.6 mm. The nozzle is positioned in the center of the GC oven at the point of minimum air velocity. The section of the column to be cooled is positioned 7 mm from the nozzle with wire eyelets. The cryogenic focusing attachment is operated so that the sample is concentrated at the head of the column as it is desorbed from the second trap in the UNACON. When LN flow to the device is turned off, the cooled section of column rapidly rises to the temperature of the oven.

The GC is connected via a direct capillary interface to a 5970B series Mass Selective Detector (MSD) equipped with a 59970A series work station (Hewlett Packard Co.). Each day prior to analyzing samples, the MSD is tuned and calibrated using perfluorotributylamine.

For qualitative analyses, the MSD is typically scanned from m/z 33 to m/z 250 with a continuous scan cycle time of approximately 0.5 sec. Identifications of compounds are made by: 1) comparing unknown spectra and retention times to spectra and retention times of known compounds analyzed using identical conditions, or 2) by comparing unknown spectra to spectra contained in the National Bureau of Standards mass spectral library[11] or in an eight-peak index[12] with confirmation by analyses of known compounds whenever possible.

For quantitative analyses, the MSD is operated to monitor multiple, individually-selected mass ions. For each compound of interest, a mass ion with high relative intensity is chosen as the quantitative ion, and a characteristic ion of equal or lower intensity is chosen as a qualifying ion for confirmation of compound identity. A typical analytical run for a sample containing multiple compounds is broken down into sequential groups of six to ten mass ions, each with dwell times chosen to give a scan cycle time for each ion group of approximately 0.25 sec.

A standard gas mixture for qualitative or quantitative analysis is prepared by injecting an aliquot of a liquid mixture of the analytes of interest into a helium-filled two-liter flask with septum cap which is then heated and maintained at 65°C.[13] Samples are withdrawn from the flask with a gas-tight syringe and are injected into a helium gas stream flowing at 100 cm^3 min^{-1} through a conditioned sampler in the direction of sample gas flow. Flow is maintained for five minutes after which the sampler is analyzed using the normal procedure. Standard gas mixtures are prepared daily.

Air sampling flow rate in both laboratory and field experiments is typically 100 cm^3 min^{-1} (20°C, 760 mm Hg). The vacuum source for replicated field samples is provided by a diaphragm pump (Air Cadet, Cole-Parmer Instrument Co., Chicago, IL). Sample flow rates are regulated with electronic mass-flow controllers placed between samplers and the pump. Typically, samples are collected simultaneously in triplicate in laboratory experiments, and samples are collected simultaneously in duplicate in field experiments. Separate replicated samples are collected for qualitative and quantitative analyses. Sample volumes are varied according to expected analyte concentrations. Typical sample volumes for urban outdoor air are 5 or 10 L. Sample volumes for indoor air may be lower.

Laboratory experiments for the direct determination of sample breakthrough volumes under simulated use conditions are conducted in a 20-m^3 environmental chamber. All interior surfaces of the chamber are clad in stainless steel. The natural ventilation rate of the chamber with the air inlet and outlet closed is 0.03 ± 0.01 h^{-1}. Prior to an experiment, the chamber is mechanically ventilated at 12 h^{-1}. The ventilation is then turned off, and the inlet and outlet are closed. Samples (5 L) for the determination of the chamber background are collected. The chamber is spiked with the analytes of interest to known concentrations by injecting microliter volumes of liquid mixtures of the analytes into a heated port connected to an air stream flowing to a mixing fan. If required, the humidity in the chamber is increased by the addition of water vapor. Air and dewpoint temperatures in the chamber are monitored throughout an experiment. Breakthrough volumes are determined from the analysis of backup samplers which are placed in series with the samplers being investigated and are exchanged at predetermined volume intervals.

Field samples were collected at the site of a five-month-old office building. The six-story building has a ventilated volume of approximately 36,700 m^3. Ventilation rates were calculated for the entire building from air velocity measurements made with a hot-wire anemometer at the building's air inlet vents on the roof. On a single day, air samples for VOC were collected outdoors on the roof (5 L) and indoors in open office spaces (3 L) at near steady-state conditions with the ventilation system operating in three different modes: 100 percent outside air, exhaust fans off; 100 percent outside air, exhaust fans on (ventilation rate = 5.2 h^{-1}); and recirculation with approximately 16 percent outside air (ventilation rate = 0.84 h^{-1}).

RESULTS AND DISCUSSION

The UNACON concentrates the sample components while eliminating much of the water retained by the sampler. However, the traps are heated over a period of approximately 21 sec, and because of heat transfer effects, they take longer to reach maximum temperature and then cool slowly. As a result, analyte vapors enter the column as a broad band. At an initial GC oven temperature of 1°C and without the use of the cryogenic focusing attachment, the peaks of low-boiling compounds are broad and poorly resolved (Figure 1a). At the same initial oven temperature, the use of the cryogenic focusing attachment effectively recondenses the analyte vapors at the inlet of the column. Peak resolution is greatly enhanced, and peak area response increases by an average of 13 percent (Figure 1b). No early emerging peaks are observed, indicating that little of the analytes escapes recondensation.[14] However, by decreasing the oven temperature to -30°C with cryogenic focusing, another similar increase in peak area response can be achieved.

The pressure drop across samplers containing Tenax, Ambersorb, and charcoal (TAC) and samplers containing only Tenax and Ambersorb (TA) was investigated as a function of sampling rate. The pressure drop

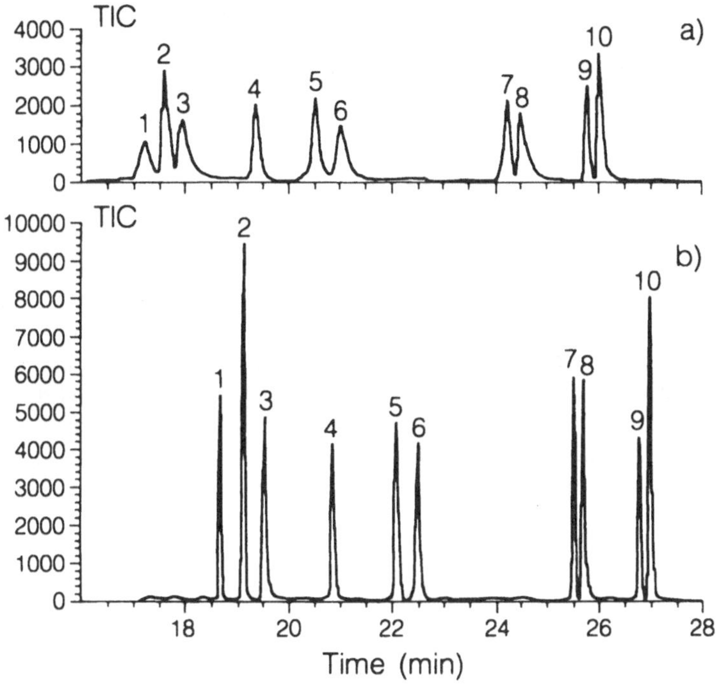

Figure 1. Total ion current chromatograms for selected ion analyses of: 1) trichlorofluoromethane, 284 ng; 2) n–pentane, 159 ng; 3) 2–propanol, 299 ng; 4) dichloromethane, 253 ng; 5) 4–methyl–1–pentene, 169 ng; 6) cyclopentane, 189 ng; 7) n–hexane, 168 ng; 8) 2-butanone, 153 ng; 9) chloroform, 188 ng; and 10) ethyl acetate, 171 ng. Oven temperature held at 1°C and then, starting between peaks 6 and 7, ramped at 10°C min^{-1} a) without cryogenic focusing and b) with cryogenic focusing.

increased substantially over a flow rate range of 50 to 250 cm^3 min^{-1} and was highly variable among samplers (Table II). Since a large pressure drop is impractical for field sampling and may also result in the vacuum desorption of trapped analytes,[15] a flow rate of 100 cm^3 min^{-1} was selected for most sampling applications. This is near the optimal rate for samplers of this geometry.[6]

Water vapor uptake by TAC and TA samplers at a sample gas flow rate of 100 cm^3 min^{-1} and a water vapor concentration of 9.52 g m^{-3} (20°C, 760 mm Hg) in the chamber air, or 55 percent relative humidity, is shown in Table III. As expected, samplers containing charcoal retain the most water. At a sample volume of 10 L, TAC and TA samplers retain 20 and 15 mg of water, respectively. This represents 21 and 16 percent, respectively, of the total amount of water passing through the samplers. Since the design of the UNACON allows much of the water to be vented while the sample is transferred to the first analytical trap, the system can tolerate up to 25 mg of water retained on a sampler before ice blockage of the capillary column occurs during trap to column transfer with cryogenic focusing. While Tenax degradation products can increase when sampling at high relative humidity,[15] the effect of water on TAC and TA sampler blanks is expected to be moderate since the samplers contain a relatively small volume of Tenax.

Breakthrough volume has been defined as the volume at which a significant amount of a constant atmosphere of an adsorbate drawn through a sorbent tube appears in the tube effluent.[6] The method of direct measurement of breakthrough volume using backup samplers collected at various volume intervals was used because it allows simulation of important field sampling parameters such as air temperature, presence of other compounds, and water vapor concentration. Breakthrough volumes were only measured for representative low-boiling compounds.

TABLE II. Pressure Drop Across Samplers Versus Flow Rate

	Pressure Drop	
Flow Rate[A] (cm^3 min^{-1})	TA Samplers (mm Hg)	TAC Samplers (mm Hg)
50	26 ± 5[B]	31 ± 6[B]
100	55 ± 11	65 ± 15
150	87 ± 19	102 ± 26
200	122 ± 29	143 ± 40
250	163 ± 41	191 ± 53

[A]Flow rate at 20°C, 760 mm Hg.
[B]Mean ± standard deviation for 5 samplers.

TABLE III. Cumulative Water Vapor Retention Versus Sample Volume

Sample Volume[A] (L)	Cumulative H_2O TA Sampler (mg)	Retention[B] TAC Sampler (mg)
2	5.1	7.5
4	7.8	11.1
6	10.1	14.2
8	12.4	17.2
10	14.9	20.3

[A]Volume at 20°C, 760 mm Hg.
[B]At H_2O vapor concentration of 9.52 g m^{-3} (55% RH).

Breakthrough volumes for higher boiling compounds can be estimated since, for most compounds, there is a good correlation between breakthrough volume and boiling point.[6] Breakthrough volume can be related to the more commonly measured retention volume, which is defined as the volume at which the peak maximum for a single injection emerges from a sorbent tube,[6] from a knowledge of sampler theoretical plates.[16]

In Table IV, breakthrough volumes for ten low-boiling compounds on TA and Tenax samplers are compared at conditions of 21°C, approximately 50 percent relative humidity, and 100 cm^3 min^{-1} sample flow rate. Significant breakthrough is defined as a loss of more than one percent of an analyte. For the ten compounds, breakthrough volumes on Tenax alone are generally small, with half of the compounds having breakthrough volumes of less than one liter. These volumes are similar to safe sampling volumes (retention volumes/2) reported for samplers of the

TABLE IV. Breakthrough Volumes for Low-boiling Compounds

Compound	Boiling Point (°C)	Breakthrough Tenax Sampler	Volume* (L) TA Sampler
Trichlorofluoromethane	24	< 0.5	< 4
n-Pentane	36	< 0.5	> 10
Dichloromethane	40	< 0.5	8–10
n-Hexane	69	1–2	> 10
2-Butanone	80	1–2	> 10
Chloroform	62	1.2	> 10
Ethyl acetate	77	> 4	> 10
1,1,1-Trichloroethane	74	< 0.5	> 10
Benzene	80	2–4	> 10
Pentanal	102	0.5–1	> 10

*Breakthrough is loss of > 1% of analyte.

same internal diameter containing 130 mg of Tenax-GC.[6] Such small volumes are inadequate for the intended application since sample volumes of 10 L may be required. The use of much larger amounts of Tenax to increase breakthrough volumes can necessitate more rigorous cleanup procedures to ensure acceptable sampler blanks. In addition, increases in breakthrough volumes may be lower than expected if sampler geometry is changed in a way that reduces theoretical plates.

With TA samplers and a sample volume of 10 L, trichlorofluoromethane (Freon-11) and, to a smaller extent, dichloromethane are the only compounds not quantitatively sampled. The charcoal layer in TAC samplers increases the breakthrough volume for trichlorofluoromethane to > 5 L while breakthrough volumes for the other compounds exceed 10 L. Thus, the addition of a backup layer of Ambersorb increases breakthrough volumes for many low-boiling compounds to at least the maximum sample volume required for the intended application. Addition of a charcoal layer may be necessary for the quantitative sampling of very low boiling compounds such as Freons; however, as noted above, it has the disadvantage of increasing water vapor retention.

Adequate retention volumes simplify the calculation of limits of quantitation. A value of ten times the standard deviation of instrumental noise and signal measurements has been recommended for the instrumental limit of quantitation.[17] By this criterion, the limit of quantitation for compounds used in this investigation is 0.1–0.5 ng. With a sample volume of 10 L and no breakthrough, the limit of quantitation for the method is expected to be 10–50 ng m^{-3} for most compounds. In practical terms, no attempt has been made to quantify concentrations much below 0.4 $\mu g\ m^{-3}$ with a 5-L sample volume.

The precision and accuracy of the method for the analysis of the ten compounds used in the study of breakthrough volume is demonstrated in Table V. Three replicate samples were collected with the TA sampler from the environmental chamber before and after the addition of known concentrations of the compounds. At background concentrations near 1 ppb, the precision of the method is better than 5 percent, with the exception of the analysis of pentanal which was more variable. At concentrations of 45–77 ppb, the precision for all compounds is better than 3.5 percent. Comparison of prepared and measured chamber concentrations indicates that the accuracy of the method for seven of the ten compounds is $\pm$ 5 percent or better. The concentration of benzene is 7 percent too high, and the concentration of pentanal is 12 percent too low for undetermined reasons. The high measured concentration of trichlorofluoromethane is probably caused by the offgassing of this compound from the

TABLE V. Precision and Accuracy for Low-boiling Compounds Using TA Sampler

	Concentration in Chamber (ppb)		
Compound	**Background**	**Prepared**	**Measured**
Trichlorofluoromethane	1.75 (0.1)[A]	49.0	58.4 (1.1)[A]
n-Pentane	1.54 (1.5)	76.5	74.5 (2.5)
Dichloromethane	0.95 (1.5)	68.8	65.3 (2.3)
n-Hexane	0.56 (2.5)	66.6	68.8 (3.2)
2-Butanone	0.74 (3.2)	72.9	72.9 (3.4)
Chloroform	—[B]	54.0	52.4 (3.1)
Ethyl acetate	0.11 (7.5)	65.9	67.3 (2.5)
1,1,1-Trichloroethane	1.11 (4.8)	44.7	46.5 (1.0)
Benzene	1.11 (4.4)	73.8	79.0 (3.0)
Pentanal	0.90 (4.0)	61.6	54.3 (3.0)

[A]Mean and (RSD) for triplicate samples.
[B]Below limit for quantitation.

polyurethane foam used to insulate the chamber. Calibration curves are typically linear with coefficients of determination (r^2) of > 0.99.

The accuracy of an analysis for VOC is strongly impacted by the potential for contamination and the formation of artifacts. For example, benzene and toluene are present in Tenax as impurities which are difficult to remove,[18] and benzaldehyde, acetophenone, and phenol are known decomposition products of Tenax.[15] On the other hand, Ambersorb XE-340 has few contaminants in the boiling point range between 100° and 300°C.[8] Blanks for TAC and TA samplers probably derive from Tenax since they typically consist of benzene (3.7 ng), toluene (2.6 ng), and benzaldehyde (> 10 ng). Water vapor and oxidants, such as ozone and oxides of nitrogen, can produce decomposition products from Tenax and from analytes adsorbed on Tenax. As examples, benzaldehyde, acetophenone and phenol have been shown to increase in magnitude when sampling air with high humidity, and cyclohexene collected on Tenax reacted with ozone to produce cyclohexadiene, three isomers of $C_6H_{10}O$ and benzene.[15] Artifacts can also be formed on carbonaceous sorbents. Terpenes, for example, have been observed to undergo degradation reactions when thermally desorbed from carbon black and a sorbent similar to Ambersorb XE-340.[9] Although artifact formation was not specifically investigated, alpha-pinene is the only compound analyzed to date which clearly undergoes some degradation.

The applicability of the method for the quantitation of VOC is demonstrated by samples taken outdoors and indoors at the site of a five-month-old, large office building. Forty-two individual compounds were detected in these samples. Approximately half of the detected compounds were identified, and many of the remaining compounds were partially classified. The identified compounds are ubiquitous components of urban outdoor and indoor air. The majority of identified compounds are hydrocarbons including 11 aromatic hydrocarbons, four n-alkanes, two branched alkanes, and one C_nH_{2n} hydrocarbon. In addition, there are four oxygenated compounds and three halocarbons. Sixteen of the compounds were quantified using multipoint calibration curves (Table VI). Precisions for average concentrations in duplicate samples collected with the ventilation system set at 100 percent outside air with exhaust fans on range from < 1 percent to 5.5 percent RSD. With this level of precision, it is possible to detect real differences in concentration with location and ventilation rate. For all compounds, except benzene and benzaldehyde, indoor concentrations consistently exceeded outdoor concentrations. When operating with 100 percent outside air, deactiva-

TABLE VI. Outdoor and Indoor Concentrations of VOC at a Large Office Building

	Outdoor Concentration	Indoor Concentration[A]		
Compound	Roof (ppb)	100% Vent. Exhaust On (ppb)	100% Vent. Exhaust Off (ppb)	16% Vent. (ppb)
n-Pentane	0.29	1.15 (2.5)[B]	2.46	0.59
Acetone	2.08[C]	6.94 (5.5)	10.2	12.6
n-Hexane	—[C]	0.56 (2.5)	1.14	—
2-Butanone	1.63	6.67 (0.8)	7.65	21.0
1,1,1-Trichloroethane	0.18	0.98 (1.4)	1.20	2.25
Benzene	1.99	2.11 (3.4)	3.65	0.87
3-Methylhexane	0.10	0.33 (2.1)	0.67	—
Trichloroethylene	—	—	—	1.30
Methylcyclohexane	0.14	—	0.63	—
Toluene	1.08	2.96 (2.6)	5.41	3.10
Ethylbenzene	0.20	0.97 (2.2)	1.39	2.28
m-, p-Xylene	0.70	3.85 (1.6)	5.38	9.67
o-Xylene	0.54	1.35 (2.6)	1.94	3.43
Benzaldehyde	1.99	3.06 (1.8)	1.80	0.82
n-Decane	0.39	0.40 (1.8)	0.47	0.60
n-Undecane	—	0.76 (0)	0.80	0.85
Total	11.3	32.1	44.8	59.4

[A]Measured in a single open office space.
[B]Mean and (RSD) for duplicate samples.
[C]Below limit of quantition.

tion of the exhaust fans decreased ventilation efficiency and increased the concentrations of all compounds except benzaldehyde. Although concentrations are blank-corrected, inconsistencies with benzaldehyde may be due to a variable sampler blank. Reduction of ventilation to approximately 16 percent outside air for 3.5 hours resulted in increased concentrations for 8 of the 16 compounds. An internal source strength was estimated for the 16 compounds using a single-equation, mass balance ventilation model.[19] The model assumes perfect air mixing. For simplicity, the penetration of the compounds into the building from outside air was taken to be unity, and removal mechanisms other than ventilation were ignored. The resulting composite source strength for the compounds was approximately 5 g h^{-1} for the entire building.

SUMMARY

The authors have assembled the instrumental components and developed procedures for an analytical method to quantify ppb concentrations of a broad range of VOC in ambient and indoor air. The method uses a multisorbent sampler in combination with a thermal desorption system for sample concentrating and inletting, a GC equipped with an on-column cryogenic focusing attachment, and a mass spectrometer. The sampler contains a series of Tenax, Ambersorb XE-340 and, in some cases, activated charcoal. Sampler breakthrough volumes for representative low-boiling compounds are $\geq$ 10 L. With a 10-L volume, limits of quantitation are $<$ 1 ppb. Overall precision is normally better than 5 percent at analyte concentrations ranging between one and several tens of ppb. Accuracy for representative low-boiling compounds is typically $\pm$ 5 percent. With this level of precision and accuracy, it is possible to detect real differences in concentrations of VOC in indoor air with location and ventilation rate. This was demonstrated by analyses made at a large office building.

ACKNOWLEDGMENTS

This work was supported by the Assistant Secretary for Conservation and Renewable Energy, Office of Building and Community Systems, Building Systems Division of the U.S. Department of Energy under Contract No. DE-AC03-76SF00098. The use of trade and product names does not imply endorsement.

REFERENCES

1. Hawthorne, A.R., R.B. Gammage, C.F. Dudney et al.: *An Indoor Air Quality Study of Forty East Tennessee Homes*. Oak Ridge National Laboratory Report ORNL-5965. ORNL, Oak Ridge, TN (1984).
2. Lebret, E., H.J. van de Wiel, H.P. Bos et al.: Volatile Hydrocarbons in Dutch Homes. *Indoor Air 4*:169. Swedish Council for Building Research, Stockholm, Sweden (1984).
3. De Bortoli, B., H. Knoppel, E. Pecchio et al.: *Measurements of Indoor Air Quality and Comparison with Ambient Air: A Study of 15 Homes in Northern Italy*. Commission of the European Communities Report, EUR 9656 EN. Directorate-General for Science, Research and Development, Joint Research Centre, Ispra, Italy (1985).
4. Wallace, L.A., E.D. Pellizzari, T.D. Hartwell et al.: Personal Exposures, Indoor-outdoor Relationships and Breath Levels for 355 Persons in New Jersey. *Atmos. Environ. 19*:1651 (1985).
5. Krost, K.J., E.D. Pellizzari, S.G. Walburn and S.A. Hubbard: Collection and Analysis of Hazardous Organic Emissions. *Anal. Chem. 54*:810 (1982).
6. Brown, R.H. and C.J. Purnell: Collection and Analysis of Trace Organic Vapor Pollutants in Ambient Atmospheres. *J. Chromatogr. 178*:79 (1979).
7. Holzer, G., H. Shanfield, A. Zlatkis et al.: Collection and Analysis of Trace Organic Emissions from Natural Sources. *J. Chromatogr. 142*:755 (1977).
8. Hunt, G. and N. Pangaro: Potential Contamination from the Use of Synthetic Adsorbents in Air Sampling Procedures. *Anal. Chem. 54*:369 (1982).
9. Isidorov, V.A., I.G. Zenkevich and B.V. Ioffe: Methods and Results of Gas Chromatographic-Mass Spectrometric Determination of Volatile Organic Substances in an Urban Atmosphere. *Atmos. Environ. 17*:1347 (1983).
10. Ogle, L.D., R.C. Hall, W.L. Crow et al.: Development of Preconcentration and Chromatographic Procedures for the Continuous and Unattended Monitoring of Hydrocarbons in Ambient Air. Preprint extended abstract. Division of Environmental Chemistry. *American Chemical Society Conference*, pp. 135–137. Kansas City, MO (September 1982).
11. Heller, S.R. and G.W.A. Milne: *EPA/NIH Mass Spectral Data Base*. U.S. Nat. Bur. Stand. Ref. Data Ser. 63. U.S. Govt. Printing Office, Washington, DC (1978).
12. *Eight Peak Index of Mass Spectra*, 3rd ed. The Mass Spectrometry Data Centre, The Royal Society of Chemistry, The University, Nottingham, UK (1983).
13. Riggin, R.M.: *Compendium of Methods for the Determination of Toxic Organic Compounds in Ambient Air*. EPA-600/4-84-041. USEPA, Environmental Monitoring Systems Laboratory, Research Triangle Park, NC (1984).
14. Graydon, J.W. and K. Grob: How Efficient are Capillary Cold Traps? *J. Chromatogr. 254*:265 (1983).

15. Pellizzari, E., B. Demain and K. Krost: Sampling of Organic Compounds in the Presence of Reactive Gases with Tenax GC. *Anal. Chem. 56*:793 (1984).
16. Sennum, G.I.: Theoretical Collection Efficiencies of Adsorbent Samplers. *Environ. Sci. Technol. 15*:1073 (1981).
17. Keith, L.H., W. Crummett, J. Deegan, Jr. et al.: Principles of Environmental Analysis. *Anal. Chem. 55*:2210 (1983).
18. Walling, J.F.: The Utility of Distributed Air Volume Sets when Sampling Ambient Air Using Solid Adsorbents. *Atmos. Environ. 18*:855 (1984).
19. Traynor, G.W., J.R. Girman, M.G. Apte et al.: Indoor Air Pollution Due to Emissions from Unvented Gas-fired Space Heaters. *J. Air Pollut. Contr. Assoc. 35*:231 (1985).

SECTION III

Special Topics

CHAPTER 12

Unstable Aerosols

SIDNEY C. SODERHOLM

University of Rochester, Medical Center, Department of Radiation Biology and Biophysics, 601 Elmwood Avenue, Rochester, New York 14642

INTRODUCTION

An aerosol consists of particles and the gas in which they are suspended. All aerosols are temporally unstable, i.e., they experience change with the passage of time. Some important aerosol characteristics which can change are: total mass concentration of a contaminant (sum of mass concentrations in the vapor and particle phases), fraction of a contaminant in the particle or vapor phase, and particle size distribution.

Consider a very general air contamination scenario: an atmospheric contaminant is emitted by a source; it may mix with and be diluted by ambient air; and it travels to a target person, object, or area where it has the potential to exert an effect, usually after being deposited. The physical dimensions involved in this process may be millimeters to thousands of kilometers and the times may be milliseconds to years. Air sampling may be performed near the source, in the ambient air, or near the target to characterize the source emission, identify the source, or predict the quality and quantity of contaminant reaching the target.

A wide variety of aerosol instabilities must be considered in air sampling because they may influence sampling strategy, occurrence of artifacts in samples, and extrapolation of measured aerosol characteristics to earlier or later times. It is often sufficient to have a qualitative understanding of aerosol instabilities in order to evaluate whether they are significant in a particular situation. The approach in this discussion is to identify a physical process, discuss aerosol characteristics which will change, present an estimate of the characteristic time of the change, and consider one or more examples. The characteristic time may be loosely defined as an estimate of the time required for the change to move one-half or two-thirds of the way toward completion. The characteristic time of one physical process may be compared to the characteristic time of another to determine whether one is much faster. The smaller the characteristic time, the faster the process occurs. Comparison of the characteristic time and the observation time will indicate whether a process will go to completion or will not have enough time to have any effect. It is assumed the reader has a standard table of aerosol properties which gives diffusion coefficient, Cunningham slip correction coefficient, and sedimentation velocity as a function of particle size for a particle density of 1 g/cm^3.[1]

Four types of aerosol instabilities will be considered: coagulation, deposition of particles on surfaces, deposition of vapor on surfaces and evaporation of vapor from surfaces, and vapor/particle interactions. Detailed equations and discussions of most of these topics are available in published literature and textbooks. This discussion will summarize some aspects thought to merit emphasis in state-of-the-art air sampling.

COAGULATION

Coagulation occurs as particles collide and stick together. This process decreases particle number concentration and increases mean particle size but does not change mass concentration. Thermal coagulation, the random collisions between particles due to Brownian motion, occurs in every aerosol. For an aerosol consisting of particles which are roughly the same size, the characteristic time of thermal coagulation is

$$T_{coag} = 56 \text{ min } \frac{10^6 \text{ cm}^{-3}}{N\, C_{slip}} \qquad (1)$$

where: N = number concentration
C_{slip} = Cunningham slip correction factor.

One practical effect of thermal coagulation is that high particle number concentrations are observed only relatively near emission sources.

The characteristic time of the loss of smaller particles due to collisions with larger ones is

$$T_{coag,2-1} = [\pi (D_{p1} + D_{p2}) (d_{p1} + d_{p2}) N_1]^{-1} \quad (2)$$

where: D_p = particle diffusion coefficient
d_p = particle diameter
N_1 = number concentration of the larger particles.

This characteristic time may be much smaller than that of coagulation among similar-sized particles and can result in relatively rapid transfer to larger particles of material originally emitted in smaller particles.

Coagulation rate is increased and the characteristic time decreased by factors which increase collision rates among particles.[1-3] This occurs due to attractive forces among particles, e.g., those due to bipolar electrical charging, or due to a phenomenon causing relative velocities between articles, e.g., sedimentation of particles with different aerodynamic diameters.

PARTICLE DEPOSITION ON SURFACES

An airborne particle tends to stick to any surface it hits, reducing the aerosol mass concentration. Gravitational and diffusional deposition, two mechanisms which are always present in the earth's atmosphere, will be briefly discussed.

Gravitational Deposition

The characteristic time of sedimentation onto horizontal surfaces is

$$T_{sed} = V/S_h v_g \quad (3)$$

where: v_g = particle sedimentation velocity
V = volume of the aerosol
S_h = total area of horizontal surfaces in the volume.

An inclined surface contributes an area to S_h which is equal to the projection of its area on the horizontal. In the idealized case of stagnant air in a chamber with vertical walls or sedimentation in laminar flow in a horizontal channel having vertical walls, the calculated characteristic time would be the time necessary for all particles having a sedimentation velocity of v_g to be removed from the aerosol. For a perfectly mixed aerosol, it is the characteristic time of the exponential decrease in concentration of particles having a sedimentation velocity equal to v_g. Since particles with larger aerodynamic diameters have larger sedimentation velocities, gravitational deposition tends to decrease the mean particle size of a polydisperse aerosol. For an aerosol of known mass median aerodynamic diameter, comparison of the characteristic time for gravitational deposition with the residence time of the aerosol in a volume would indicate whether significant gravitational deposition might occur in that volume.

Diffusional Deposition

An estimate of the characteristic time of diffusional deposition is

$$T_{diff} = (V\delta/SD_p) \tag{4}$$

where: S = total area of all surfaces in the volume
δ = the diffusion boundary layer thickness
v = volume of aerosol.

For diffusional deposition from stagnant air or from laminar flow, δ can be taken to be one-half of the smallest linear dimension of the volume, e.g., the diameter of a tube or the distance between two walls. In well-mixed aerosols, δ is much more difficult to estimate, diminishing the usefulness of the relation. A correlation has been derived from experimental data on deposition in chambers with electrically conductive walls.[3] It can be cast in the form of an estimate of the characteristic time of diffusional deposition in enclosed spaces

$$T_{diff.\ dep.} = 6.3 \text{ sec } [(V/S\ 1 \text{ cm})^{0.93} ([1 \text{ cm}^2/\text{sec}]/D_p)^{0.70}] \tag{5}$$

In the absence of other information, this correlation allows estimates of the characteristic time of diffusional deposition which can be compared with the residence time of the aerosol to indicate whether significant diffusional deposition will occur. Since the diffusion coefficient is larger for smaller particles, diffusional deposition tends to increase the mean particle size of the aerosol.

The characteristic time of particle deposition on surfaces can be significantly decreased below those predicted in equations 3–5 by attractive forces between particles and surfaces, e.g., electrical, thermal, or magnetic forces.[1,2]

SURFACE DEPOSITION/EVAPORATION OF VAPORS

Vapors are transported between surfaces and the suspending gas by diffusion through a layer of relatively stagnant gas. When this diffusion is the rate-limiting step in the transport process, the characteristic time expression is similar to equation 4

$$T_{diff} = V\delta/SD_v \tag{6}$$

where: D_v = diffusion coefficient of the vapor molecules
S = total
V = volume of aerosol.

Diffusion coefficients of vapor molecules through air are typically in the relatively narrow range of 0.05 to 0.2 cm^2/s, so it is sufficient to assume a value of 0.1 cm^2/s if the precise value is not known.[4] This estimate of the characteristic time for diffusional transport of vapor to and from surfaces suffers from the same lack of information on the value of δ as was mentioned in the case of diffusional deposition of particles.

Equation 6 does provide a basis for considering the volatilization of collected particles from filters and solid surfaces when air which is depleted in vapor is pulled through a sampler in which the volatile material has been collected. During sample collection, the contaminated air mass may move away from the sampler. Continuing to sample air which is depleted in the contaminant vapor causes some of the previously collected material to evaporate and be lost. It has been reported that in a laboratory experiment, volatilization of ammonium nitrate was less than 10 percent from the collection surfaces in an impactor and greater than 80 percent from a filter.[5]

Filters are designed for efficient collection of particles, including the collection of small particles by diffusion. It is not surprising that mass transfer of vapor from filter fibers capable of collecting small aerosol particles by diffusion would be very efficient; this is because vapor molecules have a much higher diffusion coefficient than particles. As a generalization, air which passes through a filter is likely to be saturated with vapors of all the materials collected in the filter unless the filter loading is too low. It is reasonable that volatilization losses were found to be much lower in an impactor where the ratio of the surface area covered by the sample to the effective volume filled by the airstream between stages is much smaller than the surface-to-volume ratio in a filter. Other factors which tend to decrease volatilization from the collection surface of a low pressure impactor are the short residence time of the air in the vicinity of the surface, especially at reduced pressure, and the potential for the collected material to be cooled below ambient temperature by the air which cooled during expansion through the jet. On the other hand, volatilization might be higher from an impactor collection surface than from collection surfaces in many other types of samplers because all the air passes close to the collection surface. Without experimental results, it is difficult to make definitive estimates of the degree of volatilization of material from surfaces of other samplers due to the lack of information on δ. When there is a potential for volatilization losses, appropriate experiments must be performed to determine whether they are significant.

PARTICLE/VAPOR INTERACTIONS

Airborne particles continually interact with the surrounding atmosphere through their incessant bombardment by vapor and gas molecules. For each chemical species, any imbalance in the rates with which molecules hit and leave the particle surface leads to a net mass transfer of that chemical species between the particle and vapor phases of the aerosol, i.e., evaporation or condensation. When the particle and vapor phases are in equilibrium, there is no net transfer of any chemical species between the two phases. This equilibrium can be upset by such activities as heating, cooling, diluting, and adding mass to the aerosol. Evaporation and condensation tend to change the particle size and the partitioning of each chemical species between the particle and vapor phases of an aerosol but do not change the total airborne concentration of the aerosol or its constituent chemical species. Both the characteristic time of the mass transfer and the equilibrium partitioning of materials between the particle and vapor phases will be discussed.

Characteristic Time

A characteristic time of condensation appropriate for all particles and the time of evaporation appropriate for the dominant material in a particle can be chosen as the time necessary for the mass of the particle to increase or decrease by 50 percent at the initial condensation or evaporation rate

$$T_{cond/evap} = 8.3\ \text{sec}\ \frac{\dfrac{\rho_p}{1\ \text{g/cm}^3}\quad {}^2}{\dfrac{D_v}{0.1\ \text{cm}^2/\text{sec}}\ \dfrac{V_{sat}}{1\ \text{mg/m}^3}\ (S_a - S_p)} \tag{7}$$

where: ρ_p = particle density
V_{sat} = the saturation mass concentration of the vapor over pure liquid at the ambient temperature
S_a = saturation of the ambient air by the vapor
S_p = saturation of the air at the particle surface.

Each saturation is defined as the ratio of the mass concentration of the vapor at the specified location to V_{sat}. The saturation mass concentration of the vapor V_{sat} can be calculated from the saturation vapor pressure VP using the ideal gas law

$$V_{sat} = \frac{MW\ VP}{R\ T}, \text{ in general} \tag{8}$$

$$= 0.041\ \text{mg/m}^3 \frac{MW}{1\ \text{g/mole}} \frac{VP}{1\ \text{ppm}}, \text{ at } 25°C \tag{9}$$

Vapor pressures given in units of torr or, equivalently, mm Hg can be converted to units of ppm by multiplying by the factor

$$10^6\ \text{ppm}/760\ \text{torr} \tag{10}$$

The same relations hold between the mass concentration of the vapor and the partial pressure when the gas is not saturated. Condensation occurs for $S_a > S_p$ and evaporation for $S_p > S_a$. For most materials, ρ_p is near 1 g/cm³ and D_v is near 0.1 cm²/s; the influence of these two parameters on the characteristic time will not be emphasized in the discussion. S_p is not directly observable in most situations but is related to the composition of the droplet

$$S_p = \mu\ \alpha\ K[V_{sat}(T_p)/V_{sat}(T_a)] \tag{11}$$

where: μ = mole fraction of the volatile contaminant in the droplet
α = activity coefficient
K = Kelvin correction factor for curvature of the droplet surface
T_p = temperature at the droplet surface
T_a = ambient temperature.

Raoult's law states that $S_p = \mu$ for an ideal solution. The activity coefficent is equal to unity for pure liquids in an isothermal system, by definition, and can be considered an experimental correction factor to Raoult's law for real solutions. In this discussion, its value will be taken to be unity. This value can be refined if experimental data are available for the equilibrium vapor concentration of the vapor above the liquid as a function of the solution concentration.

The Kelvin factor for the increased equilibrium vapor concentration above a curved surface for a pure material is equal to

$$K = \exp\,(4\,\sigma MW/\rho_p RTD_p) \tag{12}$$

where: σ = surface tension
R = universal gas constant.

The factor for a droplet consisting of a mixture of two materials has been published.[6] The deviation of K from unity is significant only when the droplet size becomes very small or the saturation of the ambient air is held very near unity, such as occurs above a planar surface of a pure liquid. The value of K will be taken to be unity in the examples.

The saturation at the particle surface was defined in terms of the saturation vapor concentration at ambient temperature. The factor in equation 11 which is the ratio of the saturation vapor concentration at the particle temperature over that at the ambient temperature accounts for possible temperature differences between the particle and ambient air. There is always at least a small temperature difference between the particle and ambient air during condensation or evaporation, due to the latent heat. This temperature difference is proportional to the evaporation or condensation rate and is significant only at high rates where it may decrease the condensation or evaporation rate by a factor of two or more. This temperature effect is always in the direction of moderating the condensation or evaporation rate, i.e., increasing the characteristic time. The temperature effect will be ignored in the examples discussed because decreasing the mass transfer rate by a factor of two will not affect the conclusions.

The characteristic time in equation 7 is not appropriate for the evaporation of a contaminant which exists in a particle as a dilute solution. In this case, equation 7 should be multiplied by the initial mass fraction of contaminant in the particle to give the approximate time during which the contaminant in the droplets would decrease by 50 percent at the initial rate. Thus, the characteristic time of evaporation of a contaminant from a dilute droplet is

$$T_{\text{evap, dilute}} = 8.3\text{ sec}\;\frac{\dfrac{\rho_p}{1\text{ g/cm}^3}\left(\dfrac{d_p}{1\ \mu\text{m}}\right)^2\dfrac{MW}{MW_s}}{\dfrac{D_v}{0.1\text{ cm}^2}\;\dfrac{V_{sat}}{1\text{ mg/m}^3}\;(S_a - S_p)} \tag{13}$$

where: MW_s = molecular weight of the solvent
μ = mole fraction
MW = molecular weight of the contaminant.

More detailed discussions of evaporation and condensation can be found in textbooks on mass transport as well as books on aerosols and cloud physics.[1,2,7]

During the expansion of air into a low-pressure system, i.e., a low-pressure impactor stage or a vacuum chamber, the air undergoes expansion and cooling. The expansion decreases the concentration of vapor in the air tending to cause particle evaporation and the cooling decreases the concentration of vapor at the particle surface tending to cause condensation. Whether the particles experience evaporation or condensation depends on the details of each case. However, the characteristic time of the mass transfer must be larger than that of either evaporation due to sudden expansion or condensation due to sudden cooling. Using the parameters appropriate for a system consisting of water vapor and droplets, the characteristic time of each of these two mass transfer processes is calculated to be

$$T = 0.2 \text{ msec } (D_p/1\ \mu m)^2 \tag{14}$$

For a particular low-pressure impactor stage, the transit time between the bottom of the jet and the collection surface was estimated to be 10 μs.[5] In this case, the transit time is much smaller than the characteristic times of condensation or evaporation for particles larger than about 1 μm, so no significant amount of evaporation or condensation of water would be expected for those particles which strike the plate. There might be a significant effect for submicron particles and those which are smaller than the cutoff diameter so they remain airborne for a longer time. Detailed discussions of similar phenomena have been published relating to sampling into a vacuum chamber and by impactors.[8,9]

Particle/Vapor Partitioning at Equilibrium

The particle and vapor phases of an aerosol are essentially in equilibrium when the partitioning of chemical species between the two phases is not changing significantly. This occurs when factors disturbing equilibrium have been absent for a sufficiently long time or are occurring sufficiently slowly. The particle and vapor phases are less likely to be in equilibrium when an aerosol is sampled near its source if rapid changes in temperature and dilution occur there. The particle and vapor phases of an aerosol in the general workplace or outdoor atmosphere are likely to be in equilibrium because the characteristic time of evaporation or condensation is relatively short, typically, compared to the time since the aerosol was emitted or its equilibrium was last perturbed. For aerosols whose particle and vapor phases are in equilibrium, knowledge of the partitioning of material between the two phases may be important in choosing a sampling strategy and in predicting effects. Predicting the equilibrium partitioning of material between the particle and vapor phases of an aerosol will be discussed in this section.

A single component aerosol is one in which all particles and all vapors which interact with the particles consist of a single chemical species. In this case, the partitioning can be described by

$$f_p = (C_t - KV_{sat})/C_t \quad (15)$$

where: f_p = fraction of mass in the particle phase
C_t = total airborne mass concentration, i.e., the sum of the of the mass concentrations in the vapor and particle phases.

This relation arises from the definition of the total airborne mass concentration and the requirement that the vapor and particle phases must be in equilibrium. The Kelvin correction K for curvature of the particle surface has a negligible effect if $C_t/V_{sat} >> K$. For $C_t < V_{sat}$, no material is in the particle phase of a single component system and f_p is computed to be negative, a nonphysical result.

There are few, if any, airborne systems which are truly single component. Systems which are not single component include ones in which the chemical species is soluble in other airborne particles or there are other vapors in the atmosphere which are soluble in particles of the chemical species under consideration. In either case the concentration of the chemical species in the vapor phase is not computed as that for a single component system.

Ambient air contains water vapor and hygroscopic nuclei in addition to contaminants whose partition between the vapor and particle phase we wish to predict. Therefore, a more realistic model system would consist of three components, one of which is nonvolatile. A set of equations can be written and solved to determine the equilibrium partitioning of the materials in this system if the total airborne concentration of all three components is known. Two relations arise from the definition of the total airborne concentration and the expressions for the equilibrium vapor phase concentration of each of the two volatile components

$$C_{ti} = C_{pi} + \mu_i \alpha_i K_i V_{sat,i}, \; i = 1, 2 \qquad (16)$$

where C_{pi} = concentration of the i_{th} component in the particle phase.

Assuming each particle is identical, expressions can be written for the mole fraction of each component in the particle phase

$$\mu_i = \mu_i (C_{p1}, C_{p2}, C_{p3}, MW_1, MW_2, MW_3) \qquad (17)$$

If the total airborne concentration of all three components is known, these five equations can be solved for the five unknowns: the mole fraction of each of the three components in the particles and the concentration of each of the two volatile components in the particle phase. The vapor/particle partitioning can then be described by the fraction of each volatile component in the particle phase

$$f_{pi} = C_{pi}/C_{ti} \qquad (18)$$

Consider an atmosphere which contains 0.05 mg/m^3 of sodium chloride and mevinphos, a water-soluble insecticide, at a concentration of 0.1 mg/m^3, which is equal to its threshold limit value (TLV).[10] The saturated vapor concentration of mevinphos is 26 mg/m3 as calculated from the available value of its vapor pressure at 20°C.[10] Since the airborne concentration of mevinphos is a factor of 260 times smaller than its saturation vapor concentration, one might assume that it will appear only in the vapor phase and not in the particle phase. Using the assumed values and the saturation vapor concentration of water vapor at room temperature of 17,000 mg/m^3, the five simultaneous equations 16–17 can be solved for the fraction of mevinphos in the particle phase at various relative humidities. The results are shown in Table I. Even without mevinphos, increasing relative humidity would cause each salt particle to grow into an aqueous salt solution which is in equilibrium with the ambient relative humidity. The second row of the table shows the concentration of water in the particle phase at each relative humidity. The mevinphos divides

TABLE I. Example of Equilibrium Partitioning

Relative Humidity (%)	Concentration of Water in Droplets (mg/m^3)	Fraction of Contaminant in Droplets
80	0.13	0.07
95	0.64	0.24
99	5.1	0.64

between the particle and vapor phases so as to reach equilibrium. Since there is a relatively small amount of mevinphos in the system, it does not significantly perturb the concentration of water in the particle phase. As the relative humidity increases, more particle phase water is available for mevinphos to partition into and the fraction of mevinphos in the particle phase increases until it is predicted to be greater than 50 percent at 99 percent relative humidity. This is for a material whose airborne concentration was taken to be 260 times less than its saturation vapor concentration. If the concentration of water in the particle phase were much higher, say as a mist or fog, essentially all of the mevinphos would be found in the particle phase. Any material whose total airborne concentration was a larger fraction of the saturation vapor concentration would have a larger fraction in the particle phase under each of these conditions. The requirement that the material be water soluble is not a particularly restrictive one. In this example, the solubility of mevinphos in water need only be approximately 1 percent by mass. In a fog or mist, the solubility could be orders of magnitude lower and most of the material would still be found in the particle phase.[11]

FINAL REMARKS

A qualitative understanding of aerosol instabilities can be obtained from consideration of the aerosol properties which are undergoing change and the characteristic time of the change. The concept of characteristic time allows comparisons of competing instabilities and allows predictions of whether a process which tends to change an aerosol property will go nearly to completion or have no significant effect on the aerosol during the observation time. It should be noted that accurate predictions of evaporation and condensation rates for droplet solutions are very difficult or impossible because data are not readily available on many parameters which are needed as a function of the droplet composition. Thus, it is far better to realize the possible existence of a sampling problem and avoid it than to try to produce a credible correction later based on calculations.

Consideration of the equilibrium partitioning of air contaminants between the particle and vapor phases of an aerosol leads to some conclusions which should be helpful when sampling complex aerosols. When the saturation vapor concentration of a material is much smaller than the airborne concentration, essentially all of that material will be found in the particle phase. However, material will transfer to the vapor phase if the temperature is increased or the aerosol is diluted sufficiently. If the saturation vapor concentration is on the same order as the airborne concentration, the material may be found in both the particle and vapor phases. If the saturation vapor concentration is much larger than the airborne concentration, no general statement can be made regarding the partitioning of the material between the particle and vapor phases. Such a material would be expected to be found predominantly in the vapor phase, but it may be found in the particle phase if the aerosol contains a sufficient concentration of airborne particles in which the material is soluble, even slightly. Airborne concentrations of water droplets are relatively massive in many high-humidity atmospheres. Thus, it may be quite common for many air contaminants to be found in the particle phase of such atmospheres even though their airborne concentrations may be smaller than their saturation vapor concentrations. This may occur even for materials which have a low water solubility because the droplet concentrations can be quite low. There may also be situations in which aerosols contain a sufficiently large concentration of oily droplets in which oil-soluble vapors would be readily dissolved. More definitive statements on the impact of such considerations on air sampling await additional theoretical and experimental studies.

REFERENCES

1. Hinds, W.C.: *Aerosol Technology*. John Wiley and Sons, New York (1982).
2. Mercer, T.T.: *Aerosol Technology in Hazard Evaluation*. Academic Press, New York (1973).
3. Mercer, T.T.: Brownian Coagulation: Experimental Methods and Results. *Fundamentals of Aerosol Science*, pp. 85–134. D.T. Shaw, Ed. John Wiley and Sons, New York (1978).
4. Perry, R.H. and C.H. Chilton: *Chemical Engineers' Handbook*, 5th ed., pp. 3-222. McGraw-Hill, New York (1973).
5. John, W., S.M. Wall, J.L. Ondo and H.C. Wang: Acidic Aerosol Concentrations and Size Distributions in California. *Aerosols: Formation and Reactivity*, pp. 25–28. Proceedings: Second International Aerosol Conference, September 22–26, 1986, Berlin, FRG. Pergamon Press, Oxford (1986).
6. Nair, P.V.N. and K.G. Vohra: Growth of Aqueous Sulphuric Acid Droplets as a Function of Relative Humidity. *J. Aerosol Sci. 6*:265–271 (1975).
7. Pruppacher, H.R. and J.D. Klett: *Microphysics of Clouds and Precipitation*. D. Reidell Publishing, Dordrecht, Holland (1978).
8. Dahneke, B. and D. Padliya: Nozzle-inlet Design for Aerosol Beam Instruments. Rarefied Gas Dynamics. J.L. Potter, Ed. *Prog. Astron. Aeron. 51*, Part II:1163–1172 (1977).
9. Biswas, P., C.L. Jones and R.C. Flagan: Distortion of Size Distributions by Particle Sampling Instruments. *Aerosols: Science, Technology, and Indus-*

trial Applications of Airborne Particles, pp. 191–194. B.Y.H. Liu, D.Y.H. Pui and H.J. Fissan, Eds. Elsevier, New York (1984).
10. American Conference of Governmental Industrial Hygienists: *Documentation of the Threshold Limit Values and Biological Exposure Indices*, 5th ed., p. 412. ACGIH, Cincinnati (1986).
11. Glotfelty, D.E., J.N. Seiber and L.A. Liljedahl: Pesticides in Fog. *Nature 325*:602–605 (1987).

CHAPTER 13

Sampling Reactive Materials

EUGENE R. KENNEDY

Organic Methods Development Section, Methods Research Branch, Division of Physical Sciences and Engineering, National Institute for Occupational Safety and Health, 4676 Columbia Parkway, Cincinnati, Ohio 45226

A reactive material is a material which can undergo further reaction either with itself or another material or compound under environmentally-mediated conditions. Examples of these types of reactions include polymerization, hydrolysis and oxidation. Since these reactions can occur in the workplace under ambient conditions or even on sampling media, the sampling and subsequent analysis of reactive materials collected in the work environment have been a problem for industrial hygienists and industrial hygiene chemists. Much work has been done to address this problem, but it is still far from solved as new reactive chemicals join other reactive materials in the workplace.

There have been many different approaches to the sampling and analysis of reactive compounds. These have included use of impingers containing reagent solutions,[1-13] specialized storage techniques to preserve samples,[14-23] sorbents engineered for collection of families of compounds,[24] reagent-coated sorbent tubes[25-40] and filters,[41] and direct-reading monitoring or analytical methods.[42-50] In some instances, several methods for one particular analyte have been developed using different approaches for sampling and analysis. Listed in Table I and discussed below are examples of some methods developed for the analysis of reactive materials.

TABLE I. Partial Listing of Methods for Reactive Compounds

Compound	Procedure for Sampling/Analysis
Acetone cyanohydrin	Storage of samples at 0°C; GC analysis using inert analytic column[24]
Acrolein	Reaction with sodium bisulfite; hexylresorcinol analysis[4]
	Reaction with 2-(hydroxymethyl)piperidine; GC analysis[32]
Aliphatic amines	Collection on phosphoric acid-coated sorbent; GC analysis[17]
1,3-Butadiene	Collection of charcoal and storage at −4°C; GC analysis[14]
2-Butanone	Collection on Ambersorb XE-347; GC analysis[16]
Carbonyl sulfide	Reaction with 1,3-diaminopropane; GC analysis of N,N′-trimethyleneurea derivative[39]
Chloroacetyl chloride	Reaction with 9-[(N-methylamino)methyl]anthracene; HPLC analysis[29]
Chloroformates	Reaction with dibutylamine; GC analysis[28]
Chloromethyl methyl ether	Reaction with potassium 2,4,6-trichlorophenate; GC analysis[26]
Ethanethiol, Butanethiol	Reaction with mercuric acetate-coated filter; hydrolysis with hydrochloric acid and GC analysis[20]
Ethylene oxide	Reaction with sulfuric acid; GC analysis[12]
	Reaction with sulfuric acid; oxidation and analysis by 3-methyl-2-benzothiazolinone hydrazone procedure[13]
	Reaction with hydrogen bromide; GC analysis[40]
Formaldehyde	Reaction with sodium bisulfite; chromotropic acid analysis[2]
	Reaction with N-benzylethanol amine; capillary GC analysis[27]
	Reaction with 2,4-dinitrophenylhydrazine; HPLC analysis[6,7,36,37,41]
Hexachloroacetone	Reaction with methanol; GC analysis[10]
Hydrazine	Reaction with acetone; GC analysis[11]
Hydrazines	Collection with sulfuric acid-coated silica gel; GC analysis[18]
	Reaction with benzaldehyde; HPLC analysis[38]
Methanethiol	Reaction with mercuric acetate-coated filter; hydrolysis with hydrochloric acid and GC analysis[19]
Phosgene	Reaction with 2-(hydroxymethyl)piperidine; GC analysis[30]
	Reaction with dibutylamine; GC analysis[28]
Vinyl acetate	Collection on hydroquinone-coated charcoal; GC analysis[22]
	Collection on XAD-7; GC analysis[21]

GC = Gas chromatographic.
HPLC = High-performance liquid chromatographic.

IMPINGER SAMPLING

Prior to the advent of sorbent tubes, impingers or bubblers were the primary means of sampling workplace air for volatile organic compounds. To improve efficiencies of these devices for reactive compounds, reagents were incorporated into the absorbing solutions. These reagents reacted to either stabilize the compound or prevent its volatilization from the impinger or both. One example of this type of improvement was the use of 1 percent sodium bisulfite solution instead of water in an impinger for the collection of formaldehyde.[1,2] Collection efficiencies were approximately 80 percent with water, but were improved to 95–100 percent with the use of 1 percent sodium bisulfite solution.[2] However, problems of storage of bisulfite solutions of formaldehyde have been noted.[3] Another method which used this same approach for sampling was NIOSH method P&CAM 211 for acrolein.[4] This method combined the use of reagent (sodium bisulfite) and storage at low temperature to preserve the sample with subsequent reaction for analysis of the sample. The acrolein was collected in 1 percent sodium bisulfite solution and stored at 0°C until analysis by reactions with 4-hexylresorcinol, trichloroacetic acid, and mercuric chloride. The reaction formed a blue-colored compound which was determined spectrophotometrically. Changes in the order of addition of reagents during the color-forming step have caused observed differences in the spectral absorbances of standards and samples.[5]

2,4-Dinitrophenylhydrazine in solution has been used for sampling formaldehyde as well as other aldehydes and ketones.[6-9] The aldehydes and ketones react with the 2,4-dinitrophenylhydrazine under acid conditions to form the corresponding 2,4-dinitrophenylhydrazones, which can be determined either by spectrophotometry,[8] gas chromatography,[9] or high performance liquid chromatography.[6,7] In our laboratories, when acidic, aqueous solutions of 2,4-dinitrophenylhydrazine[6] were used to sample low levels of carbonyl compounds, the levels found were substantially lower than expected. This was due to the equilibrium between the carbonyl compounds and their corresponding 2,4-dinitrophenylhydrazones. This finding limited the usefulness of this method to the measurement of higher levels of carbonyl compounds. When acetonitrile solutions of 2,4-dinitrophenylhydrazine were used,[7] collection was quantitative, even at lower carbonyl concentration levels, but evaporation of acetonitrile limited sampling times.

A method for the determination of hexachloroacetone in air has been developed using impinger sampling.[10] The interesting point about this method is that the anhydrous methanol used as the absorbing liquid in the impinger also acted as the reagent for derivatization of the compound. Methanol reacts with hexachloroacetone to form methyl trichloroacetate and chloroform. The methyl trichloroacetate is then determined by gas chromatography. One of the major disadvantages with this sampling technique is that water from the air is also collected in the methanol and reacts with hexachloroacetone at the same rate as methanol to form trichloroacetic acid and chloroform. Therefore, water reduces the amount of methyl trichloroacetate formed and acts as a negative interference. If chloroform is not present in the sampled environment, the amount of chloroform present in the sample is used to correct for any hydrolysis which may have occurred due to sampled water vapor. To attempt to minimize the collection of water in the sampling train, three impingers were used in series, with the first impinger being maintained at room temperature and the remaining two at dry ice temperature. A method using a similar concept for sampling hydrazines has also been developed.[11] Hydrazines are sampled into an impinger containing either acidified or aqueous acetone, which reacts to form the corresponding hydrazones. The major difficulty in using both of the above methods is the use of volatile organic solvents in impingers. Each method requires cooling of the impingers to maintain solvent levels.

A method for the sampling of ethylene oxide in air was developed based on the reaction of ethylene oxide with dilute sulfuric acid solution in an impinger to form ethylene glycol.[12] The ethylene glycol produced in this reaction is then determined by gas chromatography, although the sample is amenable to other analysis techniques such as colorimetric determination. This sampling approach allows the analyst the advantage of preparing standards for calibration from authentic ethylene glycol, rather than ethylene oxide gas. For those unaccustomed to working with gases and/or lacking proper equipment, the measurement of exact amounts of gas for preparation of calibration standards can be difficult and can lead to inexact standard concentrations and improper calibration of the method. A method using similar chemistry was developed using a diffusive sampling approach.[13] Ethylene oxide is sampled by diffusion into a dilute sulfuric acid solution and converted to ethylene glycol. During analysis, the ethylene glycol is oxidized to formaldehyde and analyzed spectrophotometrically after reaction of the formaldehyde with 3-methyl-2-benzothiazolinone hydrazone hydrochloride solution (MBTH). Using this analysis procedure, formaldehyde present along with ethylene oxide in the environment is an interference in the method. This interference can be corrected by analysis of an aliquot of the ethylene glycol solution before oxidation. This analysis result will give the background formaldehyde level which can be subtracted from the analysis results of the oxidized ethylene glycol solution in order to calculate the ethylene oxide level. If the method is used in this fashion, the simultaneous determination of formaldehyde and ethylene oxide can be performed.

SPECIALIZED STORAGE TECHNIQUES

Storage at low temperature is one of the most commonly-used storage techniques for preservation of reactive samples. It can reduce 1) the rate of decomposition of the compound, 2) the rate of reaction of the compound with other compounds in the sample, or 3) the rate of volatilization of the compound from the collection medium. For 1,3-butadiene, which exists as a gas at room temperature, collection of the compound on charcoal and subsequent storage at –4°C was the main method of sample storage prior to analysis by gas chromatography. This was necessary since samples stored at room temperature showed an average analyte loss of approximately 1.5 percent per day.[14] The use of a synthetic polymeric sorbent could also be considered to be a specialized storage technique. The instability of 2-butanone on charcoal has been reported.[15] Losses of as much as 20 percent of the analyte have been observed after 23 days storage. In NIOSH method 2500,[16] 2-butanone is sampled using Ambersorb XE-347, desorbed with carbon disulfide and analyzed by gas chromatography. Recovery of 2-butanone from this sorbent is good even after storage for six weeks at 25°C.

Other methods have used sorbents coated with compounds which stabilize reactive compounds collected on the sorbent. During preparation for analysis, these reagents are neutralized or removed before analyte determination. Acid-coated sorbents have been used for the sampling of low molecular weight aliphatic amines[17] and hydrazines.[18] For the aliphatic amines, the acid is neutralized with base and the amines are analyzed by gas chromatography. Detection by a nitrogen-selective detector increases sensitivity for the amines and reduces possible interferences from other organic compounds present in the samples. In the hydrazine method,[18] the hydrazines are desorbed from the sorbent with water and reacted with 2-furaldehyde to form the hydrazone derivative. The hydrazone derivative is extracted from the water solution with ethyl acetate and analyzed by gas chromatography. Methods have also been reported which use mercuric acetate-impregnated filters for the sampling of methyl-,[19] ethyl-, and butylthiols.[20] The thiols are regenerated by treatment of the filters with acid and extracted into 1,2-dichloroethane for gas chromatographic analysis, which uses flame photometric detection for increased sensitivity and specificity. An additional storage restriction for this method is that samples must be stored in the dark since storage in the light results in losses of up to 80 percent of the thiol.

Several methods for vinyl acetate have been reported[21,22,51] which have approached the stability problems of this compound differently. The major stability problems with this compound are polymerization and hydrolysis to acetaldehyde and acetic acid. The compound was not efficiently recovered when adsorbed on charcoal.[51] In NIOSH method P&CAM 278,[21] samples are collected using Chromosorb 107 sorbent tubes. Samples are thermally desorbed and analyzed by gas chromatography using flame ionization detection. This sorbent allows good recovery of the compound. The capacity of the Chromosorb 107 for vinyl acetate is limited to 3 L to prevent breakthrough. At the recommended low sampling volume and flow rate of 0.1 L/min, the usefulness of the method for long-term personal samples is limited. In the method of Kimble et al.,[22] charcoal sorbent is treated with a 5 percent solution of hydroquinone in ethanol prior to sampling to help prevent polymerization of the vinyl acetate. A drying agent is placed before the sorbent bed to remove water vapor from the air. By drying the air being sampled, hydrolysis of the vinyl acetate collected on the sorbent is prevented. Samples are desorbed with carbon disulfide containing 2 percent acetone and analyzed by gas chromatography. Samples are stable for five days at room temperature, but sample storage for longer periods requires refrigeration. With the NIOSH method, samples exhibited a storage loss of 4 percent after seven days at room temperature. After 14 days at room temperature, 14 percent losses were reported, indicating that analysis of samples should be performed as soon as possible after sampling. If this is not possible, refrigerated storage of the samples is recommended.

Acetone cyanohydrin is another compound which can decompose to several different compounds. It can undergo a rearrangement to acetone and hydrogen cyanide or dehydrate at higher temperatures to form methacrylonitrile. In a method developed for acetone cyanohydrin,[24] sampling on Poropak QS, a nonpolar silylated styrene-divinylbenzene copolymer, helped solve the problem of hydrolysis on the sorbent since it has a low affinity for water vapor. Even then, samples collected at high humidity had to be stored refrigerated before analysis. An inert analytical column made of Teflon and filled with 5 percent OV-1 on 20/40 mesh Chromosorb T is used at moderate column oven temperature (100°C) to reduce thermal degradation of the compound during analysis.

A novel approach for the sampling of polar, unsaturated compounds is presented by Williams and Sievers.[23] By chemically modifying Chromosorb 102 to allow the formation of a europium complex in the sorbent, an enhanced affinity for aldehydes, ketones, and other unsaturated compounds is incorporated into the sorbent. One percent breakthrough volumes for acetone, acetaldehyde, and acetonitrile are approximately eight to ten times the values found for unmodified Chromosorb 102. With the complexation of other metal ions into the modified polymer, the design of selective sorbents for other classes of compounds may be possible.

REAGENT-COATED SORBENTS

Of all the approaches taken for the stabilization of reactive analytes, the most intriguing is the reaction of the reactive compound with a derivatizing reagent on a sorbent. By presenting the compound of interest to a large excess of reagent coated on a sorbent, quantitative reaction is favored. In instances where the derivative has greater mass than the original compound, sensitivity of the analysis is enhanced when using mass sensitive detection such as the flame ionization detector.[25] By incorporating a halogen atom into a derivative, an increase in specificity and sensitivity can be obtained with the use of an electron capture detector.[26]

There are certain factors that require investigation when using a reaction of a compound with a derivatizing reagent as the basis of a sampling and analytical method. The reaction which will serve as the basis for the sampling method must be fast enough to allow efficient collection by the sampling medium. If the reaction is too slow, then the sampling rate and sample capacity of the medium will be limited. The reaction should give only one product in stoichiometric yield. This will simplify the analytical data interpretation and the calibration procedures. The derivative formed in the reaction should be stable and easily recovered from the sampling medium. The derivative should also be readily determined by conventional analytical means with good sensitivity. The interference of other compounds in the formation of the desired derivative should also be investigated, preferably early in the method development work. If other compounds do interfere, appropriate measures should be taken to remove the interferent before derivative formation. If a compound interferes in the analysis, different analysis conditions must be found under which there will be no interference. As a final consideration, the reagent-coated sorbent should have a reasonably long shelf life.

While these factors help define the ideal situation, reality often forces a compromise. Often kinetics are unfavorable and limitations must be placed on sampling rates and times. In NIOSH method 2502 for formaldehyde,[27] formaldehyde is reacted with 2-(benzylamino)ethanol coated on XAD-2 to form 3-benzyloxazolidine, which is determined by capillary gas chromatography. Breakthrough studies with this coated sorbent indicated that sampling rates of 80 cm^3/min or less should be used to prevent sample breakthrough. This limitation is believed to be caused by slow kinetics of the reaction. A method reported for chloromethyl methyl ether derivatizes the analyte with potassium 2,4,6-trichlorophenate coated on GLC-110 glass beads.[26] Even with the recommended low sampling rate (5–10 cm^3/min), only a moderate reaction yield of ca. 56 percent is obtained. This loss in sensitivity due to low sample volume and reaction yield is compensated by the increased sensitivity of the analytical technique (gas chromatography with an electron capture detector) for the derivative. In this particular instance, the use of the derivative for the calibration of the method eliminates the need to handle the chloromethyl methyl ether, a human carcinogen, except for derivative preparation. Attempts to expand this method to sample and analyze bis(chloromethyl)ether were unsuccessful due to the slower reaction kinetics of this compound.

The problem of hydrolysis has been addressed in methods developed for chloroformates and phosgene.[28-30] In the method developed by Hendershott,[28] chloroformates are reacted with dibutylamine coated on XAD-2 to form dibutyl carbamates and dibutylamine hydrochloride. Phosgene is also found to react with dibutylamine to form tetrabutylurea and dibutylamine hydrochloride. After desorption of the coated sorbent with hexane, excess amine, which might interfere in the analysis, is removed by extraction of the hexane with 1N hydrochloric acid. The hexane solution of the carbamates or ureas is then analyzed by gas chromatography. This method has had limited success in the sampling and analysis of other acid chlorides. An aromatic amine has also been used for the sampling of chloroacetyl chloride.[29] Chloroacetyl chloride is reacted with 9-[(N-methylamino)methyl]anthracene coated on Tenax-GC to form an amide. The amide is desorbed with toluene and determined by high-performance liquid chromatography. In a method developed for phosgene by chemists at the Occupational Safety and Health Administration Analytical Laboratory,[30] phosgene is reacted with 2-(hydroxymethyl)piperidine coated on XAD-2. The resulting derivative is a bicyclic oxazolidone which is determined by gas chromatography. 2-(Hydroxymethyl)piperidine coated on XAD-2 has also been used for the sampling of formaldehyde,[31] acrolein,[31,32] furaldehyde,[33] glutaraldehyde,[34] and pentanal.[35] The derivatives formed in these methods are oxazolidines and are determined by gas chromatography.

The reaction of aldehydes with hydrazines has been used for the sampling and analysis of both aldehydes[36,37,41] and hydrazine.[38] Several methods have been reported using 2,4-dinitrophenylhydrazine coated on sorbents for sampling formaldehyde and other aldehydes and ketones.[36,37,41] The carbonyl compounds were reacted with the 2,4-dinitrophenylhydrazine coated onto either XAD-2,[36] silica gel,[37] or filters[41] and determined by high-performance liquid chromatography. With the 2,4-dinitrophenylhydrazine-coated filters, sampling using either an active or passive sampling mode was reported.[41] In our laboratory, the potential contamination of the sorbent with formaldehyde from either the sorbents themselves or from the atmosphere was a problem which was solved by taking great care in the preparation and coating of the sorbent. In a reverse approach, XAD-2 sorbent coated with benzaldehyde and hydrochloric acid has been used for the sampling of hydrazine in workroom air.[38] The benzaldazine, resulting from the reaction of hydrazine with benzaldehyde, is desorbed with *N, N'*-dimethylformamide and determined by high-performance liquid chromatography.

In a method reported by Leiber and Berk for carbonyl sulfide,[39] conditions of the analysis provide the final step in the derivatization of the analyte. Carbonyl sulfide is reacted with 1,3-diaminopropane coated on Woelm polyimide sorbent. The product of this initial reaction is (3-aminopropyl)thiocarbamic acid. This compound is then cyclized to *N, N'*-trimethyleneurea. It was found that the cyclization took place in part during storage and in part in the injection port of the gas chromatograph. Carbon dioxide interferes with the collection of the carbonyl sulfide, but this problem is solved by using a pre-impinger containing 40 percent potassium hydroxide solution to remove the carbon dioxide.

In a method reported by Esposito et al.,[40] ethylene oxide is reacted with hydrogen-bromide-treated Ambersorb XE-347 to give 2-bromoethanol. The bromoethanol is desorbed with a mixture of acetonitrile and toluene. Excess hydrogen bromide is neutralized by addition of 100 mg of sodium carbonate to the desorbing solution. The bromoethanol is determined by gas chromatography with electron capture detection. This method allows more accurate determination of ethylene oxide by using a sampling medium which has a greater sampling capacity and gives a stable derivative.

DIRECT-READING INSTRUMENTS

The use of direct-reading instruments for the analysis of reactive analytes has the advantage of near-real-time reporting of results and the reduction of sample instability. The major disadvantages are usually complicated operation, limited sensitivity, the need for additional equipment and reagents for calibration, analytical interferences, and high cost. Also many of the devices can take only area samples and not personal samples. Currently, there are several devices for the determination of formaldehyde in near real time.[42,43] One of the devices uses a fuel cell for the measurement of formaldehyde.[42] This device is sensitive to a large number of interferences (e.g., phenol, toluene, and other oxidizable organics) which may limit usefulness. A second device uses the pararosaniline method for the determination of formaldehyde.[43] This device requires lengthy set-up and shut-down but has good sensitivity for formaldehyde at low levels. Several devices are available which use a reagent-coated paper tape through which the analyte-containing air is drawn.[44,45] The analyte reacts with the reagent coating to develop a color, which is measured spectrophotometrically. These paper tapes can be coated with a number of reagent coatings for the determination of compounds such as hydrazine, isocyanates, and phosgene. Other approaches for real-time monitoring of reactive compounds have used mass spectrometry for the determination of hydrazines[46] and analyte-specific traps with reagent-coated crystals of a piezoelectric crystal detector for the determination of toluene diisocyanate and propylene glycol dinitrate.[47,48]

The current trend in the area of direct-reading instruments is the development of portable instruments which are adaptations of conventional analytical instruments such as the gas chromatograph[49] and the infrared spectrophotometer.[50] Currently, these types of instruments work well where the workplace contamination is defined and of limited complexity. In environments where the contamination is unknown, reliance must be placed on the analytical laboratory for calibration and confirmation of results, since portable instruments lack the resolution and data handling capabilities of units found in the laboratory. As costs are reduced and performance of these devices is increased, they may play a major role in the monitoring of the workplace for contaminants, especially for reactive materials.

SUMMARY

The previously-mentioned examples cover a wide range of techniques for the sampling and analysis of reactive materials. Specialized storage techniques, such as inert sorbents, storage at low temperature or in the dark, and stabilizer coatings on sorbents, have been used to preserve the integrity of the reactive material of interest. In other cases, reagents in impingers or coated on sorbents or filters have reacted with the reactive materials and converted them to more stable and easily analyzed compounds. The potential of designing a sorbent for a particular reactive compound now exists. Along with these sampling and analysis methods, the analysis of reactive materials where they exist in the workplace air is becoming possible with portable, direct-reading instruments.

Table I represents a partial listing of the methods developed for reactive compounds. The advantages which these methods have over conventional methods, such as the charcoal tube method,[52] include better sample stability, increased sampling and analytical sensitivity, increased analyte specificity, ease of sampling, reduced need to handle toxic or carcinogenic compounds, better precision and accuracy, and simultaneous analysis of multiple compounds. Along with the advantages of these methods, the disadvantages must also be considered; these include the need for higher analyst skill level, the higher cost of specialized reagents, sorbents and analysis techniques, longer analysis times, more labor intensive sample preparation, and lack of commercially-prepared sampling devices. However, alternative sampling and analysis methods for many of these reactive materials currently do not exist.

The sampling and analysis of reactive compounds is an area which offers the analytical chemist an opportunity to use his ingenuity for the solution of difficult sampling and analysis problems. The number of reports in the literature dealing with sampling and analytical methods for reactive analytes indicates that the interest in this area is growing[53,54] and should continue to grow in the future.

DISCLAIMER

Mention of company names or products does not constitute endorsement by the National Institute for Occupational Safety and Health.

REFERENCES

1. National Institute for Occupational Safety and Health: Formaldehyde. *NIOSH Manual of Analytical Methods*, 2nd ed., Vol. 1., Method P&CAM 125. DHHS (NIOSH) Pub. No. 77-157A (1977).
2. National Institute for Occupational Safety and Health: Formaldehyde. *NIOSH Manual of Analytical Methods*, 3rd ed., Method 3500. DHHS (NIOSH) Pub. No. 84-100 (1984).
3. Daggett, D.L. and T.H. Stock: An Investigation into the Storage Stability of Environmental Formaldehyde Samples. *Am. Ind. Hyg. Assoc. J. 46*:497 (1985).
4. National Institute for Occupational Safety and Health: Acrolein. *NIOSH Manual of Analytical Methods*, 2nd ed., Vol. 1, Method P&CAM 211. DHHS (NIOSH) Pub. No. 77-157A (1977).

5. Hemenway, D.R., M.C. Costanza and S.M. MacAskill: Review of the 4-Hexylresorcinol Procedure for Acrolein Analysis. *Am. Ind. Hyg. Assoc. J.* *41*:305 (1980).
6. Kuwata, K., M. Uebori, H. Yamasaki and Y. Kuge: Determination of Aliphatic Aldehydes in Air by Liquid Chromatography. *Anal. Chem.* *55*:2013 (1983).
7. Lipari, F. and S.J. Swarin: Determination of Formaldehyde and Other Aldehydes in Automobile Exhaust with an Improved 2,4-Dinitrophenylhydrazine Method. *J. Chromatog.* *247*:297 (1982).
8. Papa, L.J.: Colorimetric Determination of Carbonyl Compounds in Automotive Exhaust as 2,4-Dinitrophenylhydrazones. *Environ. Sci. Tech.* *3*:397 (1969).
9. Fracchia, M.F., F.J. Schuette and P.K. Mueller: A Method for Sampling and Determination of Organic Carbonyl Compounds in Automobile Exhaust. *Environ. Sci. Tech.* *1*:915 (1967).
10. Kissa, E.: Determination of Hexachloroacetone in Air. *Anal. Chem.* *55*:1222 (1983).
11. Holtzclaw, J.R., S.L. Rose, J.R. Wyatt et al.: Simultaneous Determination of Hydrazine, Methylhydrazine and 1,1-Dimethylhydrazine in Air by Derivatization/Gas Chromatography. *Anal. Chem.* *56*:2952 (1984).
12. Romano, S.J. and J.A. Renner: Analysis of Ethylene Oxide – Worker Exposure. *Am. Ind. Hyg. Assoc. J.* *40*:742 (1979).
13. Kring, E.V., D.J. Damrell, A.N. Basilio, Jr. et al.: Laboratory Validation of a New Passive Air Monitoring Badge for Sampling Ethylene Oxide in Air. *Am. Ind. Hyg. Assoc. J.* *45*:697 (1984).
14. Lunsford, R.A., Y.T. Gagnon and J. Palassis: 1,3-Butadiene. *NIOSH Manual of Analytical Methods*, 3rd ed., Method 1024. DHHS (NIOSH) Pub. No. 84–100 (1984), 2nd Supplement (in press).
15. Folke, J., I. Johansen and K.H. Cohr: The Recovery of Ketones from Gas-sampling Charcoal Tubes. *Am. Ind. Hyg. Assoc. J.* *45*:231 (1984).
16. Slick, E.J., J. Posner and D. Smith: 2-Butanone. *NIOSH Manual of Analytical Methods,* 3rd ed., Method 2500. DHHS (NIOSH) Pub. No. 84–100 (1984).
17. Kuwata, K., E. Akiyama, Y. Yamazaki et al.: Trace Determination of Low Molecular Weight Aliphatic Amines in Air by Gas Chromatography. *Anal. Chem.* *55*:2199 (1983).
18. Mazur, J.F., G.E. Podolak and B.T. Heitke: Use of a GLC Concentrator to Improve Analysis of Low Levels of Airborne Hydrazine and Unsymmetrical Dimethylhydrazine. *Am. Ind. Hyg. Assoc. J.* *41*:66 (1980).
19. Knarr, R. and S.M. Rappaport: Determination of Methanethiol at Parts-per-Million Air Concentrations by Gas Chromatography. *Anal. Chem.* *52*:733 (1980).
20. Knarr, R. and S.M. Rappaport: Impregnated Filters for the Collection of Ethanethiol and Butanethiol in Air. *Am. Ind. Hyg. Assoc. J.* *42*:839 (1981).
21. Foerst, D.L.: Vinyl Acetate. *NIOSH Manual of Analytical Methods*, 2nd

ed., Vol. 4., Method P&CAM 278. DHHS (NIOSH) Pub. No. 78–175 (1978).
22. Kimble, H.J., N.H. Ketcham, W.C. Kuryla et al.: A Solid Sorbent Tube for Vinyl Acetate Monomer that Eliminates the Effect of Moisture in Environmental Sampling. *Am. Ind. Hyg. Assoc. J. 43*:137 (1982).
23. Williams, E.J. and R.E. Sievers: Synthesis and Characterization of a New Sorbent for use in the Determination of Volatile, Complex-forming Organic Compounds in Air. *Anal. Chem. 56*:2523 (1984).
24. Glaser, R.A. and P.F. O'Connor: The Analysis of Air for Acetone Cyanohydrin Using Solid Sorbent Sampling and Gas Chromatography. *Anal. Lett. 18*:217 (1985).
25. Kennedy, E.R. and R.H. Hill: Determination of Formaldehyde in Air as an Oxazolidine Derivative by Capillary Gas Chromatography. *Anal. Chem. 54*:1739 (1982).
26. Langhorst, M.L., R.G. Melcher and G.J. Kallos: Reactive Adsorbent Derivative Collection and Gas Chromatographic Determination of Chloromethyl Methyl Ether in Air. *Am. Ind. Hyg. Assoc. J. 42*:47 (1981).
27. Kennedy, E.R. and R.H. Hill, Jr.: Formaldehyde. *NIOSH Manual of Analytical Methods*, 3rd ed., Method 2502. DHHS (NIOSH) Pub. No. 84–100 (1984).
28. Hendershott, J.P.: The Simultaneous Determination of Chloroformates and Phosgene at Low Concentrations in Air Using a Solid Sorbent Sampling-Gas Chromatographic Procedure. *Am. Ind. Hyg. Assoc. J. 47*:742 (1986).
29. Klein, A.J., S.G. Morrell, O.H. Hicks and J.W. Worley: Determination of Chloroacetyl Chloride in Air by High-performance Liquid Chromatography. *Anal. Chem. 58*:753 (1986).
30. Hendricks, W.: Phosgene. *OSHA Analytical Methods Manual*. U.S. Department of Labor, Occupational Safety and Health Administration, Method 61. American Conference of Governmental Industrial Hygienists, Cincinnati, OH (1986).
31. Hendricks, W.: Acrolein and/or Formaldehyde. *Ibid.*, Method 52.
32. Kennedy, E.R., P.F. O'Connor and Y.T. Gagnon: Determination of Acrolein in Air as an Oxazolidine Derivative by Gas Chromatography. *Anal. Chem. 56*:2120 (1984).
33. Okenfuss, J.R. and E.R. Kennedy: Furfural. *NIOSH Manual of Analytical Methods*, 3rd ed., Method 2524. DHHS (NIOSH) Pub. No. 84–100 (1984), 2nd Supplement (in press).
34. Okenfuss, J.R. and E.R. Kennedy: Glutaraldehyde. *Ibid.*, Method 2531.
35. Kennedy, E.R., Y.T. Gagnon and J.R. Okenfuss: Valeraldehyde. *Ibid.*, Method 2536.
36. Andersson, G., K. Andersson, C.A. Nilsson and J.O. Levin: Chemosorption of Formaldehyde on Amberlite XAD-2 Coated with 2,4-Dinitrophenylhydrazine. *Chemosphere 10*:823 (1979).
37. Beasley, R.K., C.E. Hoffmann, M.L. Reuppel and J.W. Worley: Sampling

of Formaldehyde in Air with Coated Solid Sorbent and Determination by High Performance Liquid Chromatography. *Anal. Chem.* *52*:1110 (1980).
38. Andersson, K., C. Hallgren, J.O. Levin and C.A. Nilsson: Liquid Chromatographic Determination of Hydrazine at Sub-Parts-Per-Million Levels in Workroom Air as Benzaldazine with the Use of Chemisorption on Benzaldehyde-coated Amberlite XAD-2. *Anal. Chem.* *56*:1730 (1984).
39. Leiber, M.A. and H.C. Berk: Determination of Carbonyl Sulfide in Air by Derivatization with 1,3-Diaminopropane and Capillary Gas Chromatographic Analysis. *Anal. Chem.* *57*:2792 (1985).
40. G.G. Esposito, K. Williams and R. Bonglovanni: Determination of Ethylene Oxide in Air by Gas Chromatography. *Anal. Chem.* *56*:1950 (1984).
41. Levin, J.O., K. Andersson, R. Lindahl and C.A. Nilsson: Determination of Sub-Part-per-Million Levels of Formaldehyde in Air Using Active or Passive Sampling on 2,4-Dinitrophenylhydrazine-coated Glass Fiber Filters and High-Performance Liquid Chromatography. *Anal. Chem.* *57*:1032 (1985).
42. CEA Model 555 Formaldehyde Monitor. CEA Instruments, Inc., Westbrook, NJ.
43. Lion Formaldemeter. MDA Scientific, Inc., Glenview, IL.
44. MDA Series 7100 Continuous Toxic Gas Monitor. MDA Scientific, Inc., Glenview, IL.
45. Autostep Portable Isocyanates Monitor and Portable Hydrazines Monitor and Portable Phosgene Monitor, GMS Systems, Inc. McMurray, PA.
46. Leasure, C.S. and G.A. Eiceman: Continuous Detection of Hydrazine and Monomethylhydrazine Using Ion Mobility Spectrometry. *Anal. Chem.* *57*:1890 (1985).
47. Morrison, R.C. and G.G. Guilbault: Determination of Toluene Diisocyanate with a Silicone-Coated Piezoelectric Crystal Detector. *Anal. Chem.* *57*:2342 (1985).
48. Turnham, B.D., L.K. Yee and G.A. Luoma: Coated Piezoelelectric Quartz Crystal Monitor for Determination of Propylene Glycol Dinitrate Vapor Levels. *Anal. Chem.* 57:2120 (1985).
49. Photovac 10S Portable Photoionization Gas Chromatograph. Photovac International, Inc., Long Island, NY.
50. Miran 1B. Foxboro Co., Foxboro, MA.
51. Foerst, D.L. and A.W. Teass: A Sampling and Analytical Method for Vinyl Acetate. *Analytical Techniques in Occupational Health Chemistry*, D.D. Dollberg and A.W. Verstuyft, Eds. American Chemical Society, Washington, DC (1980).
52. Hydrocarbons, BP 36–126°C. *NIOSH Manual of Analytical Methods*, 3rd ed., Method 1500. DHHS (NIOSH) Pub. No. 84–100 (1984).
53. Melcher, R.G.: Industrial Hygiene. *Anal. Chem.* *55*:41R (1983).
54. Melcher, R.G. and M.L. Langhorst: Industrial Hygiene. *Anal. Chem.* *57*:238R (1985).

CHAPTER 14

Comparison of Number and Respirable Mass Concentration Determinations

THOMAS F. TOMBA[A] and ROBERT A. HANEY[B]

[A]Chief, Dust Division; [B]Chief Environmental Assessment and Contaminant Control Branch, Pittsburgh Health Technology Center, Mine Safety and Health Administration, 4800 Forbes Avenue, Pittsburgh, Pennsylvania 15213

INTRODUCTION

Regulations pertaining to Safety and Health Standards for Surface Metal and Nonmetal Mines and for Underground Metal and Nonmetal Mines are specified in Title 30, CFR, Parts 56 and 57, respectively. In these parts of Title 30, exposure limits for airborne contaminants are based on the Threshold Limit Values (TLVs) adopted by the American Conference of Governmental Industrial Hygienists (ACGIH) as set forth and explained in the *Threshold Limit Values for Chemical Substances in Workroom Air Adopted by ACGIH for 1973*. Exposure limits established in this edition for various mineral silicate dusts containing less than one percent quartz are based on the number of particles per cubic foot of air as determined from impinger samples counted by light-field microscopic techniques. During the past 13 years, the Mine Safety and Health Administration (MSHA) has used these limits to assess the quality of the environments of mineral processing operations.

In the 1976 edition of *Threshold Limit Values for Chemical Substances in Workroom Air*, limits based on the respirable mass of the dust per cubic meter of air that were supposedly equivalent to previously recommended limits based on the number of particles per cubic foot of air were published in Appendix G. The bases for establishing the equivalent mass concentration values were:

1. An empirical relationship, derived by Jacobson and Tomb[1] that indicated 5.65 mppcf was approximately equal to 1 mg/m^3 of respirable dust sampled with an Isleworth Gravimetric Dust Sampler, Type 113A.[2]
2. A relationship of 6 mppcf = 1 mg/m^3 developed from a calculation that assumed that the average density for silica containing dust is approximately 2.5 g/m^3 and that the mass median diameter of particles collected in midget impinger samplers (counted by the standard light field microscopic technique) and in respirable dust samplers is approximately 1.5 micrometers (μm).

Recognizing that an assessment based on a respirable mass limit would be more relevant to the health hazard, provide a simpler method of assessing the quality of an environment, and be less expensive and more reproducible than the count method, MSHA investigated the validity of the equivalent respirable mass limits recommended. This investigation was principally performed to provide documentation to support any legal actions that would result from the use of the recommended limits as equivalent "standards."

To investigate the validity of the equivalent mass concentrations recommended, the rationale published in the 1976 TLV booklet was reviewed as was the *Documentation of the Threshold Limit Values.* Empirical relationships were derived from comparative measurements obtained with a long-running midget impinger and a respirable dust sampler.

PROCEDURES

To develop the empirical relationships, comparative measurements with the midget impinger and respirable dust samplers were obtained at operations which mined or processed natural graphite, perlite, mica, diatomaceous earth, and talc (nonasbestiform).

Samples collected with the midget impinger were analyzed for number concentration using light-field microscopy following the Bureau of Mines[3] standard microprojector technique. All impinger samples were analyzed within the time constraints established by standard procedures. The results were reported as millions of particles per cubic foot. Respirable dust samples were weighed and the mass concentration of dust was determined and reported as milligrams of respirable dust per cubic meter of air sampled.

The respirable dust sampler was that typically utilized by MSHA's Metal and Nonmetal Mine enforcement personnel to assess the respirable mass concentration of dust in an environment. The sampling system consisted of a 10-mm nylon cyclone preseparator, followed by a preweighed, 37-mm diameter (5-micrometer pore size) vinyl metricel filter housed in a Millipore "Contamination Analysis Monitor" case. Airflow through the respirable dust sampling system was maintained constant at 1.7 liters per minute (lpm) using either an MSA Model G, Bendix 3900, or Bendix BDX30 pump.

The typical procedure used to obtain comparative measurements with the different sampling systems was to assemble them into a package as shown in Figure 1. Each package contained two modified midget impinger samplers, two respirable dust samplers, and a total dust sampler. The modification to the impinger consisted of replacing the standard 1-by-4.5-inch particle collection flask with a larger container that would permit extending the sampling time of the impinger from 20 minutes to four hours. Normally two packages, located at different sampling sites at a respective mineral processing operation, were used. Attempts were made to locate the package(s) so that aerosols with expected different size distributions were collected. The sampling time for comparative samples ranged from two to four hours. The number of comparative samples obtained for the respective minerals varied. The number was dependent on the type and accessability of the operations.

Filters on which the respirable dust was collected were pre- and post-weighed to 0.01 mg on an electronic, semimicroanalytical balance. The total dust samples were collected with a similar sampling system to the respirable dust samples, but without the 10-mm nylon cyclone attached. Total dust samples were also collected at a flow rate of 1.7 lpm. In addition to determining the total mass concentration of the aerosol in the environment, a representative number of the total dust samples collected were particle-sized with a Model TA II Coulter Counter. Particles from samples analyzed with the Coulter Counter were classified into 14 size intervals. The size range covered by an analysis was dependent on the diameter of the aperture tube used on the Coulter Counter during the analysis; either a 50- or 100-μm diameter aperture tube was used. The range in particles analyzed is from approximately 2 to 50 percent of the diameter of the aperture tube used.

TREATMENT OF DATA

Empirical relationships between number concentration, in mppcf, and respirable mass concentration, in mg/m^3, were derived from the compar-

Figure 1. Package of instruments used to obtain comparative midget impinger and respirable dust samples.

ative measurements for the respective minerals using the method of least squares. For each mineral, the best fit regression line relating the measurements, standard error of estimate, $S_{y/x}$, and correlation coefficient, r, were calculated. The standard error of estimate provides a quantitative measure of the variability of the data about the regression line and the correlation coefficient provides a measure of the degree of linearity between the respective variables (number and mass concentration).

Equivalent mass concentration values derived from the empirical relationships for each of the minerals were compared to the equivalent mass concentration limits specified in the 1976 TLV booklet. In addition, mass concentration equivalent values were calculated using the method given in the TLV booklet and the parameters required for that calculation, i.e., aerosol density and mass median diameter (M'_g).

Data obtained from the Coulter Counter analysis of the total dust samples were used to characterize the size distributions of the aerosols sampled. Count-versus-size data were converted to mass-versus-size data mathematically for each aerosol by using the cube of the average diameter for each interval, the material density, and the respective interval count. Cumulative mass-versus-size data were plotted on logarithmic-probability graph paper, and the mass median diameter (M'_g) and geometric standard deviation (σ_g) were determined using the graphic technique developed by Hatch and Choate.[4] The count median diameter (M_g) was then determined using the relationship:

$$\text{Log } M_g = \text{Log } M'_g - 6.9078 \text{ Log}^2 \sigma g$$

RESULTS AND DISCUSSION

Figures 2–6 graphically show the data of the comparative measurements obtained for the respective minerals, the regression lines relating the count and mass concentrations obtained, and the standard error of estimate and correlation coefficient for each of the relationships derived. The data compiled on Table I are the density of the respective aerosols, the recommended limits specified in the TLV booklet, and four count-to-mass ratios (R) derived from 1) the recommended count and mass concentration limits specified in the booklet, 2) the empirically derived regression equations, and 3) and 4) the procedure given in the TLV booklet using the M_g and M'_g valves that were determined to be representive of the respective aerosols sampled.

The significance of the correlation between the mass and number concentration relationships was statistically tested by testing the significance of the correlation coefficients (r). The hypothesis of zero correlation was rejected at the 5 percent level for all the relationships because of the large "r" values obtained.

Also shown on Figures 2–6 is the uncertainty that would be expected when using the equation to estimate the mass concentration when the number concentration measured is at the value specified in the 1973 TLV booklet. The coefficient of variation in the form of:

$$CV = \pm \; [(S_{y/x})/\text{Equivalent Mass Limit}] \times 100$$

was used to quantitate the uncertainty. For all the relationships, except that derived for mica, the uncertainty was less than 15 percent. A review

TABLE I. Comparison of Values (R) Obtained for Converting Count Concentration Data to Equivalent Mass Concentration Data

		Recommended TLV		R (mppcf/mg/m³)				Aerosol Parameters		
Aerosol	Density g/cm³	Count, mppcf	Mass mg/m³	TLV	Emp	Calc. (M_g)	Calc. (M'_g)	M_g	M'_g	σ_g
Graphite I	1.76	15	2.5	6	9.8	—	1.46	0.07	2.76	3.16
Graphite II					8.8	—	0.54	0.02	3.84	4.65
Perlite I	2.30	30	5	6	4.4	260	0.05	0.45	7.57	2.64
Perlite II					3.2	0.15	0.004	5.36	18.56	1.92
Talc I	2.75	20	3	6.6	5.5	28	0.32	0.89	3.95	2.01
Talc II					5.3	660	0.05	0.31	7.49	2.80
Diatomaceous earth	2.20	20	1.5	13	21.7	45	0.05	0.82	7.98	2.38
Mica	2.80	20	3	6.6	6.6	116	0.05	0.55	7.50	2.55

M_g = Count Median Diameter.
M'_g = Mass Median Diameter.
σ_g = Geometric Standard Deviation.
EMP = R Derived from Empirical Relationship.
Calc. (M_g) = R Calculated Using Count Median Diameter.
Calc. (M'_g) = R Calculated Using Mass Median Diameter.

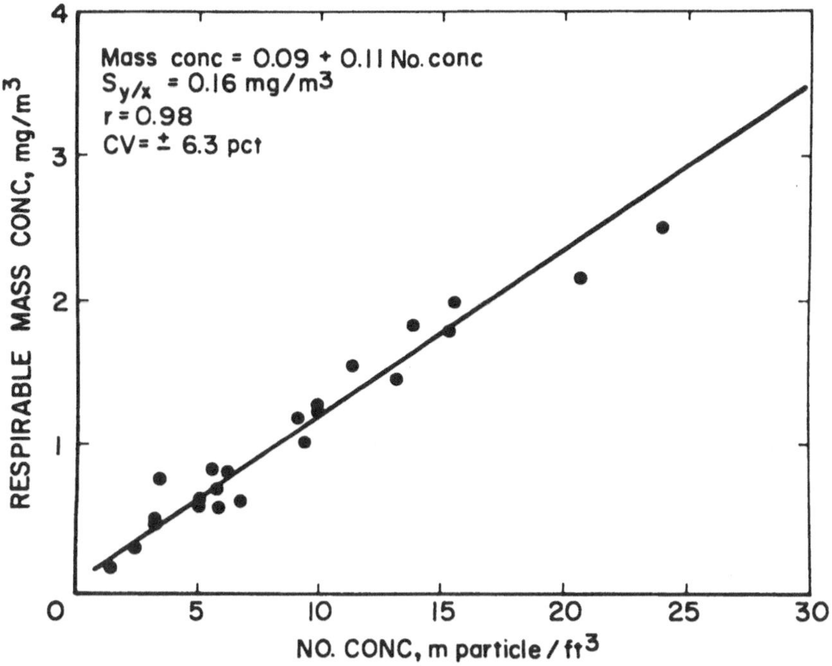

Figure 2. Comparison of dust concentrations obtained from midget impinger and respirable mass dust samples at two graphite processing operations.

of the raw mica data showed that the comparative measurements between duplicate impinger and duplicate respirable dust samples exhibited a higher degree of variability than those obtained in the other aerosols. The reason(s) for this could not be established.

A comparison (Table I) was made of the ratio (R) between the count and mass concentrations (mppcf:mg/m^3) recommended in the TLV booklet for the respective minerals and the ratios established from the empirical relationships and the calculation method using both the M'_g and M_g derived from the total dust samples. The comparison shows that only the empirically derived count to mass concentration ratio established for the mica and talc aerosols approximated the values recommended in the TLV booklet. None of the ratios established from the calculation method agreed with the values recommended in the TLV booklet or with the empirically derived values. It is apparent from the data that the M'_g or M_g established from a total dust sample measure-

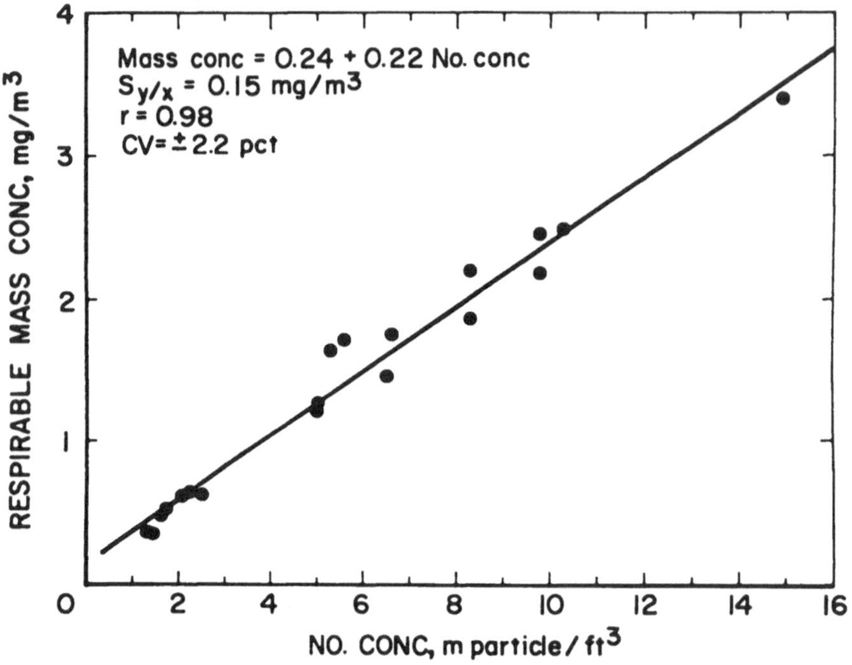

Figure 3. Comparison of dust concentrations obtained from midget impinger and respirable mass dust samples at two perlite processing operations.

ment cannot be used to derive a factor for converting number concentration determinations to equivalent mass concentrations.

Because number concentrations determined using the light-field microscopic technique only account for particles between approximately 1 μm and 10 μm in size, estimates of M'_g, M_g, and σ_g were determined for a limited number of total dust samples using only the particle size data obtained in this fraction of the sample. The parameters (M_g, M'_g, σ_g) established for the respective aerosols are shown on Table II. A comparison of these parameters with those on Table I shows that there is more uniformity in the parameters characterizing the aerosol in this fraction; i.e., the size distributions of the particles in the 1 to 10 μm fraction are more similar than those obtained when particles representative of the whole aerosol are used. These data also show that the 1.5-μm diameter specified in the TLV booklet as the M_g of particles collected by the midget impinger actually is more representative of the M_g. However, the count median diameters derived from the 1 to 10 μm fraction of the

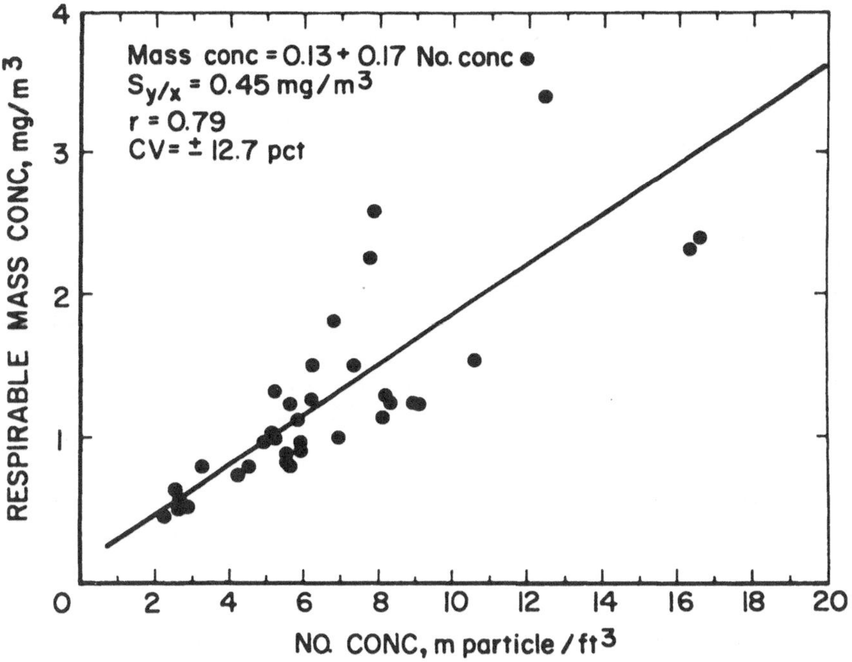

Figure 4. Comparison of dust concentrations obtained from midget impinger and respirable mass dust samples at two talc processing operations.

TABLE II. Particle Size Distribution Parameters Derived from the 1 to 10 μm Fraction of the Aerosols

	Aerosol Parameters			R(mppcf/mg/m^3)	
Aerosol	**M_g**	**M'_g**	**g**	**Calc. (M_g)**	**Calc. (M'_g)**
Graphite I	0.52	2.79	2.11	506.	1.41
Graphite II	0.36	3.52	2.39	658.	0.70
Perlite I	1.15	4.67	1.98	15.4	0.24
Perlite II	1.86	6.81	1.93	3.65	0.08
Talc I	1.31	3.42	1.76	8.74	0.49
Talc II	1.17	4.27	1.93	12.3	0.25
Diatomaceous earth	1.97	5.14	1.76	3.21	0.18
Mica	1.42	4.61	1.87	6.60	0.20

M_g = Count Median Diameter.
M'_g = Mass Median Diameter.
σ_g = Geometric Standard Deviation.
Calc. (M_g) = R Calculated Using Count Median Diameter.
Calc. (M'_g) = R Calculated Using Mass Median Diameter.

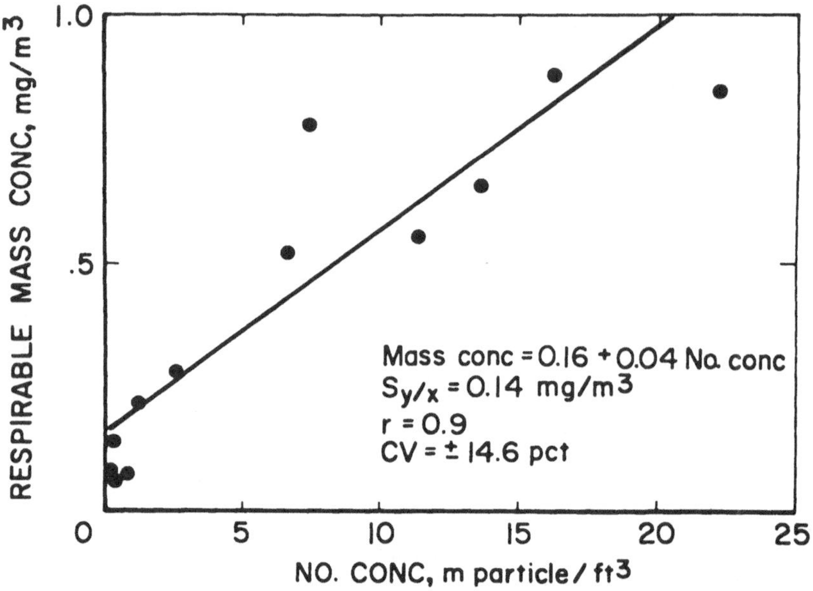

Figure 5. Comparison of dust concentrations obtained from midget impinger and respirable mass dust samples at two diatomaceous earth processing operations.

aerosols still do not provide a parameter that can be used to calculate the recommended factor for converting a number concentration to an equivalent mass concentration.

The method given in the TLV booklet for calculating a factor based on the M_g and density of the aerosol makes the implicit assumption that the size distribution of the aerosols is similar; however, as the data show, the M'_g and geometric standard deviation differ significantly for aerosols found in the same type of mineral operations as well as those established for different mineral processing operations. It should also be recognized that when using the calculation method recommended in the TLV booklet, a 15 percent difference in the diameter used to calculate an equivalency factor can result in a difference in the calculated equivalency factor of greater than 60 percent. This is due to the fact that conversion from a count to a mass concentration is a function of the cube of the particle diameter.

A review was conducted of the documentation published in the 1976 TLV booklet to arrive at, or substantiate, the value of "6" as the approximate factor used to obtain mass concentration values equivalent to previ-

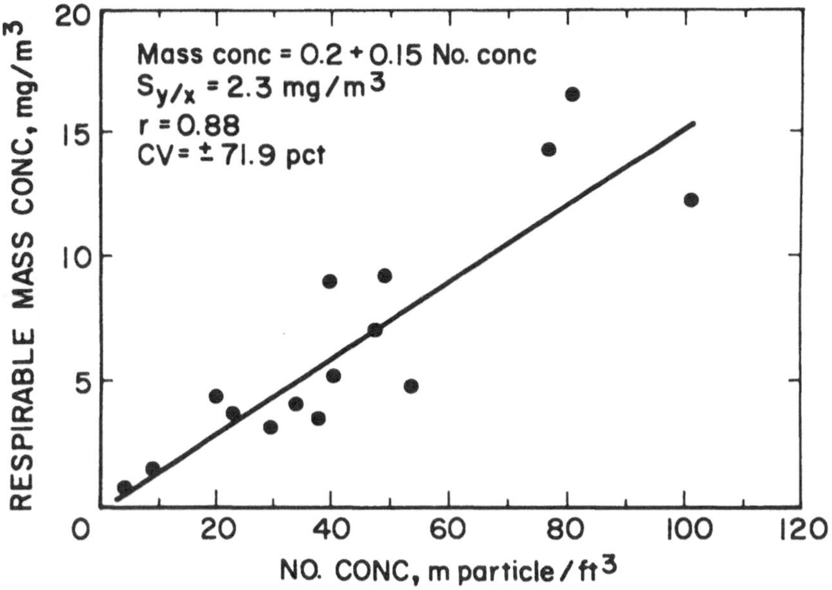

Figure 6. Comparison of dust concentrations obtained from midget impinger and respirable mass dust samples at two mica processing operations.

ously recommended number concentration values. This review showed that some of the supporting documentation is questionable. First, it is not clear which respirable dust criterion (that defined by the British Medical Research Council [BMRC] or by ACGIH) was assumed to be followed by the respirable sampler when sampling the respirable fraction of the dust. The empirical relationship of 5.6 mppcf to 1 mg/m^3 of air was derived by Jacobson and Tomb[1] from comparative measurements obtained with the midget impinger and the Isleworth Gravimetric Dust Sampler, Type 113A, an instrument that samples respirable dust according to the BMRC criteria. Mass concentration measurements obtained with a respirable mass sampler sampling respirable dust in accordance with the ACGIH criteria would be significantly lower. For coal mine dust, it has been shown[5] that the ratio between mass concentrations determined with an instrument sampling respirable dust with respect to the BMRC criteria and an instrument sampling with respect to the ACGIH criteria is 1.38.

Another questionable item deals with the statement that "the mass median diameter of particles collected in impinger samplers and counted by the standard light-field technique and collected in a respirable sampler is approximately 1.5 μm." From the size distribution data obtained from the analysis of total dust samples in the size interval from 1 to 10 μm (Table II), and from comparing size distribution data from the Coulter Counter analysis of comparative total dust samples and impinger samples collected during these studies, it would appear that 1.5 μm would be more representative of the M_g than the M'_g. This is also supported by data obtained by Cooper[6] in the Public Health Service's study of the diatomaceous earth industry. It is also highly unlikely that the M_g of the particles collected in the impinger sample would be the same as the M'_g of the particles collected in the respirable dust sampler because of the nonuniform selection process of the particle classifier on the respirable dust sampler.

The last questionable item has to do with the diameter used in the calculation method to calculate an equivalent mass concentration. The example specifies using the M'_g. From the presentation and definition of various diameters presented by Reist,[7] it appears that the diameter which should be used is the diameter of average mass which is defined as representing the diameter of a particle whose mass times the number of particles per unit volume is equal to the total mass per unit volume of the aerosol. Although by definition this would appear to theoretically be the diameter to use, the recommended limits also could not be obtained when this diameter was used in the calculation method.

Based on the review of the documentation in the TLV booklet and the empirical relationships derived from comparative measurements obtained with the midget impinger and the personal respirable dust sampler, it is concluded that: 1) "6" is not a factor that should be used universally to convert number concentration data obtained from the analysis of midget impinger samples using light-field microscopic techniques to equivalent mass concentration data; 2) because of the variability that occurs in the size distributions of the aerosols sampled (even in the 1 to 10 μm size fraction), it is unlikely that a single parameter characterizing an aerosol can be used to calculate an equivalent mass concentration; and 3) comparative measurements should be used to derive the necessary factors for converting count concentration to equivalent mass concentration data.

SUMMARY

The validity of respirable mass concentration limits for mineral dusts recommended in the 1976 ACGIH Threshold Limit Value booklet as equivalent to previously recommended number concentration limits was investigated. The investigation consisted of reviewing the documentation in the 1976 TLV booklet that was used to support the respirable mass concentration limits recommended; deriving empirical relationships from comparative measurement obtained with a midget impinger and respirable personal dust sampler at industrial operations processing graphite (natural), perlite, talc, diatomaceous earth and mica; and comparing equivalent mass concentration measurements obtained from the derived empirical relationships to those recommended in the TLV booklet.

It was concluded from the investigation conducted that the relationship, 6 mppcf = 1 mg/m^3, used to convert particle count concentration data to respirable mass concentration data was not valid. This conclusion was based on:

1. Equivalent mass concentrations established from the empirical relationships derived from comparative impinger and respirable samples did not always agree with those recommended in the TLV booklet.
2. The rationale supporting the 6 mppcf = 1 mg/m^3 relationship was questionable and could not be confirmed using data collected during this investigation.

Because there was a significant difference in the empirical relationships derived between count and mass concentration determinations and because attempts to mathematically calculate equivalent mass concentrations were unsuccessful, equivalent mass concentration limits should be empirically derived using comparative measurements obtained in the aerosol of interest.

REFERENCES

1. Jacobson, M. and T.F. Tomb: Relationships Between Gravimetric Respirable Dust Concentration and Midget Impinger Number Concentration. *Am. Ind. Hyg. Assoc. J. 28*:554 (1967).
2. Dunmore, J.H., R.J. Hamilton and D.S.G. Smith: An Instrument for the Sampling of Respirable Dust for Subsequent Gravimetric Assessment. *J. Sci. Instr. 41*:669 (1964).
3. Anderson, F.G.: A Technique for Counting and Sizing Dust Samples with a Microprojector. *Am. Ind. Hyg. Assoc. J. 23*:330 (1962).
4. Hatch, T.H. and S.P. Choate: Statistical Description of the Size and Properties of Nonuniform Particulate Substances. *J. Franklin Inst.* (1929).
5. Tomb, T.F. et al.: *Comparison of Respirable Dust Concentration Measured*

with MRE and Modified Personal Gravimetric Sampling Equipment. Bureau of Mines RI 7772 (1973).

6. Cooper, W.C., L.J. Crally et al.: *Pneumoconiosis in Diatomite Mining and Processing*. Public Health Service Pub. No. 601, G/SGPO, Washington, DC (1958).
7. Reist, P.C.: *Introduction to Aerosol Science*, Chap. 2. Macmillan Pub. Co. (1984).

SECTION IV

Real-Time Aerosol Samplers

CHAPTER 15

Modern Real-Time Aerosol Samplers

PAUL A. BARON

National Institute for Occupational Safety and Health, Robert A. Taft Laboratories, 4676 Columbia Parkway, Cincinnati, Ohio 45226

INTRODUCTION

Real-time aerosol monitoring instrument usage in the occupational health field has increased considerably in the last 15 years. Much of this increased usage is due to the availability of more sophisticated and well-characterized instruments. However, for a variety of reasons, these instruments are not used as much as they might be. Some difficulties stem from the wide range of particle sizes that can be found in industrial settings. The particle diameter can range from 0.001 μm for condensation nuclei to 200 μm and up for dust particles thrown into the air. If one is interested in the mass of these particles, the range covers 15 orders of magnitude. There are many physical mechanisms including diffusion, convection, impaction, gravitational, settling, electrostatic drift, condensation, and evaporation that can govern the behavior of particles; the importance of these various mechanisms changes in different parts of this large size range. Thus, it is rare to find instruments that can operate effectively over more than two orders of magnitude in particle size. In addition, particles in the workplace can have a wide range of chemical properties. To effectively measure workplace aerosols, it is necessary to determine the appropriate aspect of an aerosol that is to be measured. For industrial hygiene purposes, this typically means relating the aerosol measurement to the aerosol toxicity.

Quite often, more than one instrument is needed to provide measurements over a wide size range. On the other hand, in spite of all the potentially confounding variables, it is often possible to use relatively simple and limited instruments to provide very useful information. The single most common aerosol measurement in workplaces is respirable dust mass. An instrument that has a usable response in the 0.5–10 μm size range can be adequately tailored for these types of measurements. In addition, the real time and spatial information is often more important than highly accurate measurements.

Apart from the physical aspects of an aerosol, there are a number of other reasons why the instruments are often not used. First of all, there are very few aerosols present in today's industrial environment that pose an immediate danger to health. Most of the health risks are produced by chronic exposure. This means that the immediacy of information is not as critical as it is for some gases. While compliance measurements typically dictate that personal time-weighted average (TWA) results be obtained, direct reading instruments may, in some cases, be substituted for these measurements. Quite often a toxic aerosol is a component or small part of the aerosol present and a species-specific method of detection is required. There are not many direct reading aerosol instruments meeting this requirement, but some that attempt to meet this need will be mentioned below.

There are some physical aspects of the instrumentation itself that prevent its widespread use. There is no simple, rapid aerosol detector equivalent to detector tubes for gases. This means that many industrial hygienists are not as familiar with making short-term, real-time measurements for aerosols as they are for gases. Much of the instrumentation is expensive for the typical health professional and is additionally cumbersome when a wide range of contaminants must be monitored. A number of instruments have suffered reliability problems in field use and thus created a negative attitude among some users. Finally, the complexity of aerosol properties and behavior, combined with the necessity to interpret the dynamics of the workplace air, have prevented some from attempting to make real-time measurements.

The following discussion will present recent progress in addressing some of these problems with direct reading aerosol instruments. Evaluations and applications of these instruments will also be offered. The instruments discussed are primarily those developed with support of government funding. Many have become commercial products; the remainder do have certain features that might be useful in future applications.

REVIEW OF INSTRUMENT DEVELOPMENT

Following the increase in public awareness of environmental and occupational health issues in the 1960s, several governmental agencies, including the National Institute for Occupational Safety and Health (NIOSH) and the Environmental Protection Agency (EPA), were formed. Other agencies, such as the Bureau of Mines (BOM), were given new directives to address occupational health problems. Along with these new governmental initiatives came considerable amounts of money for research and development. Part of this money went toward the development of direct reading aerosol instrumentation. In addition to direct funding increases, these developments were spurred by concurrent improvements in the fields of electro-optics, electronics, and computers. Considering the availability of commercial direct reading aerosol instruments before and after this period, the last 15 years must certainly be considered revolutionary. The following discussion will present an overview of some of the developments that took place since the early 1970s.

The first direct reading dust monitor to be funded by NIOSH was the Respirable Dust Monitor (RDM-101) developed by GCA/Technology. The RDM used beta radiation attenuation to detect aerosol deposited by impaction on a greased disk. This instrument gave a result after sampling for a period of one minute or longer. The lower cutoff of particle size due to the impaction collector was about 1.0 μm. The beta radiation counting precision produced a lower detection limit of about 0.2 mg/m^3. Lower levels could be detected with longer sampling periods. It was accurate for a wide range of materials.[1,2] Offshoots of this instrument were the RDM-201 (using filtration as the collection mechanism) and the RDM-301 (same sensor as the RDM-101 with paper tape data logging).

The largest single dust monitoring program is the coal mine dust monitoring program enforced by the Mine Safety and Health Administration (MSHA). The BOM, as part of its role in supporting MSHA, funded continued efforts through the mid-1970s to develop light scattering instruments (also called photometers or nephelometers) that would measure respirable coal mine dust. In 1977, a contract with GCA/Technology resulted in the production of the Real-Time Aerosol Monitor (RAM).[3] This instrument allowed the direct measurement of respirable dust concentrations in the rugged environment and potentially explosive atmosphere of coal mines. It gave a continuous readout (allowing the operator to follow rapid changes in concentration) with a nominal measurement range of 0.001 to 200 mg/m^3, an internal calibration and zero check, a clean air sheath to keep the optics clean, a light-emitting diode to illuminate the aerosol, and an output for data logging or stripchart recording. It also had an optional drier in the sheath air system to prevent condensation on the optics under high humidity. Many of these features combined to give the instrument good long-term stability and reliability.

However, this instrument, as well as other photometers, is generally used as a supplement to, rather than a replacement for, respirable dust filter sampling. Quite often, laboratory studies have shown good correlation of instrument response with generated dust.[2,4] Good correlation in a specific study, however, does not necessarily mean good accuracy in general. Photometers and other direct reading dust instruments are sensitive to size distribution changes, humidity, and composition of the measured aerosol.

The RAM development had several offshoots. The RAM-S was a stripped-down version using an external pump with no zero and range check; it was intended for use as a control system monitor. The machine mounted respirable dust monitor (MMRDM) was funded by BOM at the request of MSHA to provide a system that would monitor the mining machine, one of the primary sources of dust in the coal mine. The MMRDM was intended for use in an alternative, less labor-intensive strategy for determining compliance with coal mine dust standards. It would have shut off the mining machine when dust concentrations reached unacceptable levels. The MMRDM was built and demonstrated but never implemented for compliance purposes.[5] Reliability problems, calibration changes with type of coal, and interference from water droplets and high humidity were some of the reasons its use was resisted by the mining operators.[6]

Another offshoot of the RAM was the development of the mini-RAM, funded jointly by BOM, NIOSH, and EPA. This instrument was a miniaturized version of the RAM that had no integral sampling pump, but it did have data logging and signal averaging capability. For the first time, it allowed the user to wear the instrument as a personal monitor. This was as important because personal exposure measurements have often shown higher and more variable concentrations than area measurements because the worker is often directly involved in the dust generating process. The passive sampling approach depended on local convection to push the aerosol through the detection region. This approach also negated some of the advantages of the RAM; namely, the clean air sheath for the optics and the zero and range check. However, when combined with a personal sampling pump and cyclone, the mini-RAM was found to give fairly reliable respirable dust readings.[7,8] Two other companies, MDA Scientific and ppm, Inc., also produced miniaturized photometers subsequent to the development of the mini-RAM, respectively designated the Personal Dust Monitor (PDM) and the Hand-held Aerosol Monitor (HAM).

In 1980, BOM and NIOSH funded the development of a personal, end-of-shift readout monitor based on the Tapered Element Oscillating Microbalance (TEOM).[9] The TEOM is a device that can be used to measure the mass collected on its vibrating tubular glass sensor. The sensor replaced the filter cassette in a standard coal mine dust personal sampler and a small filter on the sensor collected dust for an entire work shift. At the end of the shift, the sensor was placed in a special readout system that recorded and logged the collected mass. Although this system has been found to work well, the advantage of rapid and automated analysis has to be weighed against the simplicity and low cost of the current coal mine dust personal sampler. It shows some promise as an accurate, rapid readout mass sensor.[10]

One of the drawbacks of aerosol monitoring instruments is that they are typically not specific for the toxic component of an aerosol. An attempt to alleviate this drawback resulted in the development (funded by the Department of Energy, NIOSH, and EPA) of a portable X-ray fluorescence (XRF) analyzer at Columbia Scientific Industries between 1977 and 1981.[11] The detection system used radioactive sources to irradiate a filter sample with X-rays and the fluorescent X-rays of each element were selectively detected through the use of X-ray filters. While the portable XRF analyzer was not strictly a direct reading instrument, it did give the health professional field analytical capability. At the time of this development, similar types of instruments were in use for analyzing ore and metal samples. The developed instrument was capable of analyzing any one of 20 elements (between sulfur and uranium on the periodic table) in a filter sample in a period of two minutes. The sensitivity was sufficient to analyze most of these elements at their threshold limit value (TLV) on an eight-hour time-weighted average (TWA) sample. Obviously, it could also analyze shorter term, high volume filter samples as well as more rapid environmental measurements. There were a number of reliability problems with the instrument, mainly related to the computer components.[12] The instrument was never commercialized by the developer because of the availability of similar instruments and the cost of further development of an additional product. More recently, a continuous monitor for lead using a paper tape filter has been developed, using the X-ray fluorescence principle.[13]

Another instrument that was developed for a specific contaminant was the Fibrous Aerosol Monitor (FAM). This was developed at GCA/Technology (recently renamed Monitoring Instruments for the Environment [MIE], Inc.) under funding by NIOSH, BOM, and EPA in 1977.[14]

The FAM was developed to detect asbestos fibers in real time at the concentration of the then-current standard of 2 fibers/ml[3]. The FAM detected all fibers by using a combined electrostatic alignment and light scattering system. Since it detected all fibers, it was prone to interference as an asbestos monitor from some common dusts such as cotton and other organic fibers, fibrous glass, and particle chains of metal fumes. However, in the absence of other aerosols, laboratory tests indicated fairly good correlation with the standard method of phase contrast light microscopy.[15-17] Field test results were mixed, with some types of operations providing good results and others poor correlation with the reference method.[16,17] It is known that the FAM responds differently to straight fibers, such as amphiboles, than to curly fibers, such as chrysotile. The early FAM instruments also had reliability problems, mostly related to alignment of the laser beam. Many of these problems have apparently been solved.[18] Although some theoretical work has been done,[19] there is still some question as to how the instrument responds to real fibers with non-regular shapes. MIE, Inc., has recently modified the FAM with a claimed eightfold improvement in discrimination against nonfibrous dust by placing a polarizing filter in front of the detector.[18]

The Aerodynamic Particle Sizer (APS) was developed by TSI, Inc., in 1981 with the help of NIOSH funding to fill a need for direct aerodynamic sizing of aerosol particles in the respirable dust size range.[20] This instrument sizes individual particles by accelerating them through a nozzle and measuring the particle velocity with light scattering. The aerodynamic size distribution is then calculated through a calibration with unit density spheres. Although the measured size is a function of density as well as aerodynamic size, the rapid measurement and data handling capability of the APS make it a very useful laboratory field instrument.[21] The density of the aerosol particles, if known, can be factored into the calculations to determine aerodynamic size. Some other problems with APS measurements have also been noted, principally with sizing droplets with low surface tension and coincidence at higher particle concentrations. A modified version of the APS (Model 3310) has just been released by TSI. It features improved data handling capability as well as improved anticoincidence circuitry.

A very recent development is a miniaturized condensation nucleus counter (CNC), called a PORTACOUNT which is used as a field quantitative fit tester for respirators. This device was developed at TSI, Inc. with U.S. Army funding and is based on earlier work using CNCs for fit testing.[22] The PORTACOUNT can detect particles in the size range 0.02–1 μm and concentration range 0.1 to 5×10^5 particle/cm^3, allowing it to measure fit factors up to 10^5. It is small and self-contained, and has automatic, direct readout of fit factor as well as some self-diagnostic capability.

While the above discussion certainly does not present a comprehensive list of all the direct reading dust instruments that are available today, it does indicate the broad spectrum of new capability that has arisen in the last decade and a half. In the last several years, there has been a decrease in funds available to health agencies and thus less available for instrument development. However, this has allowed some efforts to be transferred to in-house research and instrument evaluation. The following discussion will present some applications of these and similar instruments in laboratory and field measurements.

DIRECT READING AEROSOL INSTRUMENT APPLICATIONS

Laboratory Studies

Many evaluations of direct reading aerosol instruments generally indicate that the instruments provide reasonably good correlation with reference methods under laboratory conditions and that this correlation degrades in field situations. There are a number of reasons for this, including that field situations are less controllable for doing accurate comparisons. In addition, the instruments themselves respond in ways not directly comparable to the reference methods. For instance, photometers respond to size distribution changes and refractive index changes in the aerosol. Thus, except for well-mixed and stable field aerosols, these instruments may not correlate well in the field with reference methods which are typically based on TWA filter samples. Therefore, the best applications of direct reading instruments are those taking advantage of the availability of immediate results and the rapid response to changes in concentration. Some instruments, such as the APS, optical particle counters, the quartz crystal microbalance (QCM) cascade impactor (California Measurements) and the PCAM (ppm, Inc.), give size distribution information as well; however, in most field instruments, this information is not available.

Various photometers have been used in our laboratories and in the NIOSH toxicology laboratories to monitor dust chambers on a routine basis. Penetration measurements of cyclones and impactors have been carried out using the APS. The use of the APS allowed the investigation of a number of variables in a short period of time.[23,24] In another analytical method evaluation system, an optical particle counter was calibrated using several impactors so that size distribution data could be taken more rapidly.[25]

A major question regarding particle size measuring instruments is about their ability to give an accurate size distribution for workplace aerosols. These aerosols can have a number of different components and cover a wide size range. An evaluation of several size distribution measurement instruments was carried out to assess their accuracy in determining the emissions from a large stand grinder.[26,27] The aim of the overall project was to improve the control ventilation system on the grinder to reduce silica exposures from casting sand. The typical use of this type of grinder was to remove risers and smooth the edges of gray iron castings. The grinding process produced an aerosol that was a combination of several materials including metal and metal oxide, hydrocarbon fume from the grinding wheel binder, and barium sulfate, a grinding wheel lubricant. The elemental data were provided by scanning electron microscope (SEM)/energy dispersive X-ray analyses of filter samples. This aerosol represented a wide range of particles from a fume to large dust particles.

Some of the larger particles were excluded by the choice of sampling location near the front of the grinder. The instrument comparison included the QCM cascade impactor, a Climet optical particle counter (OPC), a Particle Measurement Systems active scattering laser spectrometer (laser OPC), and an APS. Figure 1 shows a comparison of the number-weighted aerodynamic size distributions obtained with the various instruments as well as the size distribution obtained from filter samples analyzed with the SEM and a LeMont automated image analysis system. It can be seen that there are significant discrepancies between the various instruments. A most probable distribution was estimated by evaluating the deficiencies and strengths of each instrument and creating a composite distribution.

The QCM cascade impactor seemed to give the correct slope of the distribution but gave concentration values that were apparently too low, perhaps because of internal losses in the instrument. Its relatively low resolution also smoothed out some details of the size distribution. The laser OPC also seemed to give the correct slope of the distribution, but, because of uncertainties in estimating particle densities and refractive index, the calculated aerodynamic size distribution was also low. The laser OPC also gave spurious counts at the very small particle sizes due to noise generated by the larger particles. The APS response dropped off significantly below about 1.0 μm. It also dropped off somewhat in the 6–10 μm range due to anti-coincidence circuitry. The density correction was not factored into the aerodynamic size calculation in this figure. If

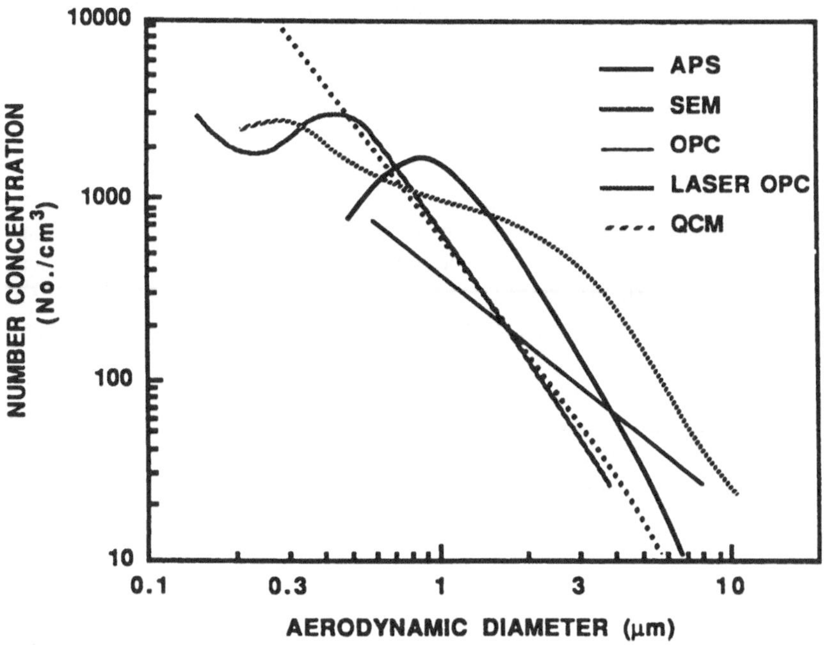

Figure 1. Size distributions of a grinding aerosol as measured by several instruments.[26]

we assume an average density of about two for the particles, then the size distribution measured by the APS comes closer to the laser OPC and QCM cascade impactor curves.

Although not a direct reading instrument, the SEM provides some interesting results for comparison. The SEM, or the transmission electron microscope, is often used as a reference method for many studies, especially in toxicology. Note that the shape of the SEM size distribution is different from the other distributions. This is apparently due to several measurement problems. The limited resolution of the image analysis system may cause problems in precise sizing of the smaller particles. Again, estimates of particle densities may cause a shift in the calculated aerodynamic sizes, although knowing the chemical composition of individual particles can help in this regard. Finally, the fact that particles are measured resting on the filter surface means that asymmetric particles may present their largest cross section for measurement. This apparently causes the size overestimation of the larger particles. This error becomes especially important when trying to estimate the mass of these particles.

Since aerosol mass is often the important parameter for health-related measurements, it is apparent that errors in the large particle size range become quite important. There is almost a sixfold difference between the lowest and highest total mass estimates from these various instruments.

From this study, it was felt that the APS, even with some limitations, gave the most useful estimate of aerodynamic size. Some recent developments may improve on this instrument further. Predictive equations have been developed to correct errors in APS size determination due to non-unit particle density.[28] Note that the correction for this density effect was not carried out in the above study and would result in about a ten percent decrease in the total mass estimated by the APS. Improvements have been made in the new APS Model 3310 to reduce the effect of coincidence in the larger particle size range as well as to extend the upper limit of the measurement range to 30 μm.

The APS has been used to investigate the properties of passive sampling devices and other direct reading instruments. The APS was used to measure penetration curves of cyclones and impactors.[22,23] However, some of these studies did not take into account the effect of density on the aerodynamic size reported by the APS. For instance, a nebulized suspension of dust was used as the challenge aerosol in several tests. This aerosol consisted of agglomerated particles, making it difficult to estimate the density of the particles. While the bulk density of the particles was high (3.5 g/cm^3), the agglomerated particles probably had a density less that 2 g/cm^3. This would decrease the corrected sizes less than ten percent.

The APS was used to determine the relative particle size-dependent response of several photometers to a volcanic ash aerosol.[29] The choice of this aerosol was largely one of convenience. Photometer response to other aerosol materials can also be evaluated by this technique. Figure 2 shows the experimental system. Aerosol was generated in a test chamber, sampled through an impactor and measured by both the APS and the instrument being tested. The impactor cut point was decreased in small increments and the difference in response of the photometer between each two cut points could be directly compared to the APS calculated mass. Figure 3 shows the size-dependent response curve for the RAM.

Field Studies

The APS is not only useful as a laboratory instrument but is sufficiently rugged to be used in the field. The items most vulnerable to workplace dust are the computer disk drive and the keyboard, so these

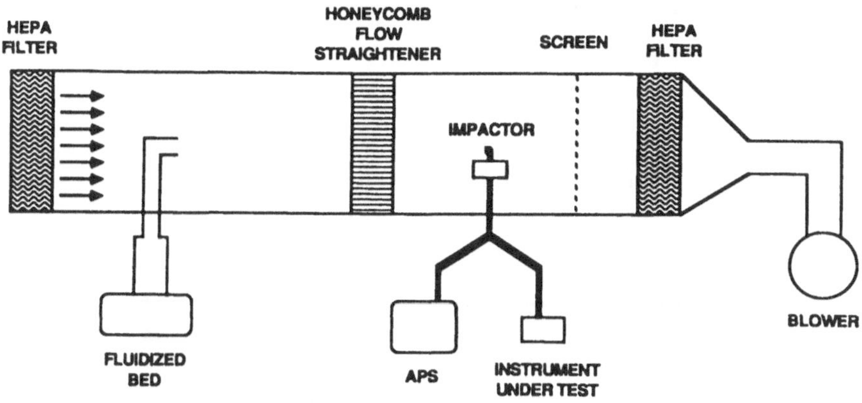

Figure 2. Test system for measuring size-dependent response of light scattering instruments.[29]

have often been placed in a glove box arrangement supplied with filtered air. Figure 4 shows a schematic of the APS and sampling system used at a limestone powder grinding operation. A Hi-Vol sampler pulled air through 4-inch plastic flexible hose to the APS inlet where it was sampled isokinetically. The flexible ducting allowed sampling from various locations as much as 15 feet from the APS cart.

Another APS application was at an outbreak of legionnaires' disease traced to the use of a health club whirlpool.[30] It was postulated that aerosol generated from the whirlpool surface carried *Legionella* bacteria from the water to people within the room. Since legionnaires' disease was not contracted by people outside that room, it was further postulated that the viable bacteria must be carried in water droplets large enough to surround the bacteria, yet small enough to become airborne and reach the respiratory tract. The APS was suspended with scaffolding over the whirlpool surface at several locations and the water droplet size distributions measured. To prevent water vapor condensation on the internal parts of the instrument, the instrument's cooling fan was connected to 2-inch diameter tubing that brought outside room air into the instrument. Figure 5 shows the number weighted size distributions at several heights in the room. All the measurements were virtually identical because poor ventilation in the room created a humid environment in which droplets generated at the water surface neither evaporated nor grew. It can be seen that a significant number of droplets were generated in the respirable size range. While these measurements were not made in an industrial

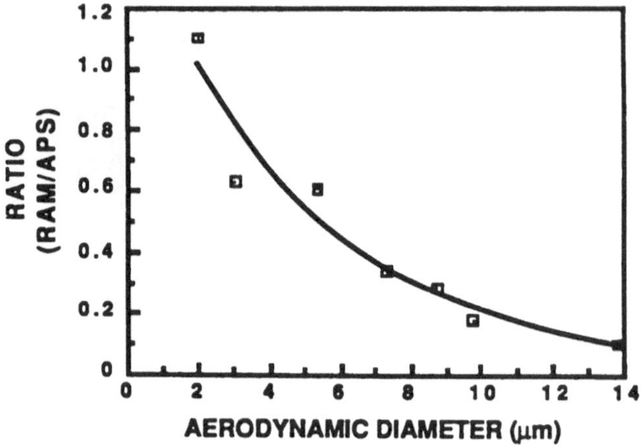

Figure 3. Size-dependent response of the RAM for a volcanic ash aerosol.[29]

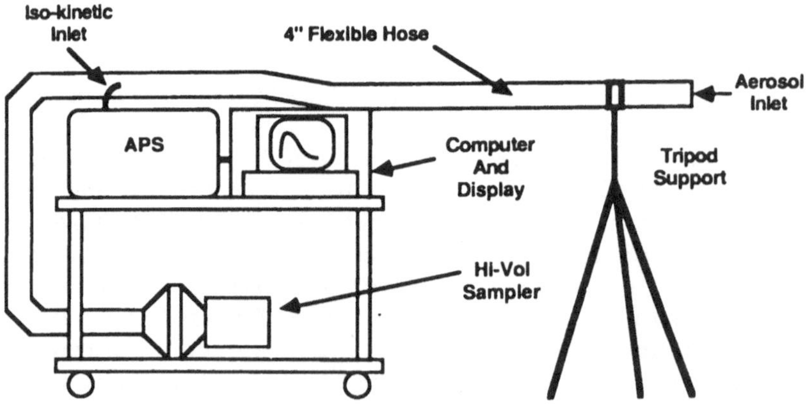

Figure 4. Field sampling arrangement for APS to allow sampling at less accessible locations.

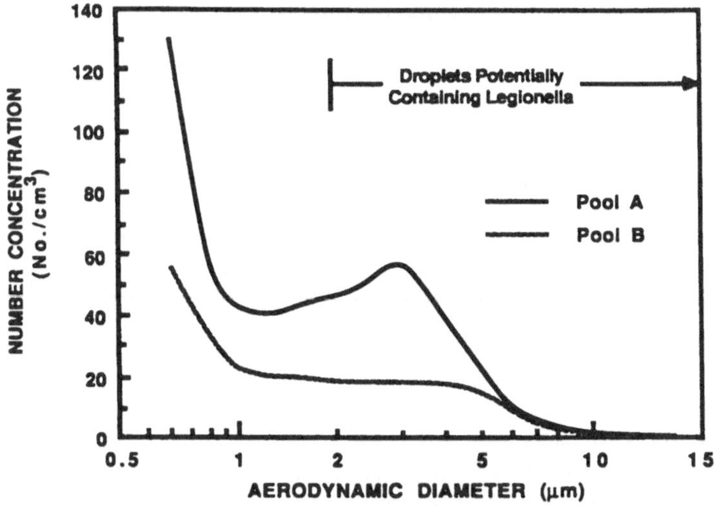

Figure 5. Size distributions of water aerosol produced by bubbling action near the surface of two different health club whirlpools.[30] Pool A had an air injection system to increase bubbling while Pool B did not.

environment, a possible application of this type of measurement may be at aeration ponds in water treatment plants and at cooling towers where bacterial generation in water droplet aerosols may cause disease.

Direct reading instruments can augment personal measurements with TWA filter samples by targeting those areas in which personal samples need to be taken and by allowing optimization of sampling volumes for the filter sample analytical method. In addition, where aerosol size distribution and concentration do not change much with time, the direct reading instrument can be calibrated to provide a direct replacement for the filter sampling.[31]

Direct reading dust instruments can be used to implement a number of human feedback mechanisms for improving workplace conditions. Documentation of the effects of work practices can aid the health professional in convincing workers to modify those practices for reduced exposure. This documentation can also be supported by direct reading instruments which provide confidence in the operation of control systems to the worker, to management, and to others who are potentially exposed in nearby locations. These feedback mechanisms are especially useful in asbestos removal or abatement operations. Asbestos abatement operations require the use of a containment system to prevent the release of asbestos into nearby work environments and into the environment. The FAM has been used to check the effectiveness of these containment systems by locating leaks and to monitor work practices within the removal operation, thus minimizing asbestos release.[16]

Recirculation of workplace air is often desirable for economic and energy conservation reasons. A direct reading control monitor measuring the performance of the filtration system in the recirculation loop is an essential component if optimum use of recirculation is to be safely achieved. The monitors themselves must have a high level of reliability and ease of maintenance. Operational parameters such as flow rate need to be self-monitored by those instruments on a routine basis.[29] The utility of these monitors was demonstrated in a foundry recirculation system. The final filter in the recirculation system was found to be torn, a fault not indicated with the simple pressure drop indicator in use. In addition, the primary cleaning system, a pulse jet air cleaner/cartridge filter, was not working correctly. The dust monitor allowed rapid repair and evaluation of the entire air cleaning system.[29,32]

The evaluation and improvement of workplace dust control measures are often difficult, if not impossible, to carry out with traditional filter sampling methods.[33,34] The Engineering Control Technology Branch at NIOSH went through a trial period with aerosol instruments to best learn how to use them. The objective was to evaluate workplace situations for optimum placement and implementation of controls. An early evaluation involved a bag opening operation at an asbestos product manufacturing plant.[35] The FAM was situated at the bag opening site; the instrument readout was manually recorded; and detailed notes were taken of the operation and the surroundings. One finding was that the principal source of asbestos dust was actually the forklift that moved pallets of bags around and brought the bags to the bag opening site. Because of the short-term activities involved, this correlation of high levels with an alternate activity would not have been brought out by traditional sampling methods. Other operations in the same plant were monitored using RAMs and the data recorded with strip chart recorders. The subsequent data for several monitors took months to reduce and analyze. Because the manual data logging and analysis processes were tedious and limited in scope, computers and data logging interfaces were assembled for subsequent evaluations. The data logging system had multiple inputs with signal conditioning that allowed data to be taken from a variety of devices including both dust instruments as well as gas/vapor measuring instruments. The data collected by the data logger were stored on disk in a format compatible with a statistical analysis package. In addition,

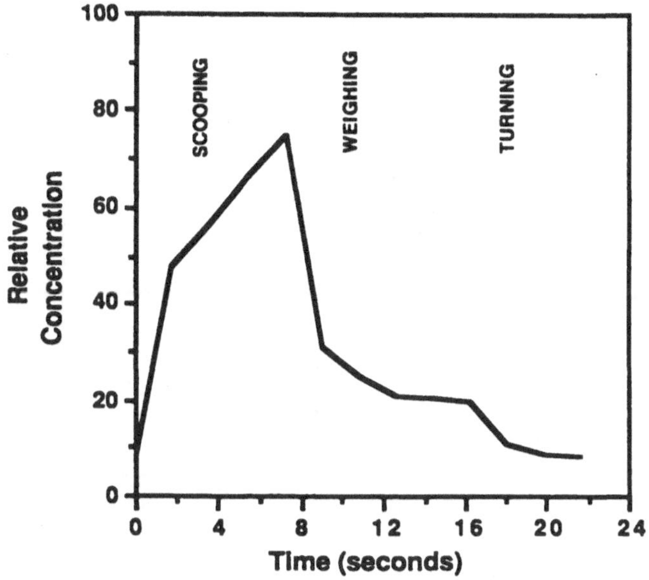

Figure 6. Predicted worker dust exposure in one cycle of a weighout and transfer operation. This curve was calculated from real-time light scattering data using time series analysis.[37]

video cameras were used to provide a permanent record of actual operations taking place in the vicinity of the monitors.

Based on experiences collecting information during a number of site surveys, integration of a variety of technologies allowed a much more complete assessment of the dynamics of various operations. For instance, in an evaluation of a manual weighout and transfer operation, the time-resolved concentrations at several points around the operation, including a Hand-held Aerosol Monitor [(HAM) ppm, Inc.] worn by the worker, allowed a number of different types of analyses to be carried out.[36] Synchronizing the video recording with the data logging allowed considerable detail to be extracted from the data. Since the operation was repetitive, a time series analysis was carried out and the work components of the operation producing the dust were detailed. Figure 6[37] shows a computer prediction of the worker dust exposure during several cycles of the operation. The time series analysis produced a breakdown of the aerosol contribution from several sources as shown in Figure 7.[37] This

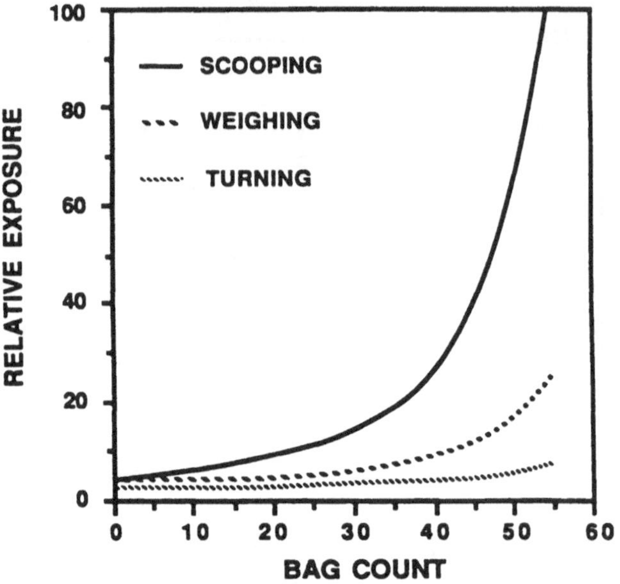

Figure 7. Predicted worker dust exposure for different components of weighout and transfer operation.[37]

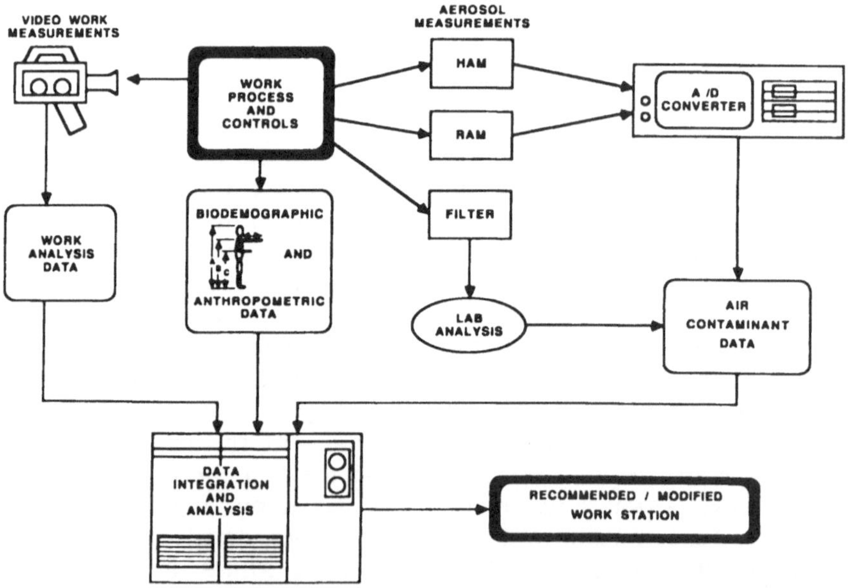

Figure 8. Overall data acquisition and analysis scheme for evaluating work processes and controls to develop recommendations for improvements.[38]

indicates that the scooping of material from a drum gave the largest contribution to the personal dust exposure.

In addition to these measurements, anthropometric and biodemographic data were also obtained to provide a more complete, quantitative picture of the operation. Figure 8 shows all of these items combined through data integration and analysis on a mainframe computer, resulting in recommendations for modified work practices and conditions.[38]

CONCLUSIONS

Over the last 15 years, a number of new instruments have been developed with the help of government funding. With the monetary cutbacks in the last several years, little further funding has gone to dust instrument development; however, work has continued in evaluating various instruments and integrating the instruments with other types of measurements. The instruments have been shown to be useful in evaluating sources of exposure and providing quantitative feedback unavailable from more traditional measurement methods. Current trends in instrument development seem to point largely in an evolutionary direction, with aerosol sensors being miniaturized and becoming more integrated with other technologies such as computers and video systems.

Increased computerization and miniaturization will hopefully produce instruments that are more reliable, self-diagnostic, and "user friendly." With these instruments, the health professional will be more able and willing to make aerosol measurements on a routine basis and solve workplace health problems.

REFERENCES

1. Almich, B., M. Solomon and G. Carson: *A Theoretical and Laboratory Evaluation of a Portable Direct-reading Particulate Mass Instrument.* DHEW (NIOSH) Report 76-114 (1975).
2. Fairchild, C.I., M.I. Tillery and H.J. Ettinger: *An Evaluation of Fast Response Aerosol Mass Monitors.* Los Alamos Scientific Laboratory Report LA-8220 (June 1980).
3. Lilienfeld, P.: *Improved Light Scattering Dust Monitor.* BOM Contract HO-377092, Bureau of Mines OFR 90-79; NTIS Pub. No. FB 299 938. Richmond, VA (1979).
4. Konishi, Y., T. Takata and K. Homma: Evaluation of the Real-Time Aerosol Monitor. *Aerosols in the Mining and Work Environment*, pp. 797-809. Marple and Liu, Eds. Ann Arbor Science, Ann Arbor, MI (1983).

5. Tomb, T.F. and A.J. Gero: Development of a Machine-Mounted Respirable Coal Mine Dust Monitor. *Ibid.*, pp. 647–663.
6. Thakur, P.C., R.W. Hatch and J.B. Riester: Performance Evaluation of Machine-Mounted Respirable Dust Monitors for U.S. Coal Mines. *Ibid.*, pp. 665–688.
7. De Garmo, S.J.: Evaluation of a Passive Real-Time Miniature Aerosol Monitor. M.S. Thesis. University of Cincinnati, Institute of Environmental Health (1984).
8. Williams, K.L. and R.J. Timko: *Performance Evaluation of a Real-Time Aerosol Monitor.* Bureau of Mines IC 8968 (1984).
9. Patashnick, H. and G. Rupprecht: *Personal Dust Exposure Monitor Based on the Tapered Element Oscillating Microbalance.* Bureau of Mines Research Contract Report, Contract HO308106; NTIS Pub. No. PB 84–173 749. Richmond, VA (April 1983).
10. Williams, K.L. and R.P. Vinson: *Evaluation of the Tapered Element Oscillating Microbalance (TEOM).* Bureau of Mines IC 9119 (1986).
11. Rhodes, J.R., J. Eckelkamp, S. Piorek et al.: *Portable XRF Survey Meter.* Final Report for Department of Energy Contract No. DE-AC05–77EV05464 (January 1983).
12. Burroughs, G.E. and P.A. Baron: *Preliminary Evaluation of a Prototype Portable X-Ray Fluorescence Analyzer.* Internal Report. Division of Physical Sciences and Engineering, NIOSH, Cincinnati, OH (December 1982).
13. Smith, W.J., D.L. Dekker and R. Greenwood-Smith: Development and Application of a Real-Time Lead-in-Air Analyzer in Controlling Lead Exposure at a Primary Lead Smelter. *Am. Ind. Hyg. Assoc. J. 47(12)*:779 (1986).
14. Lilienfeld, P., P. Elterman and P. Baron: Development of a Prototype Fibrous Aerosol Monitor. *Am. Ind. Hyg. Assoc. J. 40(4)*:270 (1979).
15. Page, S.: *Correlation of the Fibrous Aerosol Monitor with the Optical Membrane Filter Count Technique.* Bureau of Mines Report of Investigation 8467 (1981).
16. Baron, P.: *Review of Fibrous Aerosol Monitor (FAM-1) Evaluations.* NIOSH Internal Report (September 1982).
17. Iles, P.J. and T. Shenton-Taylor: Comparison of a Fibrous Monitor (FAM) with the Membrane Filter Method for Measuring Airborne Asbestos Concentration. *Ann. Occup. Hyg. 77* (1986).
18. Lilienfeld, P.: Private communication (1987).
19. Lilienfeld, P.: Light Scattering from Oscillating Fibers at Normal Incidence. *J. Aerosol Sci. 18(4)* (1987) (in press).
20. Remiarz, R.J., J.K. Agarwal, F.R. Quant and G.J. Sem: Real-Time Aerodynamic Particle Sizer. *Aerosols in the Mining and Work Environment*, pp. 879–895. Marple and Liu, Eds. Ann Arbor Science, Ann Arbor, MI (1983).
21. Baron, P.A.: Sampler Evaluation with an Aerodynamic Particle Sizer. *Ibid.* pp. 861–877.
22. Willeke, K., H.E. Ayer and J.D. Blanchard: New Methods for Quantitative

Respirator Fit Testing with Aerosols. *Am. Ind. Hyg. Assoc. J. 42(2)*:121 (1981).

23. Baron, P.: Calibration and Use of the Aerodynamic Particle Sizer (APS 3300). *Aerosol Sci. Tech. 5(1)*:55 (1986).
24. Jones, W., J. Jankovic and P. Baron: Design, Construction and Evaluation of a Multi-Stage "Cassette" Impactor. *Am. Ind. Hyg. Assoc. J. 44(6)*:409 (1986).
25. Carsey, T.P.: LISA: A New Aerosol Generation System for Sampler Evaluation. *Am. Ind. Hyg. Assoc. J. 48(8)*:710 (1987).
26. O'Brien, D.M.: Generation and Control of Respirable Dust in Industrial Stand Grinding. Ph.D. Thesis, Department of Environmental Health, University of Cincinnati, Cincinnati, OH (1985).
27. O'Brien, D., P. Baron and K. Willeke: Size and Concentration Measurement of an Industrial Aerosol. *Am. Ind. Hyg. Assoc. J. 47(7)*:386 (1986).
28. Wang, H.C. and W. John: Particle Density Correction for the Aerodynamic Particle Sizer. *Aerosol Sci. Tech. 6*:191 (1987).
29. Smith, J.P., P.A. Baron and D.J. Murdock: Response of Scattered Light Aerosol Sensors Used for Control Monitoring. *Am. Ind. Hyg. Assoc. J. 48(8)*:219 (1987).
30. Baron, P. and K. Willeke: Respirable Droplets from Whirlpools: Measurement of Size Distribution and Estimation of Disease Potential. *Environ. Res. 39*:8 (1986).
31. Baron, P.: Aerosol Photometers for Respirable Dust Measurements. *NIOSH Manual of Analytical Methods*, 3rd ed., Vol. I, 1985 Supp., pp. 43–50. NIOSH, Cincinnati, OH (1985).
32. Smith, J.P.: Use of Scattered Light Particulate Monitors with a Foundry Air Recirculation System. *Appl. Ind. Hyg. 2(2)*:74 (1987).
33. Kissell, F.N., S.K. Ruggieri and R.A. Jankowski: How to Improve the Accuracy of Coal Mine Dust Sampling. *Am. Ind. Hyg. Assoc. J. 47(10)*:602 (1986).
34. Cecala, A.B. and E.D. Thimons: *Impact of Background Sources on Dust Exposure of Bag Machine Operator*. Bureau of Mines IC 9089 (1986).
35. Heitbrink, W.A.: *Control Technology for Richard Klinger Company*. DHHS (NIOSH) In-depth Survey Report. NTIS Pub. No. PB-84-184597. Richmond, VA (1983).
36. Heitbrink, W.A., W. McKinnery and R. Rust: *An Evaluation of Control Technology for Bag Opening and Disposal—The Self Contained Filter/Bag Dump Station Manufactured by the Young Industries Inc.* DHHS (NIOSH) Report. NTIS Pub. No. PB-84-243732. Richmond, VA (1984).
37. Gressel, M.G., W.A. Heitbrink, J.D. McGlothlin and T.J. Fischback: *In-Depth Survey Report: Control Technology for Manual Transfer of Chemical Powders*. NIOSH Internal Report (ECTB No. 149-33) (December 1985).
38. McGlothlin, J.D., W.D. Heitbrink, M.G. Gressel and T.J. Fischbach: Dust Control by Ergonomic Design. To be published in *Proceedings of the IX International Conference on Production Research*. Cincinnati, OH, August 1987.

CHAPTER 16

Real-Time Portable Organic Vapor Sampling Systems: Status and Needs

JOAN M. DAISEY

Indoor Environment Program, Applied Science Division, 90–3058,
Lawrence Berkeley Laboratory, Berkeley, California 94720

INTRODUCTION

The term volatile organic compounds (VOCs) is used to designate a wide variety of organic compounds which can exist in the gaseous phase at significant concentrations at ambient temperatures. It includes aliphatic and aromatic hydrocarbons (e.g., hexane and benzene), chlorinated hydrocarbons (e.g., vinyl chloride), oxidized hydrocarbons (e.g., alcohols, ethers, ketones, aldehydes and esters), and nitrogen-containing compounds (e.g., amines and diisocyanates). The compounds in this class include toxic agents which are irritants, neurotoxins, allergens, carcinogens, and teratogens.

The VOCs have uses as industrial solvents, as raw materials for the manufacture of synthetic materials, and as components of consumer products. Widespread use of these compounds has resulted in exposure to industrial workers, to workers involved in clean-up of sites where these materials have been dumped, to office workers in buildings which are constructed of synthetic building materials, and to occupants of homes and schools. The nature of the VOC exposures to these diverse populations varies considerably. In industrial situations, workers are typically exposed to high concentrations of one or two known compounds. At toxic waste sites, a wider range of VOCs at varying concentrations is encountered and the identities of the VOCs are not always known *a priori*. In contrast to these exposures, workers in office buildings may be exposed to complex mixtures of 20 or more VOCs, each at a low concentration, but which taken in sum may be in the mg/m^3 range. Because of the potential health effects of these exposures, there is a need for instrumentation to measure the concentrations, and sometimes the identities, of VOCs in a range of environments.

Real-time portable sampling systems for VOCs sample, measure, and directly read out or record the atmospheric concentrations of these compounds in the field, often continuously. The term portable is used here to describe an instrument that can be carried or moved by one person. Portable, real-time VOC instruments currently available range from small, hand-held, relatively inexpensive ($3000 to $5000) devices that provide a generic measure of VOCs or measure a single compound to infrared monitors which measure one or a few compounds and briefcase-size portable gas chromatographs with a variety of detectors. The latter can separate and measure more individual VOCs. Prices are generally in the $12,000 to $15,000 range. Mobile gas chromatograph/mass spectrometer (GC/MS) systems have been included in this review though they are not portable as defined above. A number of these instruments are now available for field work; they provide compound identifications as well as concentrations. These instruments must be moved in a van or trailer. Prices range from $50,000 to $400,000.

There are a variety of applications for real-time portable and mobile sampling instruments. Portable monitors can be used to warn workers of dangerously high levels of toxic VOCs in industrial settings and at toxic waste sites or to monitor for compliance with regulations or commonly accepted standards. They can be used to survey toxic waste sites, to locate leaks in an industrial plant, to trace pollutant plumes, and to define safe zones in an accidental release of a VOC. These applications generally involve monitoring for a limited number of known compounds present at relatively high concentrations.

More recently, a demand had developed for instruments with greater specificity and sensitivity. Such instruments are required in situations where the identities, as well as concentrations, of the toxic compounds must be determined or where there are exposures to more complex mixtures of VOCs, such as at toxic waste sites. Instruments with greater specificity and sensitivity are also needed for research.

Applications in research include more long-term investigations of the nature, sources, and transport of VOCs at toxic waste sites and investigations of indoor air pollution.[1] Real-time instruments may be useful in efforts to determine the causes of "sick-building syndrome" in which workers in newly constructed or renovated buildings experience a variety of symptoms, including headache, nausea, and mucous membrane irritation.[2] Breath analysis for biological monitoring in the field is yet another potential research application of more specific and sensitive VOC analyzers.

This chapter reviews portable and mobile instruments that are commercially available for real-time VOC monitoring in field situations. Portable instruments for detecting explosive gases, such as methane, ethane, butane, etc., are not reviewed here but have been discussed by Nader et al.[3] A wide variety of detection systems have been used in portable VOC monitors; these include electrochemical and colorimetric detectors, thermal conductivity, flame ionization, photoionization and electron capture detectors, infrared and ultraviolet light spectrometers, and quadrupole and ion-trap mass spectrometers. The instruments reviewed here have been grouped by the type of detection system used.

SAMPLERS WITH FLAME IONIZATION DETECTORS

The flame ionization detector (FID) is an almost universal detector for organic compounds with a sensitivity in the ppm range. In the FID, a VOC with air is mixed with hydrogen and combusted to produce ions. A potential is applied across two electrodes and a current is produced which is then amplified. The current is proportional to the concentration of the VOC and the number of carbon atoms in the compound. The response is depressed by electronegative atoms such as O, S, and Cl. The detector is insensitive to water, CO, CO_2, and NO_x; however, it responds to most organic compounds and gives a linear response over orders of magnitude of concentration.

Flame ionization detectors are used in survey meters, primarily to locate areas of high concentrations or leaks, and in portable gas chromatographs. Table I presents examples of the instruments that are available. The AID Total Hydrocarbon Analyzer is a hand-held instrument which is used as a survey meter. This instrument detects VOCs and low molecular weight hydrocarbons, such as methane and ethane, which are not generally included in the VOC designation. Similar units are available from Foxboro; a gas chromatograph option available for Model OVA-128 provides greater specificity by separating the VOCs. These instruments all have internal H_2 supplies which are required for FID.

Barsky and co-workers have evaluated the effects of 0 and 90 percent relative humidity (RH) on the detection of selected and aliphatic hydrocarbons and chlorinated hydrocarbons with the Century Model OVA-128

TABLE I. Examples of Portable Real-Time VOC Samplers with Flame Ionization Detectors

Instrument	Description	Detection Range, ppm (as CH_4)
AID Portable Total Hydrocarbon Analyzer	Survey instrument, 2 units 10 × 21 × 25 cm, 15 × 16 × 10 cm, 6.4 kg, rechargeable internal H_2 supply, operates 8 hr on a battery; alarm; recorder may be connected.	Model #710: 0.1–2,000 Model #712: 1.0–20,000
Foxboro Century Organic Vapor Analyzers	Survey instrument, 22 × 30 × 11 cm, 4.1 kg, plus probe/readout assembly, 0.9 kg, rechargeable internal H_2 supply, audible alarm; operates 8 hr on battery; gas chromatograph option (OVA 108 and 128).	Model OVA 88: 1–100,000 Model OVA 108: 1–10,000 Model OVA 128: 0–10, 0–100, 0–1,000

Portable Organic Analyzer (Foxboro Analytical). They found that the linear dynamic range extended to at least 100 ppm at both 0 and 90 percent RH for most of the compounds investigated.[4]

SAMPLERS WITH PHOTOIONIZATION DETECTORS

In a photoionization detector (PID), a high energy ultraviolet lamp is used to ionize VOCs in the air sample:

$$RH + h\nu \rightarrow RH^{+} + e^{-} \tag{1}$$

where: $h\nu$ = the ionization potential of the molecule(s) of interest.

The ionized sample produces an ion current proportional to mass. The lamps used in these devices ($h\nu \simeq 10$–11 eV) do not ionize major components of air, such as O_2, N_2, CO, CO_2, and H_2O, but are energetic enough to ionize many VOCs. Interferences from more abundant hydrocarbons, such as methane and ethane, are minimized by selecting a lamp which does not emit photons of a high enough energy to ionize these compounds. By utilizing lamps with different photon energies, the detector can be modified for almost universal or more selective detection. In contrast to instruments with FIDs, an internal H_2 supply is not required for the PID detector. Barsky and co-workers have reported that 90 percent RH appears to decrease the response of the 10.2 eV lamp PID by a factor of two for most compounds tested, relative to the response under dry conditions.[4]

The sensitivity of the PID ranges from less than a ppb to ppm levels and varies with compound. Table II presents the ranges of molar responses for different classes of VOCs, relative to benzene, which have been reported for the HNU 10.2 eV lamp.[5] Molar responses for the low molecular weight alkanes and alkenes and for aldehydes, ketones and esters tend to be lower than that of benzene. In general, classes of compounds containing aromatic rings give relative molar responses greater than one and sometimes as high as three. Within a class, lower molecular weight compounds tend to give lower responses than higher molecular weight compounds.

Table III presents examples of the many portable real-time VOC samplers which utilize PIDs. The simplest instruments are hand-held, battery-operated samplers with a direct readout of concentration on a dial or LCD display. Recorders can be attached to some of these instruments, and some instruments incorporate audible alarms.

A number of portable GC instruments with PIDs are also available. These are generally the size of a briefcase or small suitcase and incorporate microprocessors for instrument control and for data handling. For most of these instruments, the GC is operated at ambient temperatures. Since this can result in variable retention times for the compounds under

TABLE II. Molar Response of the Photoionization Detector to Various Classes of Organic Compounds Relative to Benzene[A]

Organic Compound Class (n)[B]	Relative Molar Sensitivity
C_6–C_{22} Alkanes (straight, branched and cyclic) (31)	0.011–1.13
C_7–C_{19} Alkenes (16)	0.51–1.17
C_4–C_9 Aldehydes (6)	0.30–0.53
C_3–C_9 Ketones (28)	0.35–0.82
C_4–C_{12} Alcohols (10)	0.023–0.36
C_4–C_{14} Esters (20)	0.01–0.82
C_7–C_{16} Aromatic hydrocarbons (13)	1.09–1.69
Chlorobenzenes (Cl_2–Cl_6)(10)	1.20–1.44
Chlorophenols (Cl–Cl_5) (19)	1.14–1.47
Polychlorinated biphenyls (10)	2.18–2.96
Phthalates (5)	0.56–1.78
Polycyclic aromatic hydrocarbons (2–4 rings) (12)	1.88–3.08

[A]From Langhorst (1981), using HNU PID (10.2 eV lamp).
[B]Number of compounds tested.

TABLE III. Portable Real-Time VOC Samplers using Photoionization Detectors

Instrument	Description	Detection Range, as Benzene
HNU Systems Photoionization Analyzer	28 × 13 × 21 cm, 3.6 kg, with sensor (25 × 6 cm); battery operated (8-hr); meter readout; 0–10 mV recorder jacks available.	LLD:* 0.2 ppm Ranges: 0–20, 0–200, 0–2000 ppm
AID Portable Organic Vapor Analyzer	7.6 × 22.8 × 25.4 cm, 3.75 kg; battery operated (8-hr), alarm, liquid crystal display, optional strip chart recorder. Limited specificity; precision/accuracy ± 0.1 ppm, 2 sec and 5 sec response times.	Model #580 (500 ml/min): 0.1–2000 (ppm) Model #585 (50 ml/min): 0.1–10,000 ppm
AID Organic Vapor Meter Model 910	23 × 43 × 46 cm, 11.8 kg, sampling rate variable up to 41-min^{-1}, for stationary use, strip chart recorder, LED display; low and high level alarms.	10 eV lamp–0.1–2000, accuracy/precision ± 0.1 ppm; response time depends on flow rate
Photovac 10S Portable Air Analyzer	46 × 33 × 15 cm, 11.8 kg; includes a GC column operating at ambient temperature and a computer programmable for continuous monitoring, TWA, peak levels; strip chart recorder, printer/plotter and digital data displays; battery operated; internal gas cylinder for calibration; optional alarm.	0.1 ppb to ppm range
Sentex Sentor	14 × 46 × 46 cm, 13.6 kg; includes a GC column (4′ × 6′) that can be heated to 150°C, a sampling loop, microprocessor controlled; monitoring for up to 16 compounds; hard-copy print out of concentration and R.T. for 4 compounds per request, LCD display for chromatogram; automatic calibration at requested frequency.	LLD: 0.1 ppb
HNU Systems Model 301P	27 × 34 × 28 cm, 9.8 kg compact GC with 301P field pack, 12V DC power source (8-hr); isothermal operation, packed or capillary column; with line power, optional microprocessor available with printer and temperature programming upgrade.	LLD: 0.1 ppb

* LLD = Lower limit of detection.

different conditions, the instrument should be calibrated in the field. Internal calibrant gas is included in some instruments, e.g., the Photovac and Sentor. The Sentor GC has a temperature-controlled column that can be heated to 150°C and is available with electron-capture and argon ionization detectors as well as a PID.

PORTABLE GAS CHROMATOGRAPHS WITH OTHER DETECTORS

Portable GCs with other types of detectors are also available. The Sentor instrument is available with both an electron capture detector (ECD) and an argon ionization detector (Table IV). The ECD detector is particularly sensitive to chlorinated compounds with a lower limit of detection of 0.01 ppb. XonTech, Inc. also makes a portable GC with a preconcentrator and an ECD or argon ionization detector. This instrument operates on a 12-volt battery or 120/220 VAC and has both audible and visual alarms. The operating range is reported to be 10 ppb to 100 ppm.

Portable GCs with thermal conductivity detectors (TCD) are available from Bendix Corporation and Microsensor Technology, Inc., as shown in Table IV. The Micromonitor instrument is a modular instrument with up to 5 micro GC modules with TCDs.[6] The individual modules consist of a carrier gas regulation system, a sample injection system, reference and sample columns, thermal conductivity detectors, and electronics. The columns, capillary or packed, are etched on a silicon wafer and then

TABLE IV. Portable Gas Chromatographs with Thermal Conductivity Detectors

Instrument	Description
Micromonitor (Microsensor Technology, Inc.)	Up to 5 micro GCs with thermal conductivity detectors. Individual modules are 5 × 10 × 11.4 cm and consist of sample injection system, sample and reference capillary columns, carrier gas regulators, detectors etched on a silicon wafer; LLD* = 15 fg pentane from capillary column; microcomputer; correlation chromatography confirms presence of gases simultaneously detected on different columns; LCD and printer.
Unico Portable GC (Bendix Corp.)	41 × 56 × 18 cm, 22 kg, internal He carrier detector (thermistors); gas supply heated columns and detector, recorder, 115 VAC, microvolume sampling valve; LLD for benzene—89 ppm/ml, for perchloroethylene—25 ppm/ml.

* Lower limit of detection.

sealed with a flat Pyrex glass plate. Sample volumes of 80 nL can be analyzed in about 45 seconds with the capillary column. The detector is an integrated TCD on a thermally isolated Pyrex glass membrane with a thermal time constant of 200 seconds, reportedly capable of detecting 15 femtograms of pentane eluting from a capillary column. In order to minimize interferences due to co-eluting compounds, two separate columns are used and retention times on those columns are correlated using the microprocessor. If the peaks for a given compound are found at appropriate retention times on the two columns, the compound is considered to have been identified. Because of a number of technical difficulties with this system, Microsensor Technology, Inc. has temporarily halted production. They claim that the problems have been solved and that they expect to be back in production this year.[7]

Gas chromatographic instruments with FIDs, PIDs or other detectors have been developed to work best with specific analytes and must be calibrated accordingly. Although they can indicate the presence of high levels of unknown VOCs, they cannot be used to identify the compounds present nor to give more than semiquantitative information on the levels present in such instances. Nonetheless, these instruments are very useful in many field monitoring situations.

INFRARED VOC ANALYZERS

Infrared (IR) VOC analyzers detect and quantitate gases which absorb radiation in the spectral range of 2.5 to 15 μm. These instruments are based on the principle that the IR spectrum of each organic compound is unique. The IR spectrum of a compound is due to the absorption of infrared radiation to give energy transitions between vibrational and rotational energy levels of molecules in their ground electronic state. Many of the vibrational energy transitions observed in IR spectra are characteristic of certain functional groups. For example, the carbonyl functional group (C = 0) typically absorbs IR radiation of about 1700 cm^{-1}. The exact frequency will depend upon the particular compound as well as its state (gas, liquid, solid). Thus, IR spectra can provide structural information for unknown compounds or can be used to identify a compound if a spectrum has been reported for that compound.

Foxboro is the major manufacturer of portable instruments (Table V). These instruments consist of a source, a circular variable filter to select IR radiation in different frequency intervals, and a detector. Resolution with filters is typically about 10 cm^{-1}.

Detection limits depend upon the absorption coefficient of the com-

TABLE V. Miran Infrared Analyzers (Foxboro) for Volatile Organic Compounds

Instrument	Description	Application
Model 101 Specific Vapor Analyzer	47 × 14 × 15 cm, 8.2 kg; fixed cell pathlength and wavelength; factory calibrated; meter read-out in ppm and percent, AC or battery operated.	Measurement of a fixed single compound
Model 103 Specific Vapor Analyzer	10.4 kg; 13.5 m fixed cell path-length; change of filter and meter scale set permits change in monitored gas; meter readout in ppm/percent; factory cali-brated; 3 m sample hose; operated from power outlet; battery-inverted package available.	Measurement of one or more pre-selected com-pounds, one at a time
Model 1A General-Purpose Gas Analyzer	14 kg; 20 m gas cell with vari-able pathlength; variable wave-lengths; meter readout in ab-sorbance units; strip-chart recorder available; can be operated with battery or power supply; 30 m sample hose.	Scan of IR range for a general picture of con-taminants or monitoring of a single compound at a selected wavelength
Model 1B Portable Ambient Air Analyzer	12.7 kg; 20 m variable pathlength gas cell; LCD read-out of con-centration or absorbance; micro-processor with keyboard used for automatic wavelength and path-length selection, storage of calibration and analysis data, table of > 100 compounds for which analyzer is precalibrated; memory space for up to 10 user-selected compounds; audible alarm; internal battery (4 hrs).	Scan of IR range for qualitative analysis or quantitation of one com-pound at a time at selected wavelengths

pound at a given wavelength or frequency. IR absorption coefficients are typically several orders of magnitude lower than those for visible or ultraviolet absorption. In order to compensate for low IR absorption coefficients, portable instruments incorporate long (often variable) pathlength cells, typically about 1 to 20 m. The long pathlength is achieved by multiple reflections between two mirrors. Sensitivities in the ppb to ppm range can be achieved in this way. A single wavelength is selected for monitoring a given compound. The wavelength selected for monitoring should ideally be one at which no other compound absorbs, although an instrument which has a microprocessor could correct for interferences from other compounds. Table VI presents the monitoring wavelengths, pathlengths, and minimum detectable concentrations for several VOCs for the Miran 1B Portable Ambient Analyzer. The minimum detectable concentrations are typically well below the Occupational Safety and Health Administration (OSHA) maximum allowable exposure (8-hour time-weighted average). Foxboro provides similar data for over 200 compounds.

The simplest instruments, Models 101 and 103, use a fixed cell pathlength and are calibrated by the manufacturer for one (101) or a few (103) pre-selected compounds. The more complex instruments (Models 1A, 1B) provide variable pathlengths and variable wavelengths. These instruments can be used to scan the entire infrared spectrum of the air in a given location to provide a general picture of the contaminants present or can be used for quantitative monitoring at one wavelength at a time. Model 1B incorporates a microprocessor which controls instrument parameters for sequential analysis of more than one compound.

The Miran IR portable monitors have been widely used in a variety of applications, ranging from industrial hygiene and toxic waste site monitoring to research in industrial hygiene and inhalation toxicology. Their applicability for monitoring complex mixtures is limited. The infrared spectrum of a complex mixture of VOCs is the sum of the spectra of the individual compounds (assuming there are no interactions among compounds, which is generally true for gaseous mixtures). For some simple

TABLE VI. Recommended Instrument Parameters and Detection Limits for Miran Infrared 1B Portable Ambient Air Analyzer for Several VOCs

Compound	Wavelength µm	Pathlength m	Minimum Detectable Concentration ppm	Maximum Allowable Exposure, weighted (8-hr) av., ppm
Benzene	9.9	20.25	2.2	10
Ethylbenzene	9.8	20.25	3.0	100
Toluene	13.9	6.75	1.1	200
Xylene	13.1	20.25	1.9	100
Hexane	3.4	0.75	0.1	500
Octane	3.4	0.75	0.1	500
Formaldehyde	3.6	20.25	0.5	3
Acetaldehyde	9.3	20.25	1.1	200
Carbon tetrachloride	12.7	20.25	0.08	10
Chloroform	13.0	6.75	0.10	50
Vinyl chloride	11.3	20.25	0.8	1

mixtures, with a limited number of compounds, wavelengths can be found for each component at which there is little or no interference, for example, benzene and toluene (Table VI). For highly complex mixtures, such as those which might be found in a "sick building," the interferences will be overwhelming and this type of instrumentation will not be applicable. Since conventional IR detectors are not as sensitive as FID and PID detectors, it is not practical to use these as GC detectors to provide greater specificity.

MASS SPECTROMETERS

Mass spectrometers, in general, are very useful for the analysis of complex mixtures and provide great specificity and sensitivity. They are, however, very expensive instruments. In an electron impact (EI) mass spectrometer (MS), gas phase molecules at very low pressure are ionized by bombardment with accelerated electrons. The ions thus produced are accelerated by an electric field and passed through a magnetic field where they are separated according to their mass-to-charge ratio. Most commercial mobile instruments use a quadrupolar magnetic field to separate ions of different masses. The ion current at the collector is then amplified and a mass spectrum of ion current versus mass-to-charge ratio is obtained. These spectra are different for different compounds and can be used for identification and quantitation.

In chemical ionization mass spectrometry, a large excess of a second gas, such as methane or butane, is added to the ionization chamber. Most of the ions are produced from this secondary gas. These ions react with the analyte to produce secondary ions which are then separated and detected as in EI/MS. The chemical ionization (CI) mass spectrum differs from that produced in EI mode.

Several mass spectrometers have been developed for field studies and are described in Table VII. These instruments are mobile rather than portable; i.e., they are designed to be moved and operated in a van or mobile laboratory. It should be noted that there are many instruments designed for laboratory use which also can be mounted in a mobile laboratory, but the emphasis in the review is on instruments specifically designed for field use or instruments that are now being used in the field. The Bruker and VG Petra are rugged field instruments which do not include chromatographic columns. These instruments use membranes which are selectively permeable to organic compounds to provide the interface between the VOC at atmospheric pressure and the low pressure ionization chamber. Compounds are usually identified by monitoring

TABLE VII. Mobile Mass Spectrometers

Instrument	Description
Bruker Mobile Environmental Monitor	Electron-impact quadrupole MS with a 3.5 m capillary GC, programmable from 70°–190°C; 70 × 70 × 60 cm (electronics unit) and 50 × 50 × 24 cm (control unit), 145 kg; battery or generator-powered; VOCs sampled through a selectively permeable membrane; can monitor 20 compounds simultaneously, up to 8 ions per compound; series of full spectra or time curves of selected ion can be determined; video screen and printer; alarm; ppb to percent, can be remotely controlled and can transmit measurements.
VG PETRA Models SS, SM and SA	Electron-impact quadrupole MS; 80 × 72 × 57 cm (2 stacked units), 58 kg (SS), 64 kg (SM), 72 kg (SN); VOCs sampled through a selectively permeable membrane; 200 amu scan; model SS—scan and single channel monitoring; model SM —scan and multi-peak monitoring, oscilloscope output; model SA—scan and multiple peak monitoring, Apple II microcomputer, video printer output, and visual display.
Sciex TAGA 6000 E	Atmospheric pressure chemical ionization triple quadrupole MS/MS mounted in a large, air-conditioned/heated mobile laboratory; 10–1200 amu; data acquisition modes include scan-scan, parent ion scan, daughter ion scan, multiple reaction monitoring, synchronous (neutral loss/gain) scan, conventional MS scan, and multiple ion monitoring; DEC RSX-11M operating system, interactive graphics terminal and hard copy unit; can be interfaced with a GC.

selected peaks which are characteristic of the compounds of interest. These instruments are highly sensitive and can be used to detect picograms to micrograms of compounds (ppb to ppm range).

The Finnegan MAT 700 ITD is an ion trap MS detector coupled to a capillary GC. The ion trap detector is a three-dimensional quadrupole which utilizes electrodes with hyperbolic inner surfaces. In conventional mass spectrometers, ions are formed in a source region and then injected into the mass analyzer. In the ion trap, ionization and mass analyses occur in the same place. Thus, no tuning is required and the instrument is simpler, in principle, to operate. This instrument is finding use in mobile laboratories.

The Sciex Trace Atmospheric Gas Analyzer (TAGA) can analyze gases at atmospheric pressure in real time, is highly sensitive, and is designed to be used in the field. The TAGA is a tandem triple quadrupole mass spectrometer that uses sampled air as a chemical ionization reagent gas and measures either positive or negative ions.[8] In positive mode, nitrogen in air is ionized, which in turn transfers its charge to O_2. The O_2^+ ion clusters with water vapor in the air and ultimately forms the hydrated proton $H_3O^+ (H_2O)_n$ which then ionizes the VOC in the sample with little or no fragmentation. The VOC parent ions thus produced are isolated by the first filtering quadrupole and focused into a collision chamber with argon gas. Collisions between the parent ions and the argon cause fragmentation into daughter ions which are then separated by a second filtering quadrupole. The filtering quadrupoles are set to isolate the parent and daughter ions characteristic of a given compound. By sequentially monitoring for selected ion pairs, it is possible to analyze the air for several compounds in a semicontinuous mode while also reducing the risk of false positives. A full set of measurements of multiple VOCs can be obtained every few seconds. Certain isomers which give the same parent and daughter ions cannot be distinguished since there is no GC separation. Since ambient air is used in this instrument, the response factors for VOCs vary with changes in humidity and must be redetermined at regular intervals throughout the sampling period.

OTHER PORTABLE VOC MONITORING INSTRUMENTS

There are a number of inexpensive portable instruments which can be used to monitor specific compounds in situations in which there are no significant interferences from other VOCs. Some examples are given in Table VIII. MDA Scientific, Inc., markets a pocket-size monitor for formaldehyde which uses an electrochemical fuel cell with two platinum electrodes. Formaldehyde is oxidized at one electrode and an electrical current is produced. The current charges the cell to a voltage which is proportional to concentration. This instrument has been reported to be linear over the range of 0.4 to 8 ppm formaldehyde with an accuracy of 99 ± 20 percent over this range.[9] The lower limit of detection is 0.3 ppm. Positive interferences have been reported from methanol, styrene, 1,3-butadiene, and SO_2.[9] Interferences from phenol, resorcinol, formic acid, and furfuryl alcohol were reported for an earlier model.[3] A special filter is now available to minimize phenol interferences.

Sunshine Scientific Instruments markets a portable monitor which can be used for certain VOCs, such as benzene, acetone and toluene, that absorb ultraviolet light of wavelength 253.7 nm. The VOC being monitored must be known, and the atmosphere must be free of any other species which absorb or scatter (e.g., particles) ultraviolet radiation at

TABLE VIII. Other Portable VOC Monitoring Instruments

Instrument	Description
Formaldehyde Mark II (MDA Scientific, Inc.)	Pocket-size monitor with electrochemical fuel cell detector, battery operated, LCD digital read-out, 0.3 to 99.9 ppm, 20 sec response, ± 15% accuracy and precision, special filters used to eliminate interferences from phenol; interferences from methanol, ethanol, formic acid, acrolein from cigarette smoke.
Instantaneous Vapor Detector (Sunshine Scientific Instruments)	Portable monitor (13 × 10 × 43 cm, 3.6 kg) detects aniline, benzene, alcohol, acetone, pyridine, toluene, etc., by absorption of ultraviolet light; meter readout, range is about 0.01 mg/m^3; battery operated.
Toxic Gas Monitors TGM-555 (CEA Instruments)	Portable colorimetrics in which a selected VOC (formaldehyde, ethylene oxide, or phenol) is absorbed in a liquid which is then mixed with a specific chemical reagent to yield a colored product; 51 × 41 × 18 cm, 13.6 kg, with rechargeable battery (12 hrs) and constant volume adjustable air pump; digital readout, optional recorder; 0.002 to 10 ppm (formaldehyde) ± 1% reproducibility.

this wavelength. The limit of detection depends upon the absorbency index of the compound at 253.7 nm.

CEA Instruments, Inc., manufactures a portable instrument for continuously monitoring formaldehyde, which can also be converted for other VOCs such as ethylene oxide and phenol. The TGM 555 is an automated wet chemical colorimetric analyzer which uses the pararosaniline procedure for formaldehyde. Air is continuously drawn through a sodium tetrachloromercurate solution with sodium sulfite. Acid-bleached pararosaniline is mixed with this solution using a peristaltic pump, and the intensity of the resultant color is measured at 550 nm and displayed on a digital readout. A recorder output is also provided. Matthews has reported that with a series of minor modifications to the operational protocol and experimental design, the Model 555 analyzer can be used for formaldehyde monitoring at the sub-0.15 ppm levels typically found in homes.[10] Phenol can be monitored using the 4-aminoantipyrine method (0–0.1 ppm, adjustable to 0–10 ppm) and ethylene oxide with the periodate/PRA procedure (0–1.0 ppm, adjustable to 0–25 ppm).

FUTURE NEEDS

A range of portable and mobile instruments are currently available on the market for real-time monitoring of VOCs. In selecting an instrument for a particular application, the needs for specificity, accuracy, sensitivity, and response time must be kept in mind. The field conditions such as humidity, temperature extremes, and accessibility of the sampling area under which the instrument will be used must also be considered. Instruments which provide greater portability and reliability and which reduce or eliminate the influences of temperature and humidity would clearly be welcome.

There is a need to develop personal monitors which can be used to monitor one or more VOCs in industrial settings and can provide information on the time variation in concentrations as well as integrated measures of exposure. These would be useful for monitoring highly toxic VOCs and in identifying work activities that lead to high exposures. Such instruments would also have many research applications.

There is increasing recognition of situations in which VOC exposure involves complex mixtures of many compounds. In some such instances, a semi-quantitative measure of total VOCs is sufficient. However, there is often a need to identify as well as quantify the individual VOC in the mixture. At present, mass spectrometers are the only instruments which can provide identification. Truly portable GC/MS would be useful in such instances.

Gas chromatography-Fourier-transform infrared spectrometry (GC-FTIR) is being used increasingly in the laboratory for analysis of complex mixtures.[11,12] The FTIR spectrometer is based on interferometry rather than dispersion of infrared radiation. A computer is then used to transform the interferogram into an infrared spectrum. These instruments are much more sensitive than dispersion instruments and IR spectra can be obtained from nanograms to microgram quantities. Structural information can be obtained for unknown compounds. Infrared spectra can be used for compound identification, as are mass spectra, provided the spectrum of a given compound is in the library. One advantage of such instrumentation is that a high vacuum is not required. At present, the instrumentation is not portable, although such instruments can be moved on carts and possibly portable instruments will be developed.

Existing instruments for real-time monitoring of VOCs are gaining use in a variety of situations and it is likely that the need for such instruments will increase in the future. Despite their increasing use, there have been few instances in which the performance of these instruments has been critically evaluated. There is a need to systematically evaluate instrument performance and to provide guidance for appropriate applications. This is an area in which the American Conference of Governmental Industrial Hygienists can play a role.

ACKNOWLEDGMENT

This work was supported by the Assistant Secretary for Conservation and Renewable Energy, Office of Building and Community Systems, Division of the U.S. Department of Energy under contract No. DE-AC03-76SF00098. The use of trade and product names does not imply endorsement. The comments and suggestions of A.T. Hodgson, D.T. Grimsrud, and J.R. Girman on this manuscript are gratefully acknowledged.

REFERENCES

1. Blanchard, R.D. and J.K. Hardy: Continuous Monitoring Device for the Collection of 23 Volatile Organic Priority Pollutants. *Anal. Chem. 58:*1529–1532 (1986).
2. World Health Organization: *Indoor Air Pollutants, Exposure and Health Effects Assessments.* Euro Reports and Studies, No. 78, Working Group Report. Nordlinger, Copenhagen (1982).
3. Nader, J.S., J.F. Lauderdale and C.S. McCammon: Direct Reading Instruments for Analyzing Airborne Gases and Vapors. *Air Sampling Instruments for Evaluation of Atmospheric Contaminants*, 6th ed., Chap. V. P.J. Lioy and M.J.Y. Lioy, Eds. American Conference of Governmental Industrial Hygienists, Cincinnati, OH (1983).
4. Barsky, J.B., S.S. Queltee and C.S. Clark: An Evaluation of the Response of Some Portable, Direct-Reading 10.2 eV and 11.8 eV Photoionization Detectors and a Flame Ionization Gas Chromatograph for Organic Vapors in High Humidity Atmospheres. *Am. Ind. Hyg. Assoc. J. 46:*9–14 (1985).
5. Langhorst, M.L.: Photoionization Detector Sensitivity of Organic Compounds. *J. Chromatog. Sci. 19:*98–103 (1981).
6. Saadat, S. and S.C. Terry: A High-speed Chromatographic Gas Analyzer. *Am. Lab.* (May 1984).
7. Personal communication.
8. Lane, D.A., T. Sakuma and E.S.K. Quan: Real-time Analysis of Gas Phase Polycyclic Aromatic Hydrocarbons Using a Mobile Atmospheric Pressure Chemical Ionization Mass Spectrometer System. *Polynuclear Aromatic Hydrocarbons: Chemistry and Biological Effects,* pp. 199–215. A. Bjorseth and A.J. Dennis, Eds. Battelle Press, Columbus, OH (1980).
9. Coyne, L.B., R.E. Cook, J.R. Mann et al.: Formaldehyde: A Comparative

Evaluation of Four Monitoring Methods. *Am. Ind. Hyg. Assoc. J.* *46*:609–619 (1985).

10. Matthews, T.G.: Evaluation of a Modified CEA Instruments, Inc., Model 555 Analyzer for the Monitoring of Formaldehyde Vapor in Domestic Environments. *Am. Ind. Hyg. Assoc. J.* *43*:542–552 (1982).
11. Griffiths, P.R.: Fourier-Transform Infrared Spectrometry. *Science* *222*:297–302 (1983).
12. Griffiths, P.R., S.L. Pentoney, Jr., A. Giorgetti and K.H. Shafer: The Hyphenation of Chromatography and FT-IR Spectrometry. *Anal. Chem.* *58*:1349A-1364A (1986).

CHAPTER 17

Effectiveness of Real-Time Monitoring

G. EDWARD BURROUGHS and MARY LYNN WOEBKENBERG

National Institute for Occupational Safety and Health, Division of Physical Sciences and Engineering, 4676 Columbia Parkway, Cincinnati, Ohio 45226

INTRODUCTION

In the last decade, the use of direct-reading real-time instrumentation has made great strides as an industrial hygiene technique due to inherent advantages such as rapid analytical results, elimination of sample handling, and elimination of sample storage. Real-time analyzers also are able to monitor fluctuations in exposure as well as time-weighted averages (TWAs), an ability which facilitates the investigation of exposures in workplaces which have highly transient concentrations. This ability also aids the control of hazards and is a useful tool for educating both employers and employees regarding work practices and exposures. While there are negative aspects associated with the use of direct-reading instrumental methods, chiefly relating to their higher initial investment and increased complexity compared with conventional sampling techniques, the advantages in many situations clearly are greater. This is apparent through the increase in the number and variety of instruments being marketed and used by field industrial hygienists. It is anticipated that the changing nature of the workplace, as well as a need to better understand the etiology of occupationally related disease, will continue to promote both the development and the use of direct-reading instruments.

The involvement of researchers of the National Institute for Occupational Safety and Health (NIOSH) in developing and testing real-time monitoring techniques and devices has provided some examples which demonstrate the advantages of these instruments. This chapter will discuss the use of direct-reading instruments in hospitals for monitoring waste anesthetic gases, in sterilization facilities for monitoring ethylene oxide, and in grain elevators for monitoring fumigants. The discussion will include a description of the selected instrument and method, some practical concepts and limitations based on field experiences, and examples of field data to illustrate the utility of direct-reading instruments as real-time monitors.

NITROUS OXIDE MEASUREMENTS USING PORTABLE INFRARED ANALYSIS

Among the first analytes to be selectively quantitated using direct-reading instruments on-site by NIOSH industrial hygienists was nitrous oxide (N_2O). Since the mid-1970s, the lack of a reliable alternative for monitoring N_2O in the work environment of hospital personnel was the determining factor in selecting long path infrared spectrophotometry (IR) for this job.

When the problem of N_2O monitoring was originally considered, the only solution appeared to be an IR analyzer. The 1977 NIOSH criteria document on waste anesthetic gases included an on-site infrared technique for monitoring N_2O.[1] In recent years, the availability of portable gas chromatographs has solved several sampling problems; however, the high ionization potential of N_2O (12.8 eV) makes it undetectable by ionization methods, and other detectors generally lack the sensitivity required. There is no applicable conventional technique for N_2O with the exception of some passive monitoring devices which have gained wide acceptance.

The instrument used for monitoring N_2O could better be described as "transportable" rather than portable, since it weighs in the range from 20 to 50 pounds, depending on the make and model. Most available models require AC power. Some models have variable wavelengths and pathlengths, giving them the ability to analyze other substances, including other waste anesthetic gases such as halothane and enflurane. A method for monitoring N_2O using long path infrared spectrophotometry was included in the Third Edition of the NIOSH *Manual of Analytical Methods*.[2] This method was the first direct-reading instrumental technique to be included in the NIOSH Methods Manual.

Direct-reading infrared spectrophotometry has been successfully used in hospitals and in dental and veterinary operatories with no identified interferences. It covers a wide working range, from 10 ppm to more than one percent (v/v). In the back-up data supporting the published method, calibration curves were shown to be reproducible with individual data sets demonstrating statistical comparability. Experience has shown most instruments to be relatively rugged and capable of withstanding shipment. Calibration is relatively simple. Frequently, N_2O of sufficient purity can be obtained on-site for calibration, eliminating the problem of shipment of toxic materials.

The ability to measure waste anesthetic gases on a real-time basis led to the discovery that a major source of contaminant in operating rooms is leaks in the anesthesia machine. In almost every facility where levels of contaminant exceeded the recommended standard, an industrial hygienist with a real-time monitor could demonstrate to the anesthesiologist and others in the area that there was gas leaking from loose or worn connections on the anesthesia machine. Those connections could be pinpointed and frequently repaired in a matter of minutes. Following the repair, the industrial hygienist could visually demonstrate the decline in contaminant concentration.

Measurements in the area around an anesthesia machine fitting in a veterinary clinic indicated concentrations of halothane much greater than 50 ppm.[3] (This was the upper limit for which the instrument was calibrated on this occasion.) After simply tightening one fitting by hand, the concentration dropped to 6 ppm. In a dental facility, N_2O was measured at greater than 500 ppm (the limit of calibration in this instance) in a storage room containing gas cylinders supplying a manifold.[4] Within 30 minutes after the cylinder valves were closed, the concentration dropped to 10 ppm, indicating a leak in the manifold. This immediate, visible feedback had a dramatic effect on those exposed to gases and vapors undetectable by human senses.

A few practical considerations for this method are also noted. As with any analytical technique, some time is required for the operator to become familiar with the instrument. Real-time monitoring can be conducted by running a sample line from the analyzer inlet to the desired sample location. A location must be established near the area to be monitored with adequate working space and access to electrical power if necessary. This is a problem occasionally if the work space is serviced by the same ventilation system as the area to be monitored; in such instances the analyst to obtain a source of uncontaminated air (or nitrogen or oxygen) to use for zeroing the instrument.

ETHYLENE OXIDE MEASUREMENT USING PORTABLE GAS CHROMATOGRAPHY

Another environmental contaminant that we have monitored with direct-reading instruments is ethylene oxide (EtO). A few years ago, infrared analyzers were used for this work. However, recently lowered permissible exposure limits (PELs) (50 ppm to 1 ppm, TWA),[5] combined with a need to measure short duration spikes, led to the application of portable on-site gas chromatographic (GC) analysis for this job. The majority of exposures to EtO occur in sterilizing facilities (usually in hospitals) where a potentially high exposure exists for a very short time, possibly for only a few seconds, with very low concentrations generally existing for the balance of a work shift.[6,7] This exposure scenario, for substances whose exposure criteria is expressed as a TWA, is normally not a problem. For substances that cause acute toxic effects, which are potential carcinogens or mutagens, or whose PELs are expressed as ceiling values, it is desirable to measure not only the TWA but also any peak exposure since the mechanism for toxic action is not always understood.

The instrument used in these situations is a portable GC with a photoionization detector (PID).[8] Most brands and models are capable of either AC or battery operation. Carrier gas, usually ultra-pure air, can be supplied from an external supply or an integral cylinder. A four foot by one-eighth inch Carbopak BHT column at ambient temperature provides acceptable separation while keeping retention times to approximately one minute. Short retention times are advantageous in field situations since they allow for replicate analyses of samples as well as a high ratio of standard to sample analyses. Both are factors which increase the reliability of the data collected. As retention time increases, a compromise must be made between the number of times a sample can be analyzed, the number of standards that can be run, and the total number of samples analyzed per day.

The portable GC method has been used in hospitals and commercial sterilization facilities.[6,7,9] Other compounds frequently found in these facilities (e.g., Freon 12, carbon dioxide, and alcohols) are not interferences, although at high concentrations these and other compounds can increase background noise which results in decreased sensitivity. Concentrations from 0.05 ppm to several hundred ppm can be measured. A method for EtO using a portable GC has been prepared for inclusion in the Third Edition of the NIOSH *Manual of Analytical Methods*.[8] The method includes information on sampling, sample preparation, calibration and quality control, and the analytical procedure.

During a field study conducted in a facility which packages and sterilizes medical supplies, the use of on-site gas chromatographic analysis enabled the industrial hygienist to monitor worker exposure on a short-term basis.[9] This was done by collecting grab samples of the environment in 20-ml disposable syringes and then placing a rubber cap over the end port after the sample was collected. This end cap also served as a septum for the 500-μL gas-tight syringe used for injecting the sample into the GC, allowing replicate analysis of each sample. Peak EtO concentrations of up to 25 ppm were measured on occasions when an employee entered or reached into a sterilizer. Measured levels were generally in the 1 to 10 ppm range during operations such as opening the sterilizer door, pulling out a wheeled cart, or using a fork lift to remove a pallet from the sterilizer. Typically, about ten samples were grabbed during the first ten minutes (approximately) of a sterilizer down-loading process. Due to the high background caused by large amounts of isopropyl alcohol in this facility, the lower limit of quantitation for EtO was relatively high (0.3 ppm). Nonetheless, 67 of 79 samples analyzed in a two-day period were above that level, and the data enabled the industrial hygienist to formulate engineering control recommendations. The need for control had not been demonstrated by the traditional long-term sampling data.

The portable GCs seem to be slightly less rugged than the IR instruments, although they still travel well if properly packaged. On-site calibration, described in the referenced method, is relatively simple and has been shown to be very reproducible by the validation data supporting this method. The limited separation capability of these instruments has not been a problem in these situations.

PHOSPHINE MEASUREMENT USING PORTABLE GAS CHROMATOGRAPHY

Portable GCs also have been used on several occasions as real-time monitors for grain fumigants. Predominant among this work has been a series of field investigations to measure phosphine (H_3P) concentrations at grain elevators at various locations across the country.[10] A limited amount of work also has been done with carbon disulfide, carbon tetrachloride, and ethylene dibromide (discussed below).

In the grain studies, real-time monitoring techniques are beneficial because H_3P has a relatively low PEL (0.3 ppm TWA), the environment in grain elevators is relatively free from compounds with similar volatility and polarity, and the nature of the work tends to create high concentrations of short duration. A portable GC with a photoionization detector, as described in the previous section, was selected for these investigations. A six-foot by one-eighth-inch packed column (10% SE 30 on 80/100 mesh Supelcoport) provided a retention time for phosphine of approximately one minute. A limit of quantitation in the 0.02 to 0.05 ppm range was attainable during the early hours of the study.

An interesting phenomenon occurred in several series of measurements. A decrease in sensitivity developed over a day, followed by a partial recovery overnight. It was hypothesized that high molecular weight grain oils, injected onto the column with each sample, began to elute through the detector after a few hours. This created an increasing background noise over which the phosphine peak had to be measured. During the night, while the chromatograph column was flushed with carrier gas without the addition of more samples, the background diminished. A precolumn with backflush capability was installed in an attempt to eliminate this problem, but to date there has been no occasion to field test this system.

The brief nature of some jobs which create potential fumigant exposure in grain elevators makes the small sample size needed for GC analysis a major advantage. Refilling a Phostoxin tablet dispenser can take from a few seconds to a minute. Opening a sealed rail car and collecting a grain sample can take one to two minutes. During either of these jobs, an industrial hygienist can take a dozen or more grab samples if required. Each can be subsequently analyzed, in replicate, and the worker shown the magnitude of his exposure within minutes of its occurrence.

Another approach to exposure assessment is to sample the head space above grain in rail cars prior to fully opening the car lid for collection of a grain sample. In one series of field studies, data collected using this approach indicated carbon disulfide concentrations ranging from below 0.4 ppm to over 200 ppm.[11]

No written method has been prepared by NIOSH researchers for phosphine or other fumigants, although the sample collection instrument preparation, calibration and quality control, and analytical procedure sections of the EtO method can be applied with some modifications. There is no other instrument available which will provide the same combination of sensitivity and selectivity on-site as a PID GC

A very important caution should be mentioned about H_3P. At high concentrations H_3P is spontaneously flammable in air. One solution to this problem is to purchase a dilute mixture (e.g., 1% H_3P in N_2) whose concentration can be confirmed. This mixture can then be diluted to desired, known concentrations for standardization. Regulations regarding its transportation are not significantly different than for compressed air.

A limited amount of work has been done on other fumigants. The feasibility of real-time monitoring ethylene dibromide (EDB) was studied by NIOSH researchers using a portable GC with a heated column and an electron capture detector. A series of electronic problems in the GC, combined with a ban on the use of EDB as a pesticide, caused an early end to this investigation. It appeared, however, that the technique could be practical given a reasonable effort toward development. The limited use of CS_2/CCl_4 mixture for fumigation also restricted the number of field studies where these compounds were monitored with direct-reading instruments. Lab experience indicates these analyses are practical. The interesting occurrence with this combination is the ability of a PID with an 11 eV source to detect CCl_4 which has an ionization potential of 11.3 eV. While this is by no means a unique occurrence, it is noted that the sensitivity decreases rapidly as the ionization potential of the analyte exceeds the ionization energy of the detector.

SUMMARY

Many of the concepts discussed in the applications described here are viable in a wide range of real-time monitoring situations. Other concepts not discussed may have even greater application in a different field setting. For example, direct-reading real-time monitors can be used for mapping a work environment. This mapping can be used for a variety of purposes, such as source identification, determination of a subsequent sampling strategy, or ventilation design.

A trend in the development of field analytical instruments seems to be toward "smarter" units, i.e., devices with microprocessor controls, memory, automated sampling capabilities, output to computer storage and so forth. Greater portability and increased ease of operation also are strongly advertised selling points of these devices. Unfortunately, these trends also tend to increase instrument cost, which is a serious limitation for some devices. A note of caution should be sounded toward the treatment of any device as a "black box." Proper use of direct-reading instruments can provide real-time information to the industrial hygienist unobtainable by any other technique.

DISCLAIMER

Mention of company name or product does not constitute endorsement by the National Institute for Occupational Safety and Health.

REFERENCES

1. National Institute for Occupational Health and Safety: *Criteria for a Recommended Standard—Occupational Exposure to Waste Anesthetic Gases and Vapors*. DHHW (NIOSH) Pub. No. 77-140 (March 1977).
2. National Institute for Occupational Safety and Health: *NIOSH Manual of Analytical Methods*, 3rd Ed., Method No. 6600. DHHS (NIOSH) Pub. No. 84-100. Government Printing Office, Washington, DC (1984).
3. Burroughs, G.E.: *Health Hazard Evaluation Report 78-62-526, Littleton Veterinary Clinic*. DHEW, NIOSH (September 1978).
4. Burroughs, G.E.: *Health Hazard Evaluation Report 78-129-544, Alvin Jacobs, D.D.S.* DHEW, NIOSH (December 1978).
5. Occupational Exposure to Ethylene Oxide. Final Standard, 29 CFR Part 1910. *Fed. Reg. 49*:25734 (June 11, 1984).
6. Kercher, S.L. and V.D. Mortimer: Before and After: An Evaluation of Engineering Controls for Ethylene Oxide Sterilization in Hospitals. *Appl. Ind. Hyg. 2*:7-12 (1987).
7. Mortimer, V.D. and S.L. Kercher: *Control Technology for Ethylene Oxide Sterilization in Hospitals*. In-Depth Survey Report No. 146-12C. DHHS (NIOSH) (June 1986).
8. Burroughs, G.E.: Ethylene Oxide, Method 3702. A Supplement to the *NIOSH Manual of Analytical Methods*, 3rd ed. Cincinnati, OH (May 1986).
9. Gorman, R. and P. Seligman: *Health Hazard Evaluation Report 83-335-1618, Kendall Co*. DHHS, NIOSH (August 1985).
10. Zaebst, D.D., L.M. Blade, G.E. Burroughs et al.: *Assessment of Occupational Exposures of Grain Elevator Workers to Phosphine During Fumigation with Aluminum Phosphine* (in preparation 1987).
11. Ahrenholz, S.H.: *Health Hazard Evaluation Report 83-375-1521, Federal Grain Inspection Service*. DHHS, NIOSH (October 1984).

CHAPTER 18

Modern Continuous Samplers for Volatile Organics and Inorganic Gases

WILLIAM A. McCLENNY and RICHARD J. PAUR

Environmental Monitoring Systems Laboratory, U.S. Environmental Protection Agency, Research Triangle Park, North Carolina 27711

INTRODUCTION

The scientific effort within the Environmental Monitoring Systems Laboratory (EMSL), U.S. Environmental Protection Agency (EPA), for development and evaluation of new and improved methods for monitoring air quality has been driven by interest in three main topics: toxics, acidic deposition, and indoor air quality. Reliable instrumentation for EPA's monitoring networks and for special studies are needed to obtain data bases from which to distill information relevant to regulatory and scientific questions being addressed by the agency.

An example from the toxics area is the fledgling Toxic Air Monitoring System (TAMS) network. This network currently consists of two sites at each of three cities, Houston, Boston, and Chicago, with plans in 1987 for an additional site in each city and one site in a fourth city. Each site has 24-hour sampling systems for volatile organic chemicals (VOCs). Samples are collected on a 12-day schedule and then sent to EPA's central laboratories at Research Triangle Park, North Carolina, for analysis and data archival. Solid sorbent sampling using Tenax® is used at every site while two canister-based samplers are also used at one of the Houston sites. Samples are taken by sampling at constant rates over 24-hour sampling periods. Special projects in toxics monitoring include the total exposure assessment monitoring (TEAM) study in Baltimore. As part of this study, 12-hour samples were taken inside and outside selected residences using weatherized canister-based samplers outside and simpler units inside.

The central question for sampling VOCs is whether to continue using Tenax solid sorbent or to switch to the use of specially prepared canisters for sample collection and storage. The existence of sampling problems with Tenax has been documented and explored by Walling et al[1] of the EPA; he has devised a sampling approach using Tenax: the distributed air volume (DAV) approach,[2] i.e., a screening procedure to eliminate questionable results. Unfortunately, the fraction of data eliminated is very high, varying from 40 to 60 percent. Specially prepared containers for whole air collection and storage of VOCs were used in the early 1970s by air quality researchers at Washington State University. A proprietary electropolishing solution was developed for treating the interior surfaces of the canisters, and special procedures are observed in welding together hemispheres to form the container. Today the canisters can be purchased as off-the-shelf items from either of two small companies. The most common size is the six-liter spherical canister. The canisters are cleaned and evacuated to a small fraction of a torr prior to use. Typical usage includes sampling for hydrocarbons in ambient air, either for individual organics[3,4] or for nonmethane organic compounds (NMOCs) as part of a recent effort by EPA and state agencies to establish the reduction of hydrocarbon emissions necessary to reach acceptable ambient ozone concentrations.[5] The current emphasis on the use of canisters is associated with the extension of their usefulness to include collection of toxic VOCs, including substituted hydrocarbons such as vinyl chloride, ethylene oxide, methylene chloride, and tetrachloroethylene.

The development and evaluation effort in acidic deposition is mainly associated with the selection of instrumentation for sampling dry deposition components for use in the 100-station national network to be implemented through an EMSL contractual effort. Specifically, the factors influencing selection of sampling units designed for week-long, multicomponent collection are being considered. Central among these factors are accuracy, artifact-free collection, reliability, and expense. The chemical collection of inorganic gases by bonding reactions with surface coatings or selected surface materials is being studied using different sampling formats. These involve sampling targeted air components on surfaces placed perpendicular to the mass flow direction or on surfaces placed parallel to the mass flow direction. In the latter case, particles and gases can be separately collected based on the differences in diffusion, i.e., gases diffuse rapidly to the sampler wall and can be collected there.

Particles are collected downstream on size-exclusion filters. However, this convenient separation dramatically alters the phase equilibrium at the filter and causes particle-to-gas conversions. Chief among the candidate sampling units are 1) the filter pack system (usually referred to as the Canadian filter pack because of their extensive usage[6]); 2) the transition flow reactor (TFR);[7,8] and 3) the annular denuder system (ADS) as altered from the earlier Italian version[9] to meet dry deposition monitoring constraints.[10] All the units use filter packs and draw air through the collection system by use of air pumps. The addition of equipment to separate gases from particles to the simpler filter packs adds the possibility of serial and separate collection of groups of gases having different chemical bonding characteristics, e.g., acidic versus basic groups.

DISCUSSION

Volatile Organic Samplers

The use of distributed air volume (DAV) sets is the current EPA sampling approach. Four sampling channels are established, each with a different sampling rate, typically resulting in 24-hour sample volumes of 5, 10, 20, and 40 liters. The amounts of VOCs collected, divided by the respective sample volumes, should give answers that are reasonably equivalent to duplicates. If not, sample integrity is compromised and the data is invalidated. Standard operating procedures for sampling and analysis of air samples using a Tenax solid sorbent are available from EMSL.[11] Since the solid sorbent is the first element in the sampling train, contamination does not occur from upstream sampler components. However, passive sampling can occur if the cartridge channel is exposed and is a potential source of error, depending on the duration of unpumped exposure time and the effective passive sampling rates. In typical sampler designs, electronic flow controllers are used to establish a constant flow rate through the Tenax. The DAV has been used by EPA in controlled laboratory experiments and in a number of field studies to determine any problems. The DAV itself performs as expected to screen samples, but it does not identify the cause of questionable data; although some of the potential problems with Tenax are documented, their definition does not make occurrence of artifacts predictable. Indeed, results vary with location and time in an apparently random manner.

Canister-based samplers[12] allow for several sampling train configurations, since the evacuated canisters can be used to pull in sample air without need for any external power source. Subject to sampling location, a simple combination of particulate filter, critical orifice, and canister can be used to obtain a reasonably constant sampling rate until the canister pressure exceeds the level necessary to maintain a critical flow, i.e., one-half atmosphere. The substitution of an electronic flow controller for a critical orifice allows extension of the period over which a constant sampling rate can be maintained. This configuration has been used to sample for approximately 24 hours into a six-liter canister, using flow rates of less than 5 cc/min. Using the canister vacuum to produce flow results in a sample at sub-atmospheric pressure at the end of the sampling run. To supply an excess flow at the analytical system inlet, the canister can either be pressurized by a zero gas so that the diluted canister contents can be vented past the system inlet, or the canister contents can be transferred through the analytical system (reduced temperature VOC trap) into a canister at a lower pressure until a given pressure increase in the receiving canister is recorded. The canisters are also used with sampling trains that include filter, flow controller, pump, and canister, so that the canister is pressurized during collection. These components are often enclosed in a temperature-regulated, weatherized housing for operation outside. The weatherized canister systems have been used in a number of field studies.

KANAWHA VALLEY TOXICS EVALUATION STUDY (KVTE)

Canister-based samplers in weatherized housings were initially constructed, tested, and employed as part of the KVTE study in the winter of 1985–86. The sampler design schematic is shown as Figure 1. Tentative operating procedures for sampling and one for analysis of samples were prepared and were used in the KVTE study. The samplers were tested prior to deployment by ambient air sampling comparison runs. These tests showed general agreement for all compounds (within 0.3 ppbv for concentrations of 5 ppbv or less and within 10 percent for higher concentrations) except for one sampler which showed ppb losses in some cases. Five samplers were deployed in the Kanawha Valley at four sites with the extra sampler being rotated among the remaining four to establish an additional measure of quality control. This procedure identified results from one sampler as consistently low for the higher molecular weight compounds. This was the same sampler that gave consistently low readings in the earlier tests. Systematic cleaning of the sampler components has subsequently minimized losses.

Audit runs during the KVTE study using a set of reference compounds in dry nitrogen showed that higher molecular weight compounds were

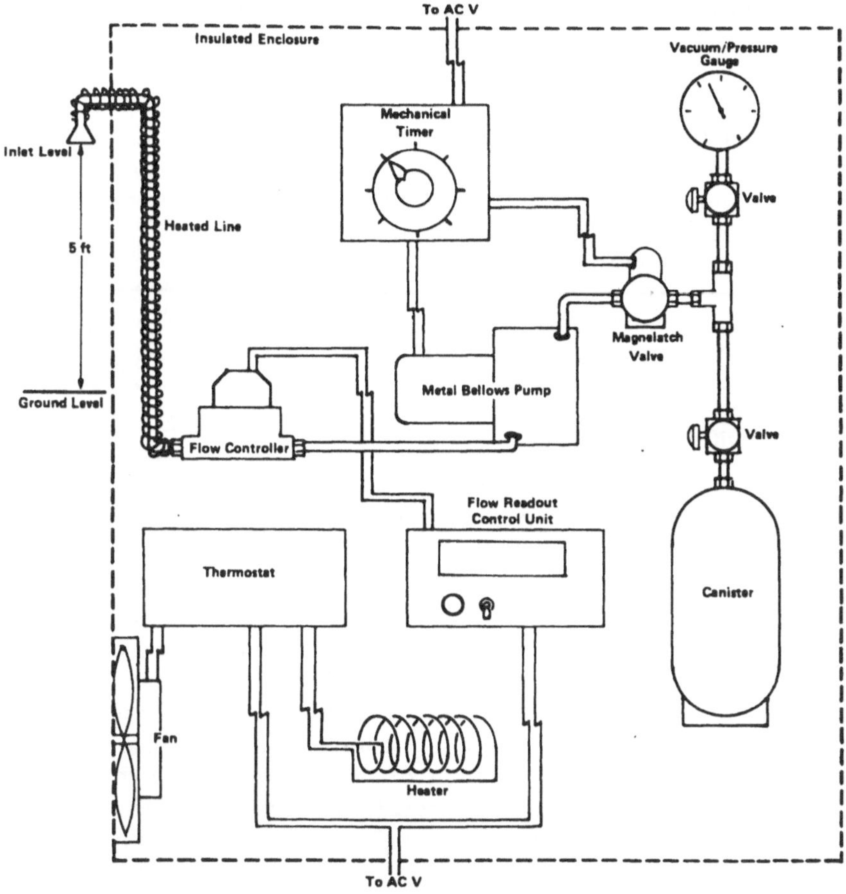

Figure 1. K-type sampler configuration.

significantly reduced in concentration upon passage through the samplers and that they reappear in subsequent samples taken from the ambient air. The same behavior was repeated after the samplers were returned to EMSL, i.e., the heavier VOCs in audit runs were held up in the sampler and subsequently returned to the airstream when a humidified sample was pulled through the sampler. No such behavior was noted when the audit sample was humidified.

The sample performance during audits suggests that water vapor usually occupies adsorption sites otherwise active with respect to some volatile organics. If water vapor is not present, volatile organics are at least delayed during passage through the sampler because of increased adsorption. When next a humidified airstream is passed through the sampler, water vapor appears to displace the adsorbed VOCs which then show up as false positive concentrations.

Although this explanation is consistent with our experimental evidence, no direct proof has been obtained. Subsequent to the KVTE, sampler component cleaning procedures have been implemented that minimize the adsorption of VOCs from dry airstreams.

Conclusions from the KVTE study are

1. Better quality control procedures are needed in selection and cleaning of components during assembly of sampler components.
2. The audit procedure should be carried out with humidified audit gas or the sampler component on which the heavier VOCs stick must be identified and replaced.

CALIFORNIA AIR RESOURCES BOARD (CARB) SUMMER 1986 STUDY

EMSL participation in the CARB summer study consisted of a Battelle Columbus task assignment to operate three samplers and one Tenax DAV at the study site (campus of Citrus College) and to analyze a portion of the samples. This portion was the 12-hour daytime samples taken over the nine-day sampling period and consisted of the Tenax DAV set and two duplicate canister samplers of the modified Kanawha Valley type (K-type). Northrop Services, the EMSL in-house contractor, analyzed the night-time samples which consisted of two duplicate canister samples from the modified K-type samplers and canisters from a sampler donated by Dr. Rei Rasmussen of the Oregon Graduate Center. The Rasmussen sampler (R-sampler) is designed differently as shown in Figure 2. Air is pumped into the sampling train at a high rate to a "dump cross" equipped with an adjustable leak. Most of the sample is dumped and the upstream volume is pressurized. A mechanical flow is adjusted to give the correct flow rate for up to 24-hour runs. The pressure difference between the upstream and downstream sides of the flow controller has to be maintained above a minimum value.

The initial results of the CARB study showed that the modified K-type samplers have contamination from Freon 113 and tetrachloroethylene, with possible deletion of trichloroethylene. Daytime samples showed higher contamination presumably because of higher temperatures causing higher emission rates from contaminated materials. Freon 113 and tetrachloroethylene contamination was traced to emissions from a sec-

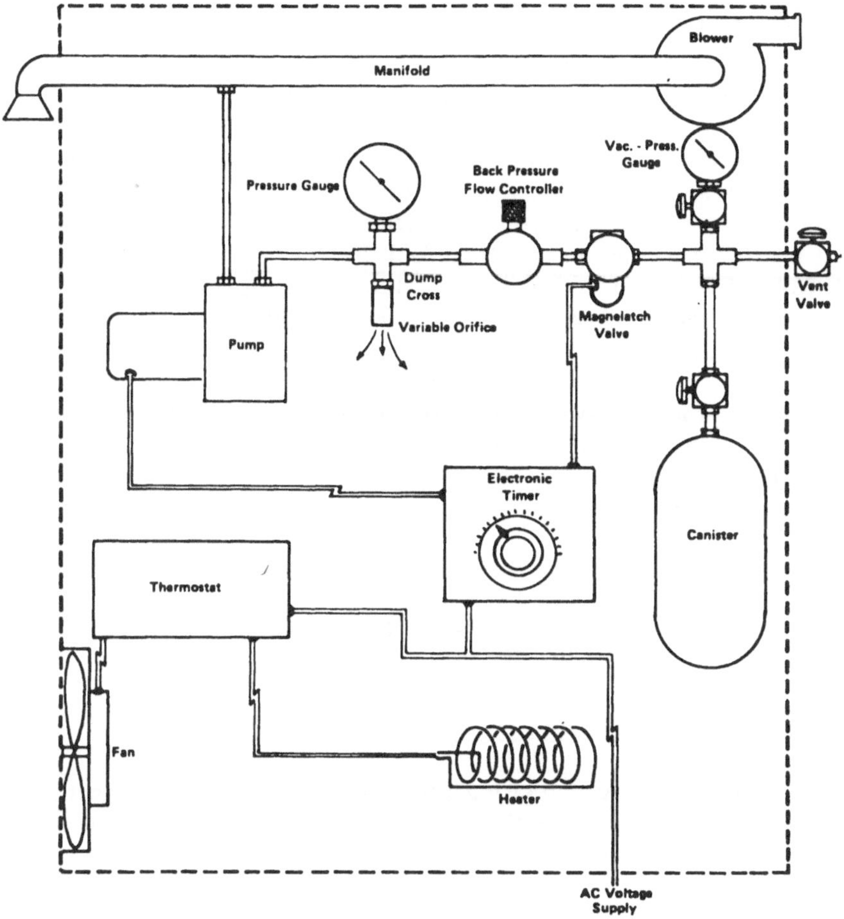

Figure 2. R-type sampler configuration.

tion of Teflon-lined stainless steel tubing. The R-type sampler was apparently free of contamination and deletions. Its only problem appears to be the collection of water in the section of high pressure between the pump and flow control element.

Conclusions from the CARB summer study are as follows:

1. High temperatures typical of summer conditions should be used to test samplers for the outgassing type of contamination. More rigid component selection and cleaning procedures are warranted.

2. R-type sampler components are apparently free of target gas contamination and of deletion problems.

HOUSTON TOXIC AIR MONITORING SYSTEM (TAMS)

Two modified K-type samplers were supplied for use in the Houston TAMS "research" site. The modification consists of adding a second pump to draw sample air through the heated inlet at a higher flow rate so that the inlet becomes an inlet manifold. Sample is drawn into the flow controller from the inlet manifold at a low rate compatible with 24-hour sampling duration. The samplers operate to collect 24-hour samples every 12 days. For the first eleven sets, nine of the duplicate canister sample analyses show generally good agreement, but also show the same problems as the CARB samples. Two of the sets showed a systematic disagreement between the duplicate canister sample analyses that has yet to be resolved.

Conclusions formed from the Houston TAMS study are

1. Unattended operation of the samplers in duplicate usually shows results equivalent to the CARB study but under Houston conditions.
2. A high degree of maintenance is apparently not required for the modified K-type samplers.

TENAX DAV AND CANISTER SAMPLER COMPARISONS

An indoor air study comparing Tenax sorbent collection with canister collection was carried out in the summer of 1985 in a furnished, unoccupied residence in Columbus, Ohio, for ten 12-hour sampling periods. The canister-based sampler consisted of an assembly of individual components in a K-type sampling train and Tenax DAV sets were used. The indoor air was spiked with a set of ten VOCs which were added to the ventilation system at the return air intake. These ten compounds are: 1) chloroform, 2) 1,1,1-trichloroethane, 3) benzene, 4) bromodichloromethane, 5) trichloroethene, 6) toluene, 7) tetrachloroethene, 8) styrene, 9) p-dichlorobenzene, and 10) hexachlorobutadiene. The concentrations for each compound varied by at least an order of magnitude in the 0–10 ppbv range. Comparison of 12 analytical results using the two sampling techniques is shown in Figure 3 as the slope of the linear regression plus and minus one standard error. The average value of the slope is 0.95 even including the lighter compounds. The sample volumes for the lighter compounds are expected to exceed the retention volume of the sorbent and thus to give a slope biased towards lower values. The methods of collection are assumed to be equivalent if: 1) a detailed statistical analysis showed that for the DAV, the slopes adequately represent the experiment and 2) the slope and intercept estimates are within two standard errors of one and zero, respectively. Under these conditions, compounds 3, 6, 7, and 9 qualify. Compound 8 violates the intercept constraint. Compounds 1, 2, and 10 do not satisfy the condition (a). Compunds 4 and 5 as well as 1 and 2 violate the slope constraint. The average ratios for the individual compounds are shown in Table I; R1 is listed to show results if only the low volume collection tube is used. A more detailed discussion is given by Spicer et al.[13]

A similar comparison study but monitoring ambient air was performed in the summer 1986 study sponsored by the California Air Resources Board, as noted above. Only 1,1,1-trichloroethane and toluene had a sufficient concentration range to warrant a linear regression analysis. Slope values R and R for Tenax results regressed on canister results are shown in Table I for the nine sampling periods. Toluene concentrations ranged from 3–7 ppbv, while 1,1,1-trichloroethane concentrations ranged from 1–4 ppbv. Intercept values for toluene plus and minus one standard error were 0.86 ± 0.46 ppbv for the full DAV set

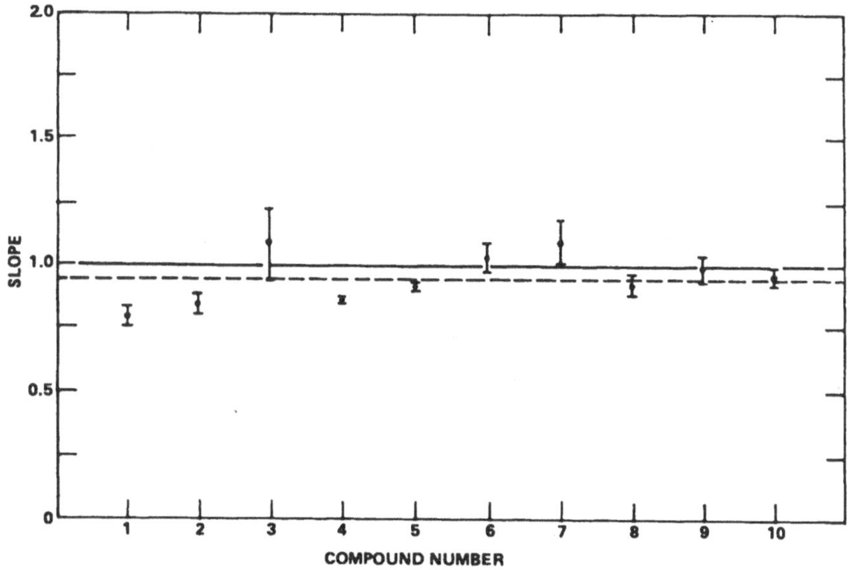

Figure 3. Tenax distributed air volume and canister slope plus and minus one standard error from the linear regression of Tenax mean vs. canister mean.

and 1.80 ± 0.90 ppbv for single-tube Tenax. For 1,1,1-trichloroethane, the intercept values were 0.02 ± 0.15 ppbv and -0.09 ± 0.18 ppbv.

Discussion—Inorganic Samplers

EMSL has the mandate within EPA to identify and/or develop a multicomponent sampler for use in a nationwide deposition network. The latest efforts have been directed towards the comparison of a number of samplers in a two-week study at one of the six stations (the Research Triangle Park location) currently operating in the network. The comparison involved 13 consecutive 24-hour collection periods. Approximately 95 percent successful data capture occurred. Of particular importance to EPA was the comparison between the filter pack (FP), annular denuder system (ADS), and transition flow reactor (TFR), since each is a candidate for providing the multicomponent sampling in the dry deposition network. To facilitate a comparison, analyses of samples from these three methods were performed by the same laboratory. Field blanks, duplicate samples, periodic flow measurements, and other quality assurance features were included as part of the tests. Four of the methods operated during the study and the species monitored are given in Table II. Also, a tunable diode laser system (TDLAS) provided reference measurements for nitric acid and nitrogen dioxide. Since the TDLAS measurements consist of monitoring absorption features of these gases in

TABLE I. Comparison of Tenax DAV Results and Canister Results

	Indoor Air		Outdoor Air	
Compound Name	R*	R_1**	R*	R_1**
Chloroform	0.80 ± 0.03	1.02 ± 0.07	—	—
1,1,1-Trichloroethane	0.86 ± 0.03	1.06 ± 0.06	0.82 ± 0.05	1.02 ± 0.07
Tetrachloroethene	1.10 ± 0.09	1.17 ± 0.12	—	—
Bromodichloromethane	0.86 ± 0.02	0.92 ± 0.04	—	—
Trichloroethene	0.92 ± 0.02	0.95 ± 0.04	—	—
Benzene	1.08 ± 0.14	0.97 ± 0.16	—	—
Toluene	1.03 ± 0.06	1.10 ± 0.009	0.62 ± 0.09	0.50 ± 0.18
Styrene	0.93 ± 0.04	0.96 ± 0.05	—	—
p-Dichlorobenzene	0.99 ± 0.05	1.03 ± 0.05	—	—
Hexachlorobutadiene	0.97 ± 0.04	0.91 ± 0.04	—	—

* R = Slope of linear regression analysis, Tenax vs. canister, using average result of DAV set, ± one standard error.

** R_1 = Slope of linear regression analysis, Tenax vs. canister, using low volume collection tube in the DAV set, ± one standard error.

the infrared portion of the spectrum, the TDLAS is considered to provide unambiguous identifications.

Research Triangle Institute personnel performed sampling and analysis for the ADS and TFR systems as well as the analysis for the FP system. Atmospheric Environment Service (Canada) personnel sampled using the FP. Unisearch Associates (Canada) personnel provided the TDLAS measurements. Atmospheric Sciences Research Laboratory (EPA) personnel provided additional TFR comparisons.

Although the data analyses for the study are only partially done, some preliminary results are now available. These results can be summarized as follows. The designation RTI-TFR refers to a nonstandard interpretation of the TFR data. The characterization of two methods as being equivalent means that on a statistical basis no difference could be shown at a 95 percent confidence level. The word "total" describing nitrate denotes nitrate extracted from the cyclone inlet on either the ADS or TFR plus the sum of gaseous and particle nitrate found elsewhere in the sampler. The word "total" describing ammonium denotes the sum of ammonium found on the cyclone inlet to either the ADS or TFR plus the ammonium found on the Teflon filter collection surfaces.

Daily Samples

1. Total nitrate and total particulate sulfate results from samples taken with the three multicomponent samplers are statistically equivalent.
2. Nitric acid results from samples taken with FP and ADS systems are equivalent to the TDLAS reference measurements. One of the TFR systems was consistently much too high, although a nonstandard interpretation of the TFR data gave better agreement. A second independent TFR system was close to but slightly higher than the

TABLE II. Species Monitored at the EMSL 1986 Methods Comparison Study

Species	ADS	TFR	FP	TDLAS
NO_2		X		X
SO_2	X	X	X	
HNO_3	X	X	X	X
NH_3	X	X		
SO_4^{2-}	X	X	X	
NO_3^-	X	X	X	
NH_4^+	X	X	X	
HNO_2	X			

consensus, although it was used in only five of the thirteen sampling periods.
3. Nitrogen dioxide measurements from TFR samples are equivalent to TDLAS reference measurements.
4. Sulfur dioxide results from ADS and TFR samples are not statistically equivalent; ADS > TFR.
5. Total ammonium results from TFR samples are equivalent to those from FP samples. ADS sample analyses for total ammonium are not equivalent to FP sample results (ADS < FP).
6. Total ammonium plus ammonium results are equivalent for the ADS and FP samples.

Overall, the study showed good agreement among the three multicomponent sampler results and the TDLAS results for daily samples. However, a revised method for calculation of TFR, HNO_3 measurements was necessary to obtain agreement. Additional comparisons of the sum of daily samples with the weekly sample, done only for the ADS, indicated a change of nitrite to nitrate over time with resultant potential underestimation of HNO_2 and over-estimation of HNO_3 and total nitrate. Similar TFR and FP comparison data were not available.

DISCLAIMER

The research described in this article has not been subjected to Agency review. Therefore, it does not necessarily reflect the views of the Agency and no official endorsement should be inferred. Mention of trade names or commercial products does not constitute endorsement or recommendation for use.

REFERENCES

1. Walling, J.F.: The Utility of Distributed Air Volume Sets when Sampling Ambient Air Using Solid Adsorbents. *Atmos. Environ. 20*:51–57 (1986).
2. Walling, J.F., J.E. Bumgarner, J.D. Driscoll et al.: Apparent Reaction Products Desorbed from Tenax Used to Sample Ambient Air. *Atmos. Environ. 18*:855–859 (1984).
3. Seila, R.L.: *Non-urban Hydrocarbon Concentrations in Ambient Air North of Houston, Texas*. U.S. Environmental Protection Agency, EPA-600/3-79-010 (February 1979).
4. Rasmussen, R.A.: Personal communication.
5. McElroy, F.F., V.L. Thompson, D.M. Holland et al.: Cryogenic Preconcentration-Direct FID Method for Measurement of Ambient NMOC:

Refinement and Comparison with GC Speciation. *JAPCA 36*:710–714 (1986).

6. Anlauf, K.G., P. Fellin, H.A. Wiebe et al.: A Comparison of Three Methods for Measurement of Atmospheric Nitric Acid and Aerosol Nitrate and Ammonium. *Atmos. Environ. 19(2)*:325–333 (1985).
7. Knapp, K.T., J.L. Durham and T.G. Ellestad: Pollutant Sampler for Measurements of Atmospheric Acidic Dry Deposition. *Environ. Sci. Technol. 20*:633–637 (1986).
8. Durham, J.L., T.G. Ellestad, L. Stockburger et al.: A Transition-Flow Reactor Tube for Measuring Trace Gas Concentrations. *JAPCA 36*:1228–1232 (1986).
9. Possanzini, M., A. Febo and A. Liberti: New Design of High Performance Denuder for the Sampling of Atmospheric Pollutants. *Atmos. Environ. 17*:2605–2610 (1983).
10. Sickles, J.E., II, L.L. Hodson, E.E. Rickman, Jr. et al.: *Sampling and Analytical Methods Development for Dry Deposition Monitoring*. Final Report on EPA Contract Number 68-02-4079 (in preparation).
11. Director, Methods Development and Analysis Division, MD-784. U.S. Environmental Protection Agency, Research Triangle Park, NC 27711.
12. McClenny, W.A., T.A. Lumpkin, J.D. Pleil et al.: Canister-based VOC Samplers. *Proceedings of the EPA/APCA Symposium on Measurement of Toxic Pollutants*, pp. 27–30. Raleigh, NC (April 1986).
13. Spicer, C.W., M.W. Holdren, L.E. Slivon et al.: Intercomparison of Sampling Techniques for Toxic Organic Compounds in Indoor Air. *Proceedings of EPA/APCA Symposium on Measurement of Toxic Pollutants*, pp. 27–30. Raleigh, NC (April 1986).

CHAPTER 19

Simultaneous Spectroscopic Determination of Gaseous Nitrous Acid and Nitrogen Dioxide in Polluted Indoor and Outdoor Air Environments

HEINZ W. BIERMANN, JAMES N. PITTS, JR. and ARTHUR M. WINER

Statewide Air Pollution Research Center, University of California, Riverside, California 92521

INTRODUCTION

During the last 20 to 30 years, concern over the possible health implications of human exposure to certain major gaseous pollutants has led to widespread, systematic measurements of their concentrations in ambient outdoor air. Additionally, a great deal of information now exists on the emissions, atmospheric transformations, and ultimate fates of such "criteria pollutants" as nitrogen dioxide.[1] Until recently, however, relatively little attention was focused on the levels and potential health impacts of NO_2 in indoor air environments.

Since circa 1980, the situation has changed dramatically. Extensive research has established that under certain conditions commonly experienced by a significant fraction of the world's population, indoor levels of NO_2 can equal or exceed federal and World Health Organization (WHO) air quality standards, as well as the NO_2 concentrations typically measured in outdoor urban/suburban air.[2-8]

This problem has been exacerbated by 1) the current trend toward energy-efficient homes and occupational environments in which ventilation rates are significantly reduced and 2) the increased use of unvented space heaters burning such fuels as kerosene.[9-11] In a home with a gas stove without external ventilation, peak NO_2 concentrations up to 1 ppm and one-hour averages of approximately 0.25 to 0.50 ppm have been observed.[2] These levels can be compared with 1) typical outdoor values in polluted air of 0.02 to 0.05 ppm annual average and 2) hourly averages that occasionally reach 0.25 ppm or higher in some heavily polluted airsheds (e.g., the Los Angeles area). Other examples of relative personal exposures to indoor vs. outdoor NO_2 concentrations for a variety of conditions are given by Spengler and Soczek.[4]

STATEMENT OF THE PROBLEM

Epidemiologists have employed the expanding data base on indoor vs. outdoor concentrations of NO_2, and associated personal exposures in residential and occupational settings, to estimate possible health impacts of this combustion-generated pollutant.[8,12,13] While such information is clearly important, there is an associated problem which is often overlooked. Conventional techniques for measuring NO_2 do not permit the unequivocal characterization and real-time measurement, in indoor or outdoor ambient air, of trace levels (ppb) of its labile co-pollutant, gaseous nitrous acid (HONO).

The possible health impact of nitrous acid was of scientific and societal concern to Congress a decade ago. Thus, Section 104b of the 1977 Amendments to the U.S. Clean Air Act

> requires the Administrator of the Environmental Protection Agency to review and critique all available information on nitric and nitrous acids, nitrites, nitrates, nitrosamines and other carcinogenic and potentially carcinogenic derivatives of oxides of nitrogen.

As atmospheric chemists, the authors' primary interest in this trace species is based on its importance for the production of ozone in photochemical air pollution. Thus photolysis of gaseous HONO in sunlight is an early morning source of the hydroxyl radical which "drives" the smog cycle[1]

$$\mathrm{HONO} + h\nu(\lambda < 400\ \mathrm{nm}) \rightarrow \mathrm{HO} + \mathrm{NO} \tag{1}$$

However, its health implications are also of interest. For example, as well as being a chemical mutagen,[14,15] HONO is an effective nitrosating agent, reacting rapidly with gaseous secondary amines in air[16] to form carcinogenic nitrosamines.[17,18] Could inhalation of this gaseous nitrite lead to *in vivo* formation of nitrosamines?[19]

This chapter will first briefly review the past research by the authors leading to the unequivocal detection and measurement of parts per billion (parts/10^9, ppb) levels of gaseous HONO in 1) heavily polluted ambient outdoor air, 2) dilute auto exhaust, and 3) synthetic indoor atmospheres containing ppm levels of NO_2. These measurements were conducted with a rapid scanning, long pathlength UV/visible differential optical absorption spectrometer (DOAS), which is described below.

This chapter will then discuss in detail the authors' current research utilizing the DOAS technique to address the following key questions: Is gaseous HONO formed indoors from NO_2 emitted by a residential gas cooking stove located in a mobile office/home? What levels of the copollutant formaldehyde (H_2CO) occur simultaneously when the stove is operating? How do indoor levels of HONO compare with its concentrations in outdoor ambient air?

OVERVIEW OF THE DOAS TECHNIQUE

The reliable identification and measurement of both stable and labile pollutants at ppb, or even parts per trillion (parts/10^{12}, ppt) concentrations in complex atmospheric systems, are challenging analytical problems.[1] The goal of providing both sensitivity and specificity, as well as capabilities for real-time measurements, demands new levels of instrumental sophistication and power.

Over the past 15 years several laboratories, including the Statewide Air Pollution Research Center (SAPRC) at University of California – Riverside, have elected to develop long pathlength spectroscopic methods which have the potential for providing ppb to ppt detection limits in the infrared, visible, and near-ultraviolet (UV) spectral regions.[1,20,21] Achieving such sensitivities requires optical paths from several hundred meters to 10 kilometers or more, depending upon the extinction coefficients (i.e., optical cross sections) and spectral features of the particular molecule being measured.

Beginning in 1979, initially in collaboration with Drs. Perner and Platt from Julich, Germany, who developed the first operational instrument and applied it to ambient air studies,[20] a UV/visible DOAS system was used to identify and measure gaseous HONO in heavily polluted urban ambient air at various sites and under various atmospheric conditions.[22-25]

The DOAS system measures the concentrations of trace components in air by their absorption of light in the visible or near-UV spectral regions. Specifically, it is used to detect those molecules with pronounced vibronic structures, such as NO_2, HONO and H_2CO; their differential absorption spectra are shown in Figure 1.

The original spectrometer first used to measure HONO in outdoor urban/suburban air is shown in Figure 2. Briefly, it consists of a broad-band 450-W high-pressure xenon arc lamp whose light is focused and transmitted parallel to the ground through up to 17 km of the atmosphere.[26]

Selected 40-nm segments of the transmitted intensity are dispersed with a grating monochromator and the wavelength-intensity profiles of these segments are repetitively monitored with a rapid scanning device and a photomultiplier tube. This scanner enables the system to record the

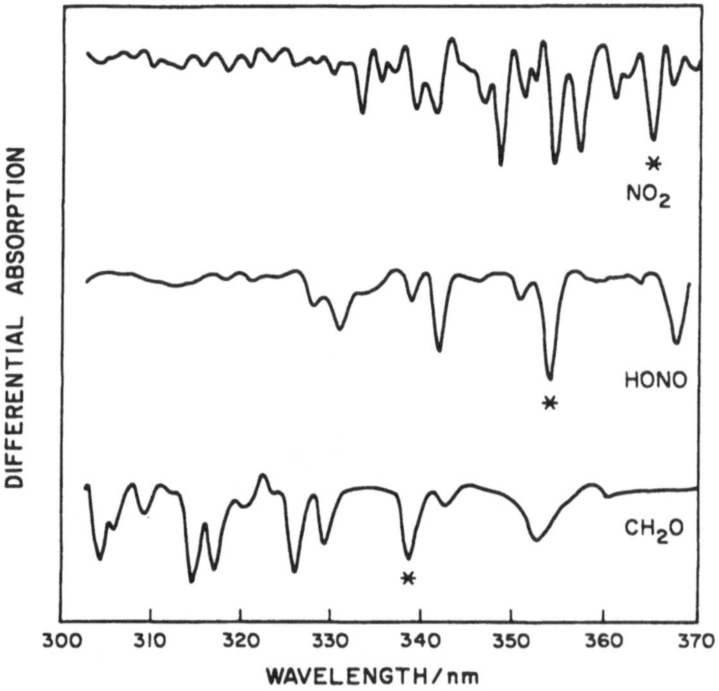

Figure 1. Differential absorption spectra of NO_2, HONO, and CH_2O. The absorption bands marked by asterisks are used for quantitative analysis. Their molar absorption coefficients in L mol^{-1} cm^{-1} are 26 for NO_2 at 365 nm, 110 for HONO at 354 nm, and 20 for CH_2O at 339 nm.

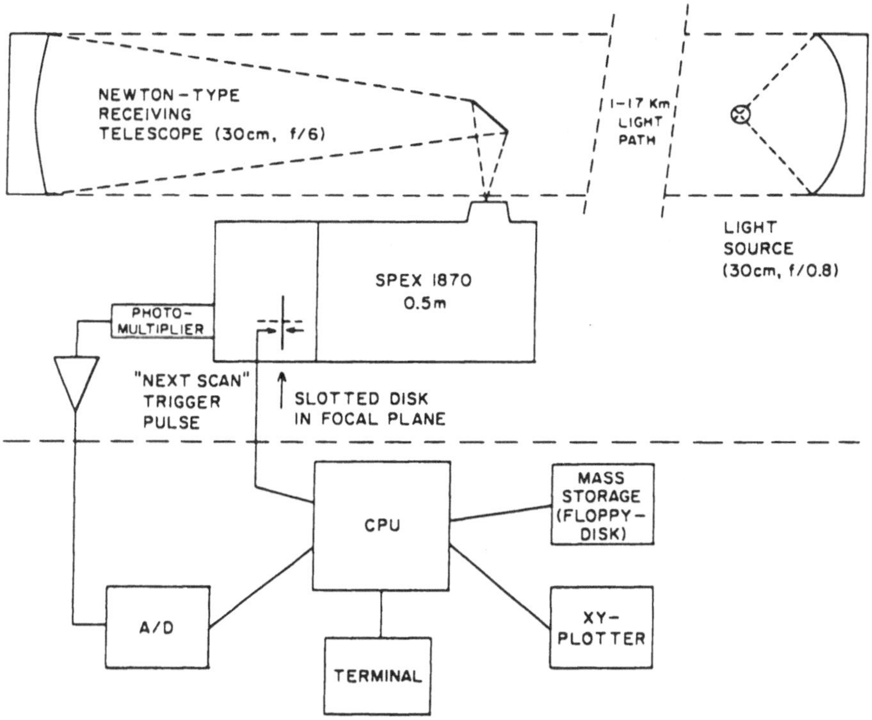

Figure 2. Schematics of the original differential optical absorption spectrometer designed for single-pass light paths of 1 to 17 km length.

same 40-nm spectrum every 10 msec, a rate which is well above the peak of the frequency distribution for atmospheric turbulence near the ground (0.1 to 1 Hz).

The inherent sensitivity of the DOAS technique is seen in Table I; this gives the detection limits at an optical path of 1 km for several compounds which are important species in polluted ambient air. The limits are calculated for normal operating conditions, i.e., a "detectable" signal of 0.0005 absorbance units ($\log_{10}I_0/I$); this is five times the minimum detectable signal under optimum conditions, 0.0001.

As indicated above, species with structured absorption spectra present in the light path (e.g., NO_2, HONO, and H_2CO) can be identified and their concentrations then deduced from differences in the optical densities on and off the key absorption peaks. Absorption lines with optical densities as low as 10^{-4} (base 10) can be detected. Data accumulation and processing are performed using a DEC MINC-11/23 computer system.

TABLE I. DOAS Detection Limits* for Selected Compounds

Compound	Detection Limit (ppb)
Anthracene	0.02
Benzaldehyde	2
Benzene	1
Formaldehyde	6
Naphthalene	0.1
Nitrogen dioxide	5
Nitrous acid	1
Sulfur dioxide	0.5

* Calculated for a detectable signal of 0.0005 absorbance units ($\log_{10}[I_0/I]$); minimum detectable signal under optimum conditions 0.0001; optical path = 1000 m.

NITROUS ACID IN OUTDOOR AIR

In a collaborative program with Drs. Platt and Perner, using their original single-pass version of the DOAS spectrometer, the first ambient HONO was identified at Riverside, California in the fall of 1979. Its gradual buildup during the night and its rapid decay after sunrise are shown in the series of spectra in Figure 3.[22] The optical path was 0.86 km and the beam was 2 to 20 m above ground; 3.3 ppb was the maximum concentration observed at 0526 hours PDT. Following sunrise the concentrations dropped off rapidly due to the efficient photodecomposition of HONO (reaction 1).

Subsequently, during the summer of 1980, in an experiment designed to place the spectrometer at a site characterized by much higher emissions of NO_x from mobile sources, HONO measurements were performed near downtown Los Angeles (DTLA) across the intersection of two major freeways (U.S. 710 and U.S. 10.)[23] Figure 4 shows the time-concentration dependence of HONO on the morning of August 8, 1980. Qualitatively, this time dependence is typical of that for the 25 days on which measurements were taken, the HONO concentration slowly building up to a broad maximum before dawn and then decreasing to levels below our detection limit (approximately 0.2 ppb) after sunrise when photolysis occurs.

Maximum HONO levels observed ranged from a low of $<$ 0.5 ppb to approximately 8 ppb at 0600 hours (PDT) on August 8, 1980. Most maxima fell in the range of 2 to 6 ppb and were observed between 0300 and 0600 hours (PDT).

In a recent wintertime study near the Los Angeles International Airport (SW of DTLA), a modified DOAS system was employed. Instead of

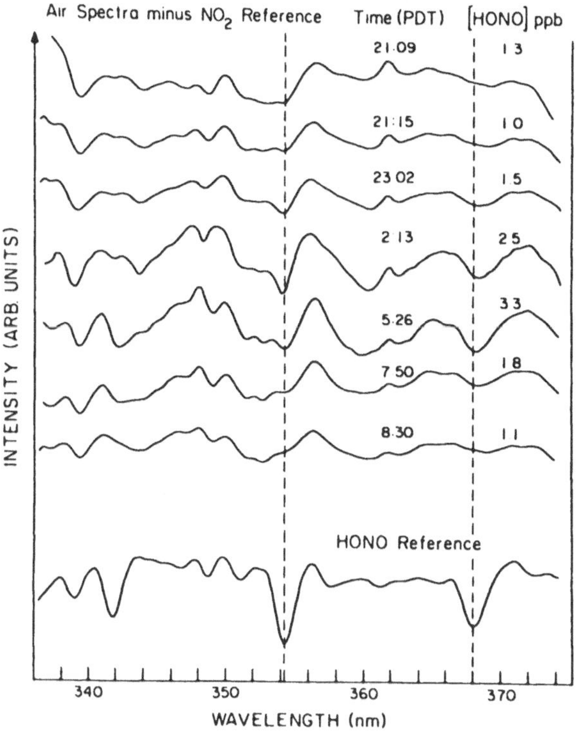

Figure 3. Detection of HONO in DOAS spectra acquired during the night of August 4–5, 1979, at Riverside, CA.

a single-pass beam 1 to 17 km in length, a new, 25-m base path, open multiple reflection optical system was designed and operated. It is similar to the multiple reflection DOAS system used in our mobile office/home studies discussed below. Time-concentration profiles of NO_2 and HONO are shown in Figure 5.

To summarize, levels of gaseous HONO ranging from less than 1 to as high as 10 ppb have been measured at night by the DOAS technique in ambient outdoor air in southern California and West Germany under various conditions. Thus, there is reason to believe gaseous HONO will build up at night in most, if not all, major urban centers throughout the world where significant quantities of combustion-generated NO_x are present.

This observation has an impact on predictions made by air quality models because early morning photolysis of HONO produces hydroxyl radicals (reaction 1) which drive photochemical smog. These ambient

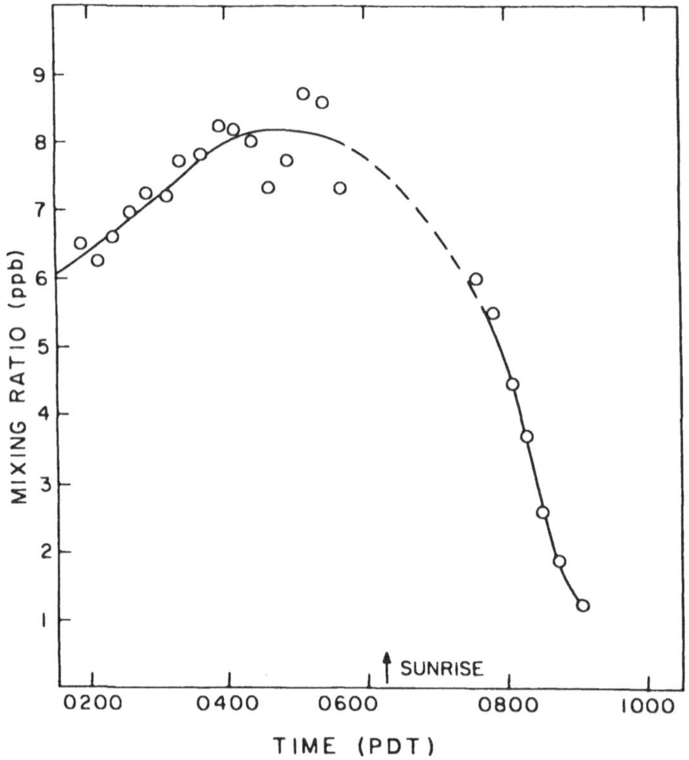

Figure 4. Time-concentration profile of HONO monitored by DOAS near downtown Los Angeles on August 8, 1980.

HONO data may also have implications for health assessments of oxides of nitrogen. They suggest that attention should be focussed not only on the *criteria* pollutant NO_2, as is often the case, but also on its night-time co-pollutant, the noncriteria species gaseous nitrous acid. As noted earlier, it is a gaseous mutagen and effective nitrosating agent. This suggests the possible *in vivo* formation of nitrosamines.

While the latter is highly speculative, it is perhaps relevant that one can estimate that an average person breathing air containing 4 ppb of HONO (approximately 8 $\mu g/m^{-3}$) for four hours will inhale approximately 16 μg of the substance. Since HONO exposures of this magnitude have been seen under several common conditions in ambient air (*vide supra*), the authors caution that the term "trace" pollutant should be used with care when describing the ambient levels of gaseous mutagens and/or possible carcinogens. Thus, for comparison's sake, ambient levels of the carcino-

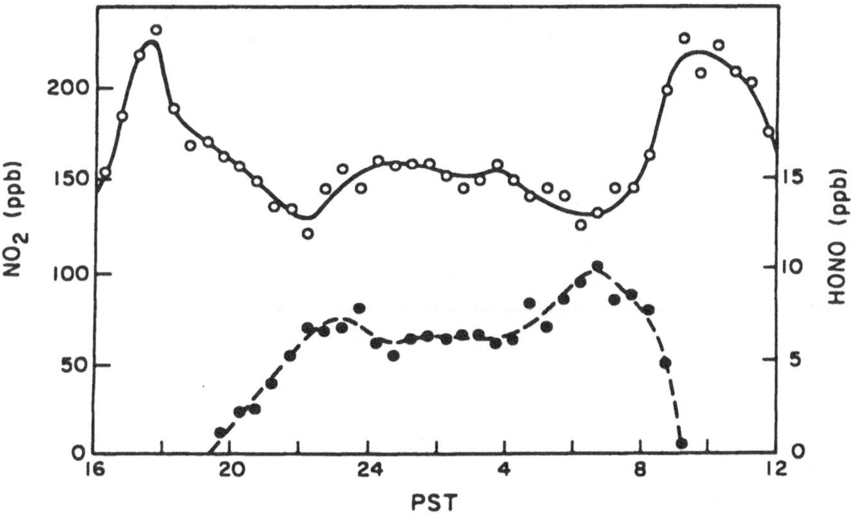

Figure 5. Time-concentration profiles of NO_2 and HONO monitored by DOAS southwest of downtown Los Angeles during the night of January 27–28, 1986.

gen benzo(a)pyrene in Los Angeles air are of the order of 2 nanograms m^{-3}, about one thousand times less than the observed night-time levels of the HONO.

SOURCES OF NITROUS ACID: "SHORT PATH" DOAS STUDIES

At the time the first DOAS measurements were made, circa 1980, the sources of ambient HONO were not known. However, the authors suggested that "heterogeneous processes appear to be involved because the known nighttime homogeneous chemistry is too slow to account for the concentrations we observed."[19]

Auto Exhaust

Proceeding with the hypothesis that HONO formation may occur immediately after exhaust emissions of NO_x, e.g., from light duty motor vehicles (LDMV) or diesel trucks, are injected into the atmosphere, a short-path DOAS system suitable for monitoring such emissions was developed.[27] This instrument had multiple reflection optics and a base path of only 1.3 m but a total optical path of ca. 30 m. In this configuration, the entire assembly of mirrors and the associated spectrometer was mounted on a movable platform. This permitted its positioning in the exhaust plume in open air, i.e., with no "cell walls" on which surface reactions could occur. It was positioned approximately 2 m behind the tailpipe of an LDMV being run on a dynamometer.

While we could not survey a large selection of cars for statistical evaluation, we observed HONO levels in the diluted exhaust ranging from nondetectable (< 12 ppb) for a 1982 California car equipped with an effective three-way catalyst and associated low NO_x emissions to 340 ppb for a heavily used 1974 station wagon having high NO_x emissions and run on leaded gasoline.

These experiments showed that some high NO_x-emitting LDMV could produce direct emissions of HONO; however, the emissions did not appear to be high enough to account for the levels we observed in ambient southern California air.

Hydrolysis of Nitrogen Dioxide

A second possible source of HONO, with widespread implications, is the heterogeneous disproportionation of NO_2 with water.

$$2\,NO_2 + H_2O \xrightarrow{\text{surface}} HONO + HNO_3 \qquad (2)$$

This process also involves the conversion of NO_2 to nitric acid and is another sink for NO_x.

This reaction was studied at ppm levels in environmental chambers in Japan by Sakamaki et al. using FT-IR spectroscopy and multipass optics[28] and at ppb levels by the authors using primarily the SAPRC evacuable chamber (Figure 6).[29] Nitrous acid and NO_2 was monitored with a DOAS instrument using a multiple-pass optical system with 3.77 m base path. Forty passes through the chamber gave a total optical path length of 151 m; this provided a sensitivity of ca. 0.5 ppb for HONO and ca. 20 ppb for NO_2.

As seen in Figure 7, gaseous nitrous acid was indeed formed in significant amounts (up to 80 ppb in 24 hours) when air at 50 percent relative humidity (RH) containing 840 ppb of NO_2 was introduced into our environmental chambers.

The results of this study are consistent with those of Sakamaki et al. at much higher initial NO_2 concentrations, and they support the conclusion

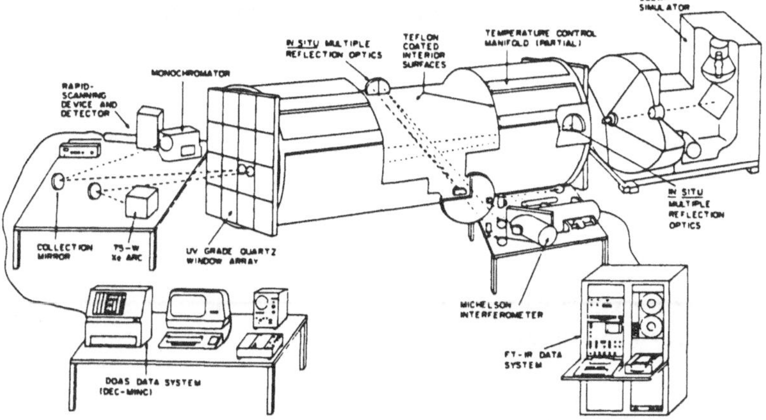

Figure 6. Schematic diagram of our evacuable environmental chamber and the attached Fourier transform infrared and DOAS UV/visible spectrometers.

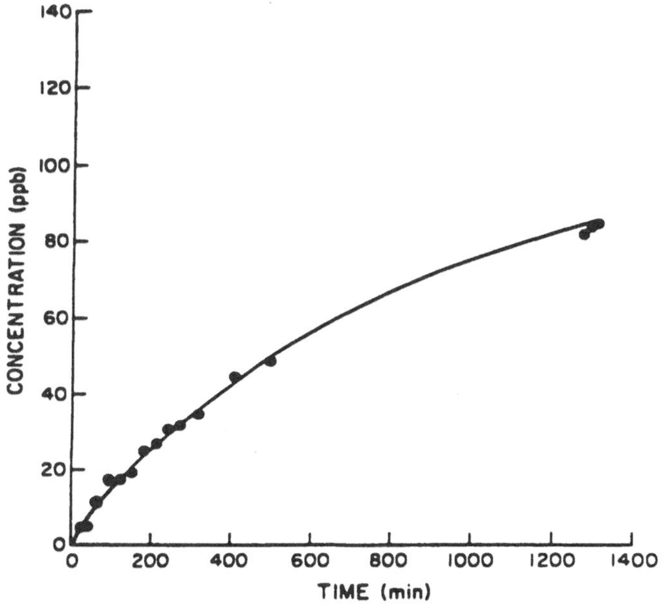

Figure 7. Time-concentration profile of HONO generated in our environmental chamber from synthetic air containing 840 ppb NO_2.

that the overall process in the ppb range of atmospheric interest is the heterogeneous reaction 2.

Interestingly, the other product, HNO_3, was not observed at the concentrations expected from reaction 2 in the gas phase by either group. One explanation is that since gaseous HNO_3 is known to be a "sticky" compound; it may be formed in reaction 2, then rapidly migrate to the walls where it deposits out.

HONO FORMATION IN SIMULATED POLLUTED INDOOR ATMOSPHERES

With these demonstrations of the utility of the compact, multipass DOAS optical system to detect and measure HONO, the authors were encouraged to employ a similar arrangement in exploratory experiments addressing a major problem of indoor air pollution cited above—the characterization and measurement in real time of labile trace species within indoor environments. Specifically, the following questions were asked:

- Is HONO present indoors under certain conditions?
- If so, how do its average level and short-term peak concentrations compare with those that were observed in outside ambient air?

For the preliminary experiments, the focus was on the potential formation of HONO within indoor environments containing clean air to which ppm levels of NO_2 were added (i.e., *simulated* polluted indoor air). Thus, a 2.2-m base path (90-m total optical path) indoor DOAS system was installed in the SAPRC-California Air Resources Board mobile laboratory (Figure 8). With this the authors demonstrated the indoor formation of HONO from concentrations of NO_2 in air ranging from 5.2 to 11.5 ppm.[30] A set of DOAS spectra of HONO, as a function of time for an initial concentration of 11.5 ppm of NO_2 and 43 percent RH, is shown in Figure 9.

These observations also raise the possibility of substantially higher rates of HONO formation within typical indoor environments which possess a greater surface area and variety than the rather sparsely furnished mobile laboratory used for our preliminary experiments. On the other hand, it is also possible that the increased surface area and variety of furnishings could lead to an increased rate of heterogeneous *removal* of HONO.

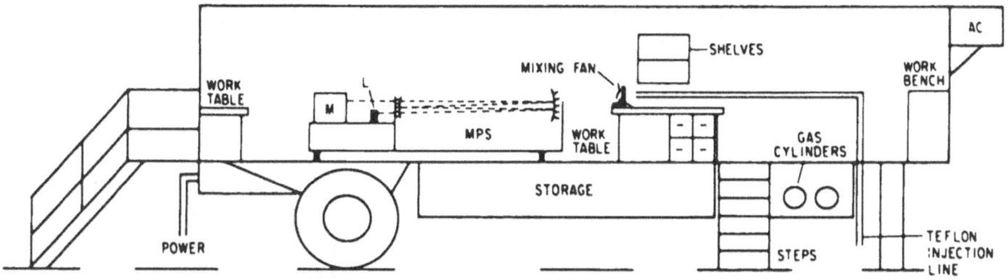

Figure 8. Setup of DOAS system in our mobile laboratory. L marks the light source, MPS the multi-pass system, and M the monochromator.

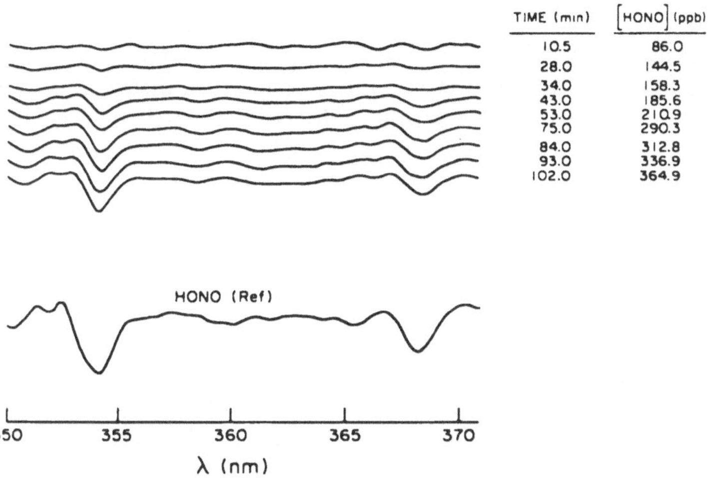

Figure 9. Detection of HONO in DOAS spectra obtained in our mobile laboratory when the air inside was spiked with 11.5 ppm NO_2.

FORMATION OF HONO IN A MOBILE OFFICE/HOME WITH A GAS COOKING STOVE

Under U.S. Environmental Protection Agency (EPA) and University of California support, the exploratory DOAS studies were extended to more complex, typical indoor environments and to lower, more realistic pollutant levels.

In this study, the following specific questions were addressed:

- Is HONO formed indoors under realistic conditions, e.g., from NO_2 emitted by a residential gas stove located in a mobile office/home?
- If so, what are the concentration-time profiles for NO_2 and HONO?
- What levels of the co-pollutant H_2CO occur when the gas stove is operating?

Toward this end, a new, more sensitive DOAS system was developed which permitted *in situ*, unambiguous determination of the concentration of trace pollutants in the ppb or ppt range (Table I). Time resolutions down to approximately five minutes are achieved.

EXPERIMENTAL

The light source employed in the new indoor DOAS system is a 75-W, high pressure xenon arc lamp whose radiation is focussed into a multiple reflection system with the Horn and Pimentel modifications[31] (Figure 10). The mirrors are coated with a multilayer dielectric coating to minimize reflection losses. The base path of the indoor mirror system is 10 m; after transiting the mirror system up to 102 times, the total optical pathlength is 1020 m. Finally, the exiting light is focussed onto the entrance slit of a 0.5-m spectrograph (SPEX 1870).

As discussed at the outset, a special scanning device is attached to the spectrometer which enables it to record a spectrum of approximately 70 nm width within about 10 msec. The central part of the rapid scanning device is a thin metal disk with a series of narrow slits etched into it radially. The disk spins in the exit focal plane of the spectrograph and at any given time one particular slit is used as the exit slit (Figure 11). The width of the scanned spectrum depends on the length of the arc that a slit covers and on the dispersion of the grating. The light passing through the exit slit is measured by a photomultiplier tube; the output signal is digitized by a high speed analog-to-digital converter and read by a minicomputer (PDP 11/23).

During one scan, about 500 digitized data points are taken. Consecutive scans, at a rate of approximately 80 scans per second, are signal averaged by the computer software. Appropriate software is also used for spectral deconvolution and for the calculation of optical densities.

As stated earlier, molecules with pronounced banded structures can be measured quantitatively from the difference between a suitable absorption band minimum and its adjacent maxima (ignoring any superimposed continuum). The procedure for identifying and measuring gaseous pollutants involves determination of the differential absorption coefficients of the pure species in air in a 10-cm pathlength cell using known

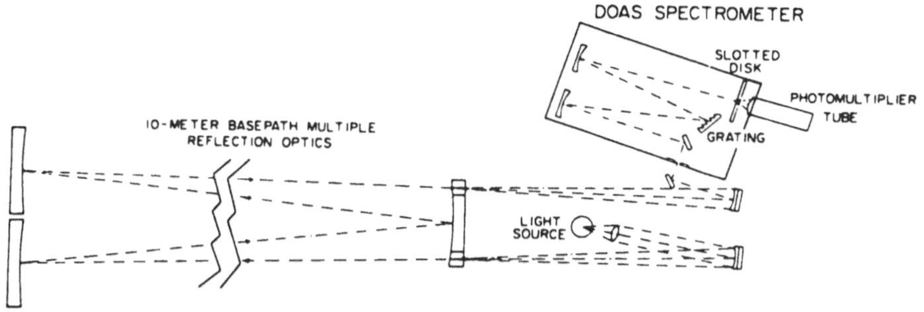

Figure 10. Schematic diagram of the new differential optical absorption spectrometer interfaced to a multiple reflection optical system.

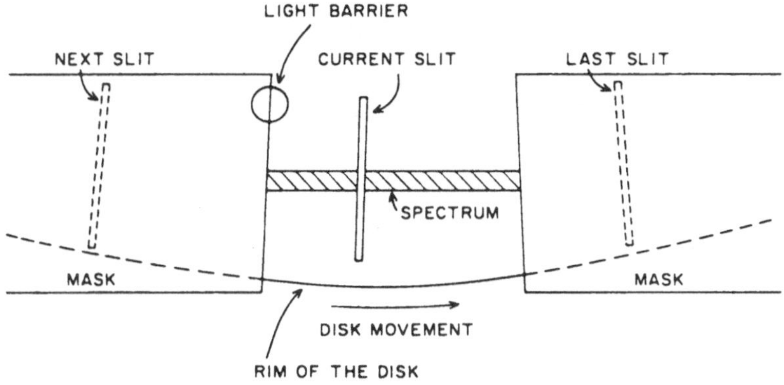

Figure 11. Schematic diagram of a portion of the rapid scanning device placed in the exit focal plane of the DOAS spectrometer.

vapor pressures of the gases. For labile species like HONO, the differential optical absorption coefficients are derived from literature data. Similar spectra of the pure species (e.g., the spectra in Figure 1) are then used to deconvolute the complex ambient air spectra. Details of this procedure are described elsewhere.[32]

The quoted detection limits of 5 ppb for NO_2, 1 ppb for HONO, and 6 ppb for H_2CO are based on the absorption cross sections (cm^2 $molecule^{-1}$) of the bands (marked by asterisks in Figure 1) at 365 nm for NO_2 (1.0×10^{-19}), 354 nm for HONO (4.2×10^{-19}), and 339 nm for H_2CO (7.8×10^{-20}). The respective values for ϵ, in units of L mol^{-1} cm^{-1} ($\log_{10}[I_0/I] = \epsilon cL$), are 26 for NO_2, 110 for HONO, and 20 for H_2CO.

A comparison of the absorption band positions of the three compounds in Figure 1 shows that their features partly overlap. NO_2 and H_2CO are easily separated, but a quantitative analysis for HONO first requires a careful subtraction of the overlapping bands from both NO_2 and H_2CO. Since in air pollution studies HONO is generally produced from NO_2 in air, there is usually a large excess of NO_2 present. In addition, the emissions from our gas stove reported below contained large amounts of H_2CO. Thus, while NO_2 and H_2CO concentrations can usually be calculated in real time during data acquisition, the analysis for HONO requires a data reduction procedure involving "in house," interactive computer processing routines. The need to subtract significant contributions from NO_2 and H_2CO also decreases the accuracy of HONO measurements.

Figure 12 is a schematic drawing of the new DOAS system set up in a 19-year-old mobile home, previously used as an office facility on the UC Riverside campus. The office/home is 50 ft long, 10 ft wide, and 8 ft high; it is subdivided into four rooms of 1000 ft^3 (26.5 m^3) each. One end room (#4) is sealed off from the other three and houses the DOAS instrumentation and computer. The beam of white light from the xenon source passes through the center of the other three rooms (1, 2, and 3) through pre-existing openings in the partitioning walls.

During experiments, the inner doors are kept open and box fans induce rapid mixing throughout the three rooms. The office/home is also equipped with a central air conditioning/heating system. When the heating or A/C unit is used in an experiment, its fan is kept running continuously.

In room 2, a used, conventional kitchen stove was installed and connected to a natural gas line. The rooms were otherwise unfurnished.

To characterize the air exchange rates with and without the air conditioning (A/C) unit operating, CH_4 was used as an "inert" tracer. About

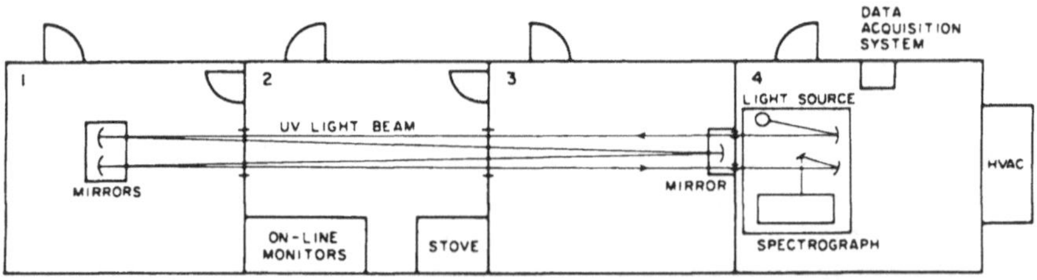

Figure 12. Schematic drawing of the mobile office/home showing the placement of the DOAS system and the associated multiple reflection optics.

20 ppm of CH_4 was injected into one of the rooms and gas chromatography samples taken repeatedly to determine its rate of loss over time. From these measurements, an air exchange rate of approximately 0.5 h^{-1} without the central ventilation unit running was calculated. When the ventilation system (heating-A/C) was turned on, the exchange rate increased to 1.5 h^{-1}.

During each observation period, spectra were acquired continuously with an averaging time of either six minutes or fifteen minutes. Spectra were stored on floppy disks for subsequent analysis. Before each run, background spectra were taken. After each run, the mobile home was vented with outside air and another set of background spectra recorded. Temperature and humidity data were derived from wet bulb/dry bulb thermometer readings.

RESULTS AND DISCUSSION

The first objective was to confirm that reactions of NO_2 in a simulated polluted atmosphere would produce significant levels of HONO. Therefore, sufficient NO_2 was injected into three connecting rooms (1, 2, and 3) to achieve an initial uniform concentration of 6 ppm throughout the three rooms at an initial relative humidity of 34 percent. This is the same order of magnitude as the initial levels of NO_2 studied in the experiments in the ARB trailer (*vide supra*).

The decrease of NO_2 and the buildup of HONO are shown as a function of time in Figure 13. Within approximately 1.5 hours, HONO reached a level of about 30 ppb.

From this experiment, we conclude that in a simulated polluted indoor atmosphere in a mobile office/home, reactions of NO_2, possibly reaction 2, produce significant concentrations of gaseous nitrous acid.

The second objective was to see if HONO could be detected and measured in this indoor environment under more realistic conditions, i.e., at much lower concentrations of NO_2 in complex mixtures of combustion-generated pollutants. Therefore, in the second experiment, the time-concentration profiles of NO_2, HONO, and H_2CO was determined in the connecting rooms when three top burners of the kitchen stove in room 2 were operating on "medium high." The stove was run on natural gas at a flow rate of 0.9 L^{-1} min^{-1} to each burner for six hours. The flames were exposed to the open air, i.e., they were not covered by cooking utensils. Experiments were conducted with the air conditioning off and on.

The time-concentration profiles for NO_2 and HONO are shown in Figures 14 and 15. The solid line corresponds to the levels present with

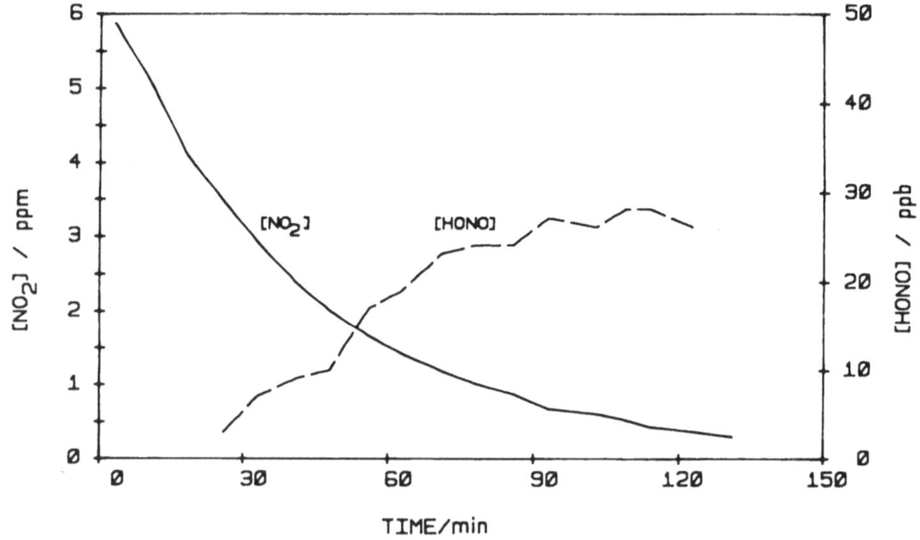

Figure 13. Time-concentration profiles of NO_2 and HONO showing the buildup of HONO after injection of NO_2 into the room air.

the central A/C system turned off. During the measurement period, the temperature inside the three rooms rose from 28°C to 39°C (82°–102°F). In a second run, under otherwise identical conditions, the thermostat of the air conditioner was set to 28°C (82°F), the starting temperature of the previous run; the data from this experiment are represented by the dashed lines.

With the A/C unit off, the NO_2 concentrations, shown in Figure 14, reach a plateau of ca. 1200 ppb after about three hours; with the A/C on, they level off at ca. 500 ppb. After turning the burners off (shown by arrows in the figures), the NO_2 decays with a half-life of approximately 45 minutes.

The ratio of the equilibrium concentrations of NO_2 with the A/C off and on is approximately 2.3, close to the corresponding ratio of approximately 3 of the concentrations of the "inert" tracer methane with the A/C off and on. This implies that there is no major change in the chemistry of NO_2 when the A/C is switched on; it is primarily an effect of increased dilution with outside air.

As shown in Figure 15, HONO begins to appear shortly after the burners are turned on. It reaches equilibrium values of ca. 30 ppb and 10

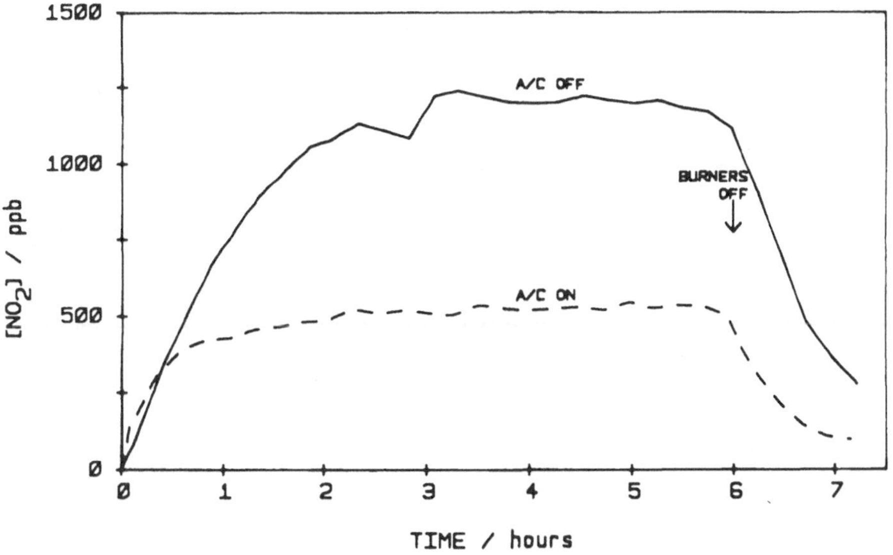

Figure 14. Time-concentration profiles of NO_2 inside our mobile office/home with three top burners of the kitchen stove in room 2 turned on. The solid line represents data taken with the A/C unit off, the dashed line with the A/C unit on. The arrow indicates the time the burners were turned off.

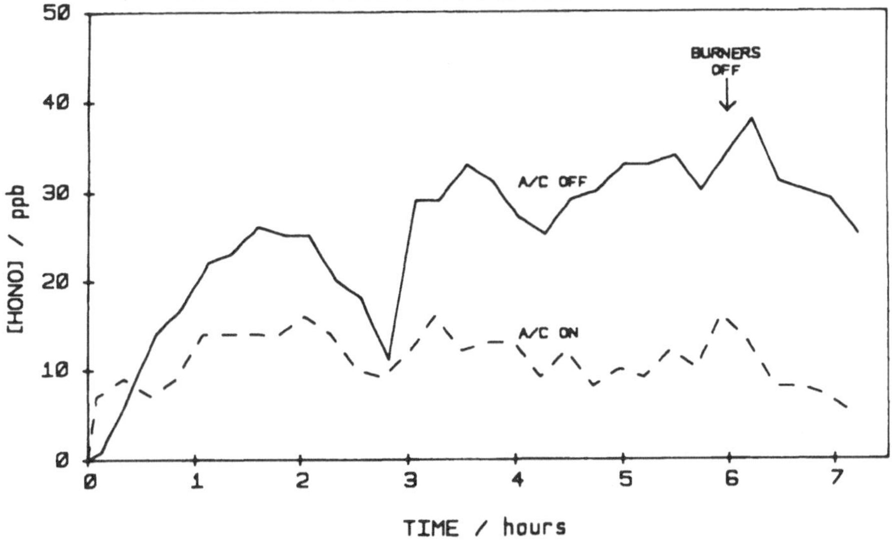

Figure 15. Time-concentration profiles of HONO inside our mobile office/home with three top burners of the kitchen stove in room 2 turned on. The solid line represents data taken with the A/C unit off, the dashed line with the A/C unit on. The arrow indicates the time the burners were turned off. [NOTE: see footnote p. 284.]

ppb for the A/C unit off and on, respectively.* Clearly, HONO is produced indoors when NO_2 is present from the combustion of natural gas in a common kitchen stove.

The observed ratio of HONO/NO_2 (Figure 16) is approximately 2.5 percent. To date, there is no simple, accurate, and inexpensive analytical technique available for HONO. Thus, if no measurements of the important co-pollutant HONO can be made in health effects studies, an "educated guess" would be to assume a level of approximately 2 to 3 percent of the NO_2 present. However, since the formation of HONO from NO_2 by reaction 2 is most likely heterogeneous, this ratio probably depends on the nature and area of the surfaces in the indoor environment. Therefore, the preliminary value of the HONO/NO_2 ratio should be used with caution.

In ambient air, HONO/NO_2 ratios of up to 5 percent (*vide supra*) were observed. The lower value for this ratio (2 to 3%) inside our mobile office/home may indicate a larger sink for HONO in an indoor environment (e.g., loss on surfaces).* Studies are currently underway to characterize this phenomenon.

FORMALDEHYDE

In order to measure the formaldehyde emitted in the gas flame, its background level in our mobile office/home had to be determined. Although it is 19 years old, a significant background seemed possible since the interior walls were lined with laminated paneling.

In fact, as shown in Figure 17, there was a significant background concentration of H_2CO. A temperature increase from 25°C to 37°C produced a corresponding rise in H_2CO from ca. 30 to 80 ppb. When the central heating system was turned off, the H_2CO concentration dropped off, with a slight hysteresis effect.

With the three burners of the stove "on" and the A/C turned off, a significant increase in H_2CO levels was observed (Figure 18) concurrently with the buildup of NO_2 and HONO (Figures 14 and 15). However, in contrast to NO_2, the H_2CO concentration did not plateau to an equilibrium value but continued to rise to ca. 750 ppb; when corrected for the

---

*Subsequent to the preparation of this chapter, additional experiments have been performed recently with a new optical system optimized for detection of HONO. Under similar burning conditions, 50–60 ppb of HONO have been detected in the presence of approximately 1200 ppb of NO_2. This HONO/NO_2 ratio of approximately 5% inside the mobile office/home agrees well with the ones observed in ambient air.

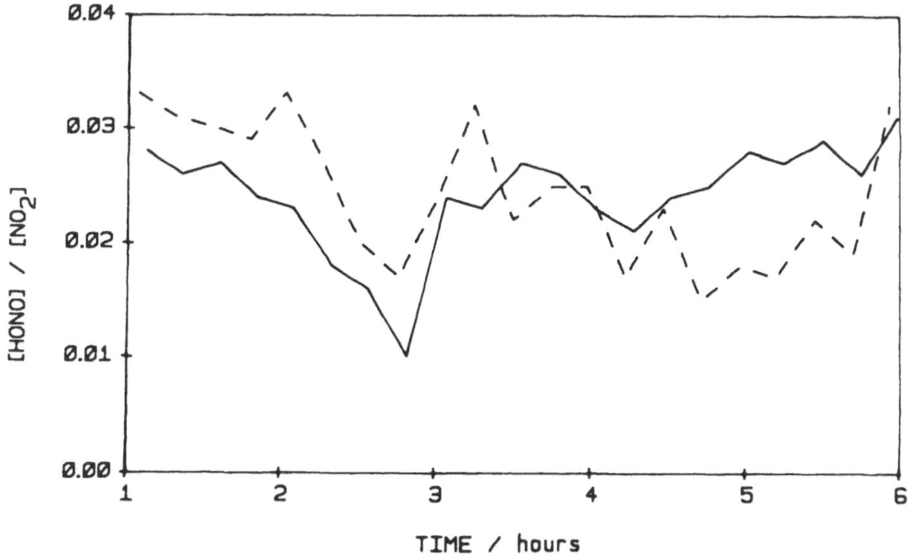

Figure 16. Ratio of $HONO/NO_2$ calculated from the observed HONO and NO_2 concentrations while the kitchen stove was turned on. [NOTE: see footnote p. 284.]

background level (Figure 17), the H_2CO contribution from combustion was ca. 670 ppb.

CONCLUSIONS

The authors have demonstrated that a long path DOAS system with multiple pass optics is a useful tool to investigate the time-concentration profiles of the labile pollutant HONO, as well as its more stable co-pollutants NO_2 and H_2CO, in an indoor air environment. While it is a sophisticated and expensive instrument, it performs unique functions that cannot be achieved by conventional analytical techniques. It permits unambiguous, *in situ* measurements of labile trace species with high time resolution and detection limits in the low ppb range.

The detection of significant quantities of HONO in our mobile office/home has interesting health implications for indoor air pollution. Thus, in the atmospheric systems we have studied to date, the concentrations of HONO indoors can exceed those observed in polluted outdoor urban air.

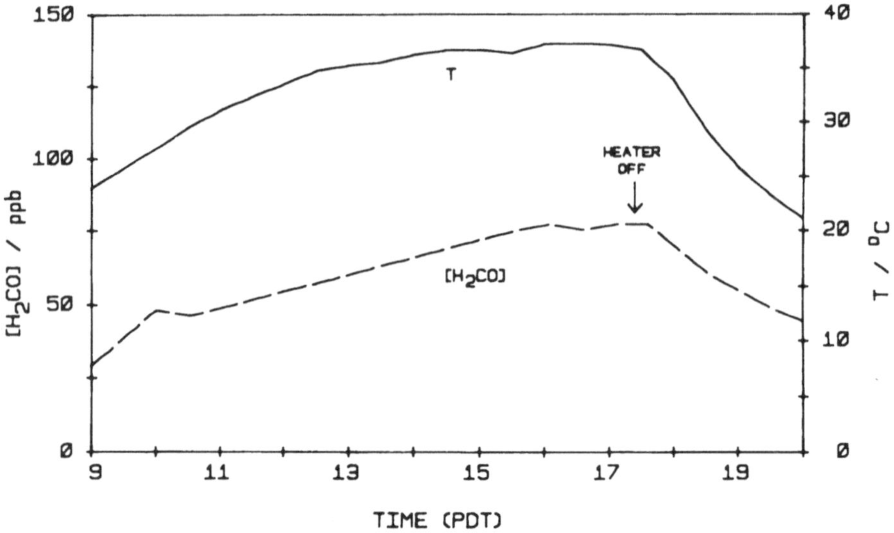

Figure 17. Change in the H_2CO background level in our mobile office/home (dashed line) during a temperature rise from 25°C to 37°C (solid line) by the central heating system. The relative humidity dropped from 34 percent to 26 percent during the heating period.

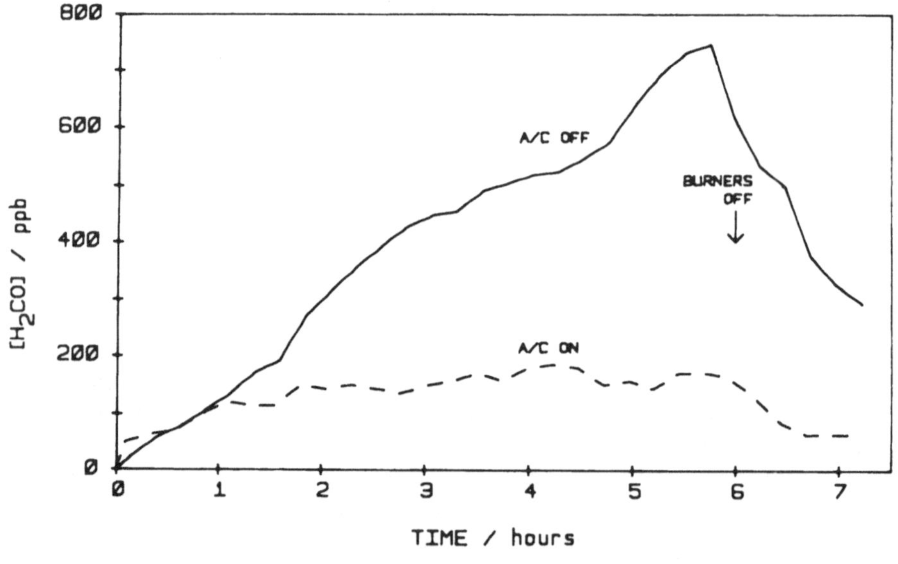

Figure 18. Time-concentration profiles of H_2CO inside our mobile office/home with three top burners of the kitchen stove in room 2 turned on. The solid line represents data taken with the A/C unit off, the dashed line with the A/C unit on. The arrow indicates the time the burners were turned off.

RECOMMENDATIONS

When developing risk assessments of personal exposure to NO_2 in indoor residential and occupational environments, the presence and possible health effects of its co-pollutant, HONO, should also be evaluated. Furthermore, the potential impact of HONO on the chemistry of indoor atmospheres should also be considered.

ACKNOWLEDGMENTS

Although the research described in this article has been funded wholly or in part by the U.S. Environmental Protection Agency under assistance agreement R-812263-01 to Dr. James N. Pitts, Jr., it has not been subjected to the Agency's peer and administrative review and therefore may not necessarily reflect the views of the Agency, and no official endorsement should be inferred. We also acknowledge support from the University of California–Riverside.

We wish to thank Mr. William D. Long for assisting in these experiments, Ms. Minn P. Poe for assistance in processing the data and Ms. I.M. Minnich for preparing the manuscript. We also appreciate helpful discussion with Dr. B.J. Finlayson-Pitts.

REFERENCES

1. Finlayson-Pitts, B.J. and J.N. Pitts, Jr., Eds.: *Atmospheric Chemistry: Fundamentals and Experimental Techniques*. John Wiley & Sons, New York (1986).
2. Yocum, J.E.: Indoor-outdoor Air Quality Relationships. A Critical Review. *J. Air Pollut. Control Assoc. 32*:500 (1982).
3. Spengler, J.D., C.P. Duffy, R. Letz et al.: Nitrogen Dioxide Inside and Outside 137 Homes and Implications for Ambient Air Quality Standards and Health Effects Research. *Environ. Sci. Technol. 17*:164 (1983).
4. Spengler, J.D. and M.L. Soczek: Evidence for Improved Ambient Air Quality and the Need for Personal Exposure Research. *Environ. Sci. Technol. 18*:268A (1984).
5. Wesolowski, J.J.: An Overview of Indoor Air Quality. *J. Environ. Health 46*:311 (1984).
6. Gammage, R.B. and S.V. Kaye, Eds.: *Indoor Air and Human Health*. Lewis Publishers, Inc., Chelsea, MI (1984).
7. Quackenboss, J.J., J.D. Spengler, M.S. Kanarek et al.: Personal Exposure to Nitrogen Dioxide. Relationship to Indoor-outdoor Air Quality and Activity Patterns. *Environ. Sci. Technol. 20*:775 (1986).

8. Papers presented at the Third International Conference on Indoor Air Quality and Climate. Stockholm, Sweden (1984). *Environ. Int. 12* (1986).
9. Yamanaka, S., H. Hirose and S. Takada: Nitrogen Oxides Emissions from Domestic Kerosene Fired and Gas Fired Appliances. *Atmos. Environ. 13*:407 (1979).
10. Traynor, G.W., J.R. Allen, M.G. Apte et al.: Pollutant Emissions from Portable Kerosene-Fired Space Heaters. *Environ. Sci. Technol. 17*:369 (1983).
11. Yamanaka, S.: Decay Rates of Nitrogen Oxides in a Typical Japanese Living Room. *Environ. Sci. Technol. 18*:566 (1984).
12. Fischer, P., B. Remijn, B. Brunekreef et al.: Indoor Air Pollution and Its Effect on Pulmonary Function of Adult Non-smoking Women. 2. Associations Between Nitrogen Dioxide and Pulmonary Function. *Int. J. Epidemiol. 14*:221 (1985).
13. Yokoyama, Y., H. Nitta and K. Maeda: Nitrogen Dioxide Concentration Indoors and Its Implication for Health Effects for School Children. First International Conference on Atmospheric Sciences and Applications to Air Quality, Part II, Seoul, Korea, 1985. *Atmos. Environ. 20*:2077 (1986).
14. Mahler, H.P. and E.H. Cordes: *Biological Chemistry*, p. 862. Harper International Editions, New York, London (1971).
15. Watson, J.D.: *Molecular Biology of the Gene*, p. 255. W.A. Benjamin, Inc., Publishers, London-Amsterdam (1976).
16. Pitts, J. N., Jr., D. Grosjean, K. Van Cauwenberge et al.: Photooxidation of Aliphatic Amines Under Simulated Atmospheric Conditions: Formation of Nitrosamines, Nitramines, Amides and Photochemical Oxidant. *Environ. Sci. Technol. 12*:946 (1978).
17. Shapley, D.: Nitrosamines: Scientists on the Trail of a Prime Suspect in Urban Cancer. *Science 191:*268 (1976).
18. Magee, P.N.: Nitrosamines and Human Cancer. *Bansberry Report 12*. Cold Spring Harbor Laboratory (1982).
19. Pitts, J.N., Jr.: Formation and Fate of Gaseous and Particulate Mutagens and Carcinogens in Real and Simulated Atmospheres. *Environ. Health Perspect. 47*:115 (1983).
20. Perner, D. and U. Platt: Detection of Nitrous Acid in the Atmosphere by Differential Optical Absorption. *Geophys. Res. Lett. 6:*917 (1979).
21. Tuazon, E.C., A.M. Winer and J.N. Pitts, Jr.: Trace Pollutant Concentrations in a Multiday Smog Episode in the California South Coast Air Basin by Long Pathlength Fourier Transform Infrared Spectroscopy. *Environ. Sci. Technol. 15:*1232 (1981).
22. Platt, U., D. Perner, G.W. Harris et al.: Observations of Nitrous Acid in an Urban Atmosphere by Differential Optical Absorption. *Nature 285:*312 (1980).
23. Harris, G.W., W.P.L. Carter, A.M. Winer et al.: Observations of Nitrous Acid in the Los Angeles Atmosphere and Implications for Predictions of Ozone-Precursor Relationships. *Environ. Sci. Technol. 16*:414 (1982).

24. Pitts, J.N., Jr., H.W. Biermann, R. Atkinson and A.M. Winer: Atmospheric Implications of Simultaneous Nighttime Measurements of NO_3 Radicals and HONO. *Geophys. Res. Lett. 11:*557 (1984).
25. Biermann, H.W., E.C. Tuazon, A.M. Winer et al.: Simultaneous Absolute Measurements of Gaseous Nitrogen Species in Urban Ambient Air by Long Pathlength Infrared and Ultraviolet-Visible Spectroscopy. *Atmos. Environ.* (in press, 1987).
26. Platt, U., A.M. Winer, H.W. Biermann et al.: Measurement of Nitrate Radical Concentrations in Continental Air. *Environ. Sci. Technol. 18:*365 (1984).
27. Pitts, J.N., Jr., H.W. Biermann, A.M. Winer and E.C. Tuazon: Spectroscopic Identification and Measurement of Gaseous Nitrous Acid in Dilute Auto Exhaust. *Atmos. Environ. 18*:847 (1984).
28. Sakamaki, F., S. Hatakeyama and H. Akimoto: Formation of Nitrous Acid and Nitric Oxide in the Heterogeneous Dark Reaction of Nitrogen Dioxide and Water Vapor in a Smog Chamber. *Int. J. Chem. Kinet. 15:*1013 (1983).
29. Pitts, J.N., Jr., E. Sanhueza, R. Atkinson et al.: An Investigation of the Dark Formation of Nitrous Acid in Environmental Chambers. *Int. J. Chem. Kinet. 16:*919 (1984).
30. Pitts, J.N., Jr., T.J. Wallington, H.W. Bierman and A.M. Winer: Identification and Measurement of Nitrous Acid in an Indoor Environment. *Atmos. Environ. 19:*763 (1985).
31. Horn, D. and G.C. Pimentel: 2.5 km Low-Temperature Multiple Reflection Cell. *Appl. Opt. 10:*1892 (1971).
32. Platt, U. and D. Perner: Measurements of Atmospheric Trace Gases by Long Path Differential UV/Visible Absorption Spectroscopy. *Optical and Laser Remote Sensing*, pp. 97–105. D.V. Killinger and A. Mooradian, Eds. Springer Series in Optical Sciences, Vol. 39. Springer Verlag, New York (1983).

SECTION V

Sampling Strategy

CHAPTER 20

Sampling Strategies for Prospective Surveillance: Overview and Future Directions

MORTON CORN

CSP Division of Environmental Health Engineering, Department of Environmental Health Sciences, School of Hygiene and Public Health, Johns Hopkins University, 615 North Wolfe Street, Baltimore, Maryland 21205

INTRODUCTION

This chapter will focus on two distinct topics. First, the current status of air sampling in the occupational environment in the United States and possible future trends will be addressed; second, two examples will be presented which illustrate different sampling strategies, one of them for prospective surveillance. Each of the examples is an actual case study of airborne contaminant concentration data collection in production facilities. Both cases are active and ongoing.

Unfortunately, the majority of the effort for occupational air sampling in the United States is compliance-related; i.e., legally enforceable standards or Permissible Exposure Limits (PELs) have been established and employers must demonstrate compliance with these standards. Documentation of meeting the standard has consumed an inordinate amount of professional effort. The administrative practicalities of enforcement resulted in relating compliance to "any eight-hour sampling period" under the Occupational Safety and Health Administration (OSHA) statute. Therefore, the statistical aspects of samples taken during an eight-hour period have received a great deal of attention,[1-3] but they represent only one need addressed by sampling strategies.[4] Adherence to the PELs has been referred to as an aspect of "game theory."[5-6] The inordinate focus on statistical aspects of compliance sampling suggests that the long-term meaning of adherence to the standard has become a secondary consideration.[7-10]

The philosophical meaning of adoption of a PEL or a Threshold Limit Value (TLV) is a form of risk assessment in the sense of the National Academy of Sciences definition of risk assessment.[11] Thus, it is a very serious game, indeed, when one considers adherence or nonadherence to PELs.

A major theme of work undertaken by this author, initially in collaboration with N.A. Esmen, has been the approach to ongoing environmental sampling for possible utilization in prospective cohort epidemiological studies, while also using results to demonstrate compliance with all applicable regulations.[12,13] Utilizing air sampling data in historical cohort studies can be considered a jigsaw puzzle in its own right. One must piece together from limited samples a portrait of the cumulative exposure of groups of workers or individuals. Substantial progress has not been made in this area.[14]

Air sampling and biological monitoring of those exposed supplement one another in the ongoing effort to maintain healthful work environments. The science of biological monitoring is progressing very rapidly. It can be stated with certainty that air sampling in the 1990s will be juxtaposed with biological monitoring for a large number of potentially toxic contaminants. Because exposure to most potentially toxic contaminants in air is not at the gross levels of decades ago, the emphasis is on detection of subtle physiological changes which can be considered predictors of disease. This is where biological monitoring is most useful. Although these early indicators of incipient disease are presently controversial, with time their full meaning will be understood and controversy will subside. Thus, a prospective air sampling program should be created in close coordination with medical personnel who are developing biological monitoring systems. It is also a truism that as attempts are made to detect lower levels of response to a toxic agent, host factors become more significant; this is another reason for intensive development of indicators of early biological change in those most susceptible to the agent in question.

CASE STUDY OF PROSPECTIVE SURVEILLANCE SYSTEM IN A SEMICONDUCTOR MANUFACTURING FACILITY

The zone theory proposed by Corn and Esmen[12] was utilized in a facility of some 10,000 employees, about 3000 of whom were exposed to upwards of a total of about 3000 chemicals. It required one full-time hygienist for six months to establish the system. The first step in application of the zone method was to determine the inventory of chemicals and who was exposed where. Table I summarizes questions asked department managers or technicians in the process of setting up the methodology. After completion of these interviews in over 100 departments employing the approximately 3000 persons referred to above, an inventory of chemicals could be associated with each department. Because many industries are experiencing very rapid technological change, such lists should be updated on a fairly frequent basis (perhaps every six months).

The second step in application of the strategy was to toxicologically code materials used (Table II). It was necessary to then make some decisions about the frequency of monitoring chemicals. The toxicological nature of the chemical was the basis for these judgments. This is perhaps the most controversial part of the strategy. *The strategy does not address those chemicals which act acutely*. The reason for this is that with a statistical sampling scheme it is highly unlikely that one would detect brief excursions of concentrations of chemicals that can affect

TABLE I. Questions for Department Managers or Technicians

1. What are the general processes going on in your department?
   a. Flow chart of processes
2. How often do processes change—how often are chemicals added or dropped from your operations?
   a. Major vs. minor changes
3. Are there distinct groups of workers within your department who could be described as:
   a. Chemical users? b. Nonchemical users?
4. Do workers consistently have the same tasks, or do responsibilities vary?
   a. Do workers work in the same area all the time or in different locations?
5. How is the department divided physically—do everyone and all chemicals exist in one room, many rooms, different buildings, etc.?
6. To what extent do workers from other departments handle your chemicals?
   a. Chemical control, etc.
   [Refer to list of chemicals authorized for that department]
7. Can you give a brief description of how each chemical is used (e.g., plating tank, degreaser, rinse, etc.)?
   a. With each description, rate possible exposures (closed system, open system, etc.)
8. If possible, can you group together chemicals which are used in similar ways or have similar exposure conditions?
9. Do you have now or do you plan in the future on having 2nd and 3rd shifts?
   a. How do the above questions apply to the 2nd and 3rd shifts?
10. Do you know of operations in other departments which are similar to yours?

TABLE II. Toxicity Codes

Acute	Chronic
Asphyxiation (X)	Suspect carcinogen (C)
Skin irritant/acute dermatitis (S)	Systemic poison (S)
Respiratory irritant (R)	Pneumoconiosis (P)
Mucous membrane irritant (M)	Immune-allergenic (A)
Gastrointestinal tract effects (G)	Suspect genetic-mutagen (G)
Skin burns (B)	Suspect reproductive-teratogen (T)
Central nervous system (N)	Chronic dermatitis (D)
Eye irritant (E)	Respiratory (R)
Skin absorption (P)	Anemia (M)
Heart (H)	Central nervous system (N)
General irritant (M,R,S) (I)	Eyes (N)

those exposed in a short period of time. Real-time instrument monitoring, preferably with large display units that employees can view, is the air sampling solution for chemical exposures to acutely acting toxicants. Therefore, the basis for monitoring was developed for chronic materials and a hierarchy of classification was utilized (Table III).

The frequency with which chemicals are monitored permits a statistical base to be built with each iteration of the randomized sampling schedule.

TABLE III. Routine-Scheduled Chemical Sampling for Manufacturing Areas*

Priority	Characteristics	Tentative Sampling Frequency**
A1	Documented chronic toxicity, existing standard, frequent usage	Quarterly
A2	Documented chronic toxicity, existing standard, infrequent usage	Semi-annually
A3	Documented acute toxicity, existing standard, frequent usage	Annually (negative documentation)
A4	Documented acute toxicity, existing standard, infrequent usage	Annually or less (document that procedure was measured)
A5	Chronic toxicity suspected, no standard, frequent or infrequent usage	None (monitor toxicity data closely—considered A1, A2 Categories,
B1	No chronic toxicity, no standard, frequently used	None (monitor toxicity data)
B2	No chronic toxicity, no standard, infrequently used	None (monitor toxicity data)

* Long-term goal is to have on-site continuous monitoring where possible.

** Various factors, such as employee/management concerns or industrial hygienist judgment, may alter individual chemical sampling frequency.

Amongst the employees in any given zone a number of employees are randomly selected for each quarterly, semi-annual or annual iteration, as shown in Table IV. This table is offered as an example of the type of scheduling which can be performed using this approach. Approximately 20 percent of employees in each zone are sampled for a given chemical on an annual basis. During any given year, employees inquiring about their exposure will be told that they may or may not have been measured that year, but their exposure is the exposure of the zone members.[12] Over a

TABLE IV. Estimates of Work Load for Industrial Hygiene Air Sampling Using Exposure Zone Sampling Strategy

	Time Allotment (manual) hour	Time Needed (manual) per year	Time Needed (w/computer) per year*
Scheduling			
Quarterly/semi-annual/ annual list	2/quarter	8 hr	4 hr
Daily scheduling	4 ± 1/week	208 ± 52 hr	208 ± 52 hr
Updating Sampling Strategy (I.H. Engineer)			
Toxicity/ingredients information	2 ± 1/week	104 ± 52 hr	52 ± 26 hr
Chemical lists/zones	2 ± 1/week	104 ± 52 hr	52 ± 26 hr
Sampling (technical)			
Pump calibration	0.5/exposure	62 hr	62 hr
Expeditions	2 ± 0.5/exposure	248 ± 62 hr	248 ± 62 hr
Pickup, cap, and ship	1 ± 0.25/exposure	124 ± 31 hr	124 ± 31 hr
Input/analysis of results	0.15 ± 0.075	128 ± 64 hr	64 ± 32 hr
Reporting and Interpreting (I.H. Engineer)			
Employee/management interaction	0.15 ± 0.075 per sample	128 ± 64 hr	64 ± 32 hr
Resource direction/ analysis	4 ± 1/week	208 ± 52 hr	104 ± 26 hr
TOTALS		1322 ± 429 hr	982 ± 287 hr
I.H. Engineer		544 ± 220 hr	272 ± 110 hr
Technician		562 ± 157 hr	498 ± 125 hr
Scheduling		216 ± 52 hr	212 ± 52 hr

* Manual refers to the use of programs such as Lotus 1-2-3 on the PC; w/computer refers to the use of an integrated series of programs used for all aspects of this project.

Number of samples = 853; number of expeditions = 124.

period of years, the statistical power of this system increases as the number of samples for each zone increases.

Before any sampling expeditions into the plant by the air sampling technician, he/she must contact the department manager to be sure the sampling can be performed on the day selected and that the appropriate operations are taking place. If sampling cannot be performed on that day, the technician will reach agreement with the manager on a suitable day. In other words, the selection through a random sampling scheme is adhered to, but the time or day of sampling may shift somewhat.

The scheme permits the hygienist to estimate the work load for sampling. Table IV indicates the time allotment for different persons implementing the schedule. This particular table is based on 124 expeditions into the plant by the technicians with 853 samples planned. The entire planning of expeditions and samples, as well as treatment of data resulting from sampling and analyses, has been programmed into a personal computer. Special software was developed for the system. There is a significant managerial advantage in being able to plan an entire year's work for budgeting and planning purposes. Table V indicates the zones by key, the number of expeditions, and the number of samples collected in manufacturing. In the process of starting the program, the initial estimate of expeditions was revised downward as indicated.

Results from the method can be reported in terms of individual results or accumulation of results according to a frequency distribution. The log normal distribution is a very good fit to zone data. Data are reported as geometric mean, geometric standard deviation (GSD), and the phi parameter 12 used to predict the number of excessive exposures in the entire population zone, if all employees were sampled (Table VI).

The exposure zone method appears to be working satisfactorily in perhaps a dozen facilities that have adopted it; these facilities range from heavy manufacturing to semiconductor operations. The method can be utilized to predict adherence to regulatory requirements and also for prospective and retrospective cohort studies.

ANALYSIS OF RETROSPECTIVE DATA IN A THERMOMETER MANUFACTURING FACILITY

The purpose of this case study is to indicate that sampling strategies developed for one purpose may subsequently prove to be useful for another purpose, although this is certainly not a common occurrence. This case study originated in 1967 during a survey for mercury-in-air in a thermometer production facility in a tropical location. The survey indi-

TABLE V. Expeditions Required in Manufacturing

Zone	# Expeditions Initial Plan	# Expeditions Final Plan	# Samples
C1	14	8	84
C4	15	7	90
M1	71	40	426
M2	16	8	96
K2	10	5	60
C3	18	12	108
K1	3	2	18
K4	10	5	60
K5	5	3	30
T3	4	2	24
K3	5	3	30
T1	4	2	24
L1	5	3	30
X1	16	9	96
C2	4	3	24
T2	1	1	6
Chem			
Serv	33	26	198
TOTALS	234	139 (234 × 6) = 1404	

Revised estimates after initial implementation:
Number of expeditions (139) × 40% = 56 expedition/yr.
Number of samples (1404) × 32% = 450 samples/yr.

TABLE VI. Past Sampling Data by Work Zone

Zone	Population at Risk	Sample Size	Agent Concentration Low	Agent Concentration High	Internal Action Level	Sample Parameters Geo Mean	Sample Parameters GSD	Sample Parameters Phi
AGENT: A								
M1	21	5	0.03	1.0	1.25 ppm	0.15	5.70	1.2
AGENT: B								
M1	21	4	0.1	5.5	25 $\mu g/m^3$	0.72	5.14	2.2
AGENT: C								
M1	21	6	0.3	21.0	5.0 ppm	2.7	5.40	0.36
AGENT: D								
C4	54	17	0.7	7.9	12.5 ppm	1.9	2.5	2.0

cated excessive concentrations of airborne mercury and recommendations, subsequently implemented, were made for both environmental and medical surveillance and engineering controls. Intensive plant activities subsequent to the 1967 survey resulted in substantial progress in all categories of recommendations by the end of 1967; these measures have been continued to the present day. Contact was made with the facility in the 1970s, in 1981, and again recently, when a legal action was filed against the corporation for damages to the health of workers employed in the facility. This action stimulated a search for records related to the programs established in 1967 and revealed that air sampling and associated records were assiduously pursued during this period.

Mercury thermometer manufacturing is a complicated production process with over 125 individual hand operations. The flow of the material, as shown in Figure 1, is essentially related to filling hollow glass tubing with mercury and creating all of the characteristics of the clinical thermometer. The step of Opening and Cavity is a process which involves expanding the glass tubing with heat and forming a glass cavity for mercury. It is followed with sealing and bulbing of the rod. Mercury is not used in either of these two steps. The formed tubes are sent to Degas and Fill. In the Degas process, mercury is introduced into the tube, a vacuum is formed and this is then used to draw the mercury into the chamber. These two processes are accomplished by machines. All remaining steps involve handling of filled thermometers, and there is potential exposure to mercury. A combination machine follows Degas and Fill; thermometers are sealed, bulbed on the opposite end, and

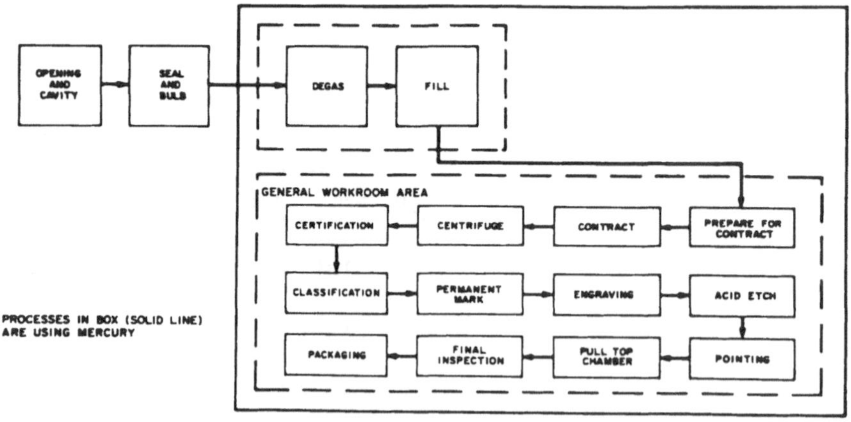

Figure 1. Thermometer production: process flow chart.

graded. From Combination, the thermometer is in the main workroom where the following processes are conducted: Prepare for Contract, Contracting, Centrifuge, Certification, Pull Top, Let Out for a Split, Pointing, Permanent Mark, Engraving, Acid Etch, and Final Inspection. Although each of these steps is not described in this chapter, it should be noted that there is opportunity for mercury exposure at every step.

In 1967 a system was established whereby an employee trained to measure mercury vapor in air using instantaneous detection instruments would make rounds of the facility, initially once during the day (1967 to 1970) and after 1970 in the morning and afternoon at different nonscheduled times. Different individuals were selected and trained for the task over the years. Measurements were expressed as the concentration of mercury vapor in milligrams per cubic meter of air. Instruments utilized changed as technology progressed. At first, the Beckman ultraviolet spectrophotometer, Models K21 and K23, was used; subsequently, a Bachrach MV-2 (J-W) Mercury Vapor Sniffer and a Jerome Mercury Meter were used. Instruments were calibrated periodically at the factory and checked against recently calibrated instruments in the field. All data were reported to one significant figure, the limit of reliability of these instruments. These data were analyzed recently in preparation for litigation.

The data base consisted of approximately 105,990 airborne mercury concentration measurements representing 2334 measurement days. The data from 1971 to 1974 represented air samples collected from 21 work categories. The work categories varied somewhat over the years as the process improved with respect to the introduction of machinery. From 1975 to 1980, for example, 23 work categories were needed and there were shifts in numbers of categories, both up and down, during the intervening years. Samples were collected during the morning and afternoon.

Figure 2 is an example of each of the graphs of air concentrations prepared for the different areas of the plant. Figure 2 results are for Degas and Fill; a total of 4800 measurements are represented in the figure. Figure 2 shows the improvement in the facility and also permits management to demonstrate general compliance with the 0.10 mg/m^3 eight-hour standard for mercury-in-air, although eight-hour samples were not obtained. The brief increase during the period represented by 1972 in this particular area was associated with major plant redesign. The figure suggests that controls were successful after the original 1967 strategy; however, during major plant redesign and installation in the 1970–72 period some increase in concentrations occurred. The improvement in conditions continued throughout the 1970s.

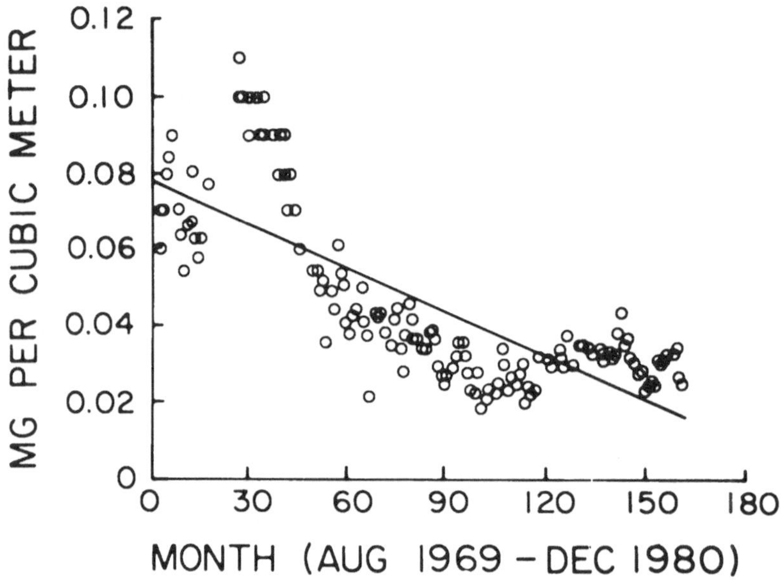

Figure 2. Least squares regression of monthly mean mercury vapor concentrations in the Degas Area.

A data base of this size is statistically very powerful to make the point of adherence to a longer-term standard, although only randomized short-term measurements were made.

The data can also be plotted in other ways. For example, the results have been plotted as moving averages for each area to demonstrate and highlight trends.

The reason for presenting this case is that while strategies of air sampling are usually developed for immediately perceived purposes, they can have other value and should be periodically reviewed with this in mind. The strategy described was developed to control individual source releases of mercury; it subsequently proved valuable for inferring exposures to the contaminant and for verifying trends of reduction in airborne contamination. This case also suggests that records of short-term air sampling for source identification require the same care and should be associated with the same long-term retention times as personal sampling records.

In summary, the exposure zone method has been described as applied to a complex chemical utilization facility, and a sampling strategy directed at source identification and control has been described as useful for inferring exposure concentrations. Both involve sampling strategies and approaches to obtaining air samples.

ACKNOWLEDGMENT

I am indebted to Ernest Timlin and Thomas Hall for their contributions to the case studies presented here.

REFERENCES

1. Leidel, N.A., K.A. Busch and W.E. Crouse: *Exposure Measurement Action Level and Occupational Environmental Variability*. DHEW (NIOSH) Pub. No. 76-131 (1975).
2. Leidel, N.A., K.A. Busch and J.R. Lynch: *NIOSH Occupational Exposure Sampling Strategy Manual*. DHEW (NIOSH) Pub. No. 77-173 (1977).
3. Brief, R.S. and R.A. Scala: A Statistical Technique for Determining Compliance With Dual Hygienic Standards. *Am. Ind. Hyg. Assoc. J. 37*:474-478 (1976).
4. Corn, M.: Strategies of Air Sampling. *Recent Advances in Occupational Health*. J.C. McDonald, Ed. Churchill-Livingstone, New York (1981).
5. Rappaport, S.M.: The Risks of the Game: An Analysis of OSHA's Enforcement Strategy. *Am. J. Ind. Med. 6*:291-303 (1984).
6. Corn, M.: Editorial—Air Sampling Strategies in the Work Environment. *Am. J. Ind. Med. 6*:251-252 (1984).
7. Tuggle, R.M.: Assessment of Occupational Exposure Using One-Sided Tolerance Limits. *Am. Ind. Hyg. Assoc. J. 43*:338-346 (1982).
8. Rock, J.C.: A Comparison Between OSHA-Compliance Criteria and Action-Level Decision Criteria. *Am. Ind. Hyg. Assoc. J. 43*:297-313 (1982).
9. Tuggle, R.M.: The NIOSH Decision Scheme. *Am. Ind. Hyg. Assoc. J. 42*:493-498 (1981).
10. Corn, M., P. Breysse, T. Hall et al.: A Critique of MSHA Procedures for Determination of Permissible Respirable Coal Mine Dust Containing Free Silica. *Am. Ind. Hyg. Assoc. J. 46*:4-8 (1985).
11. National Academy of Sciences: *Risk Assessment in the Federal Government: Managing the Process*, p. 21. Washington, DC (1983).
12. Corn, M. and N.A. Esmen: Workplace Exposure Zones for Classification of Employee Exposures to Physical and Chemical Agents. *Am. Ind. Hyg. Assoc. J. 40*:47-57 (1979).

13. Corn, M.: Strategies of Air Sampling. *Scand. J. Work Environ. Health* *11*:173–180 (1985).
14. Esmen, N.A.: Retrospective Industrial Hygiene Surveys. *Am. Ind. Hyg. Assoc. J.* *40*:58–65 (1979).

CHAPTER 21

A Cost-Effective Air Sampling Strategy

HOWARD E. AYER

University of Cincinnati, Cincinnati, Ohio 45267–0056

INTRODUCTION

Air sampling is only part of a total health and safety program. Major objectives for any company or facility must be to prevent disasters inside the facility and to avoid contributing to a disaster outside the facility. Another important goal is to prevent traumatic injury from falls, burns, cuts, and striking or being struck by other objects. Workers and others must also be protected from physical agents including noise, heat, and radiation, both ionizing and non-ionizing.

This chapter describes strategies for air sampling by operators of private or governmental facilities. Other chapters address the strategies for epidemiological studies and enforcement of governmental regulations.

Before sampling takes place, it is important to identify the materials which are potential air contaminants and for which air sampling would be appropriate. The Occupational Safety and Health Administration (OSHA) Hazard Communication Standard has had value in informing workers of their potential exposure. It has perhaps had even greater value in informing manufacturers of the potential hazards to which their workers might be exposed and which might require control. As part of the total worker health program, these potential contaminants should be examined for their degree of potential hazard. Alternatives to their use in the manner proposed should be sought if they appear to present an unacceptable degree of risk. If not, any additional controls necessary to reduce risks to worker health should be installed.

PURPOSES OF AIR SAMPLING

There are five major reasons for air sampling. These are health protection, environmental protection, compliance with governmental regulations, protection from unwarranted worker compensation claims or tort actions, and process protection.

Air sampling can assist in health protection by assuring the effectiveness of process controls or other engineering controls. It can also be used to assure that worker exposures to toxic or irritating materials are within acceptable limits.

In environmental protection, air sampling is used to monitor emissions which might be damaging to paint, plants, animals, or human health of people outside the facility. To a lesser extent, it is used to determine contributions to concentrations in ambient air outside the facility.

A major reason for air sampling is to comply with governmental regulations, principally those of the OSHA and the Environmental Protection Agency, but also those of the Nuclear Regulatory Commission, the Coast Guard, and others such as state and local authorities. Where these regulations govern, the strategies outlined here can be used to supplement them. Although the purpose of these regulations is usually protection of health or the environment, their inflexibility will often mean that samples must be taken which are perceived by the facility to be unnecessary for health protection or, conversely, that some real hazard in, or related to, the particular installation has not been regulated.

Generic inhalation hazard regulations could address sampling strategy, but political considerations make them unlikely. Regulations meeting the letter of the OSHAct would be so restrictive and unwieldy as to constitute a burden upon users which would cause either general disregard for the standards or further flight into foreign territories.

A purpose for air sampling which has arisen in these litigious times is documentation of healthful working conditions to prevent future successful lawsuits from workers alleging health damage from their work. Although most samples taken for this purpose will have no direct value in the protection of worker health, they may occasionally uncover some unsuspected health hazard. In any event, they may prove invaluable in preserving the economic viability of the organization.

A final purpose of air sampling is to ensure the protection of a product or process from corrosion, abrasion, contamination, or some other deleterious effect. That purpose will have priority in the facility and need not be covered here.

SAMPLING AIR CONTAMINANTS FOR HEALTH PROTECTION

The best strategy for health protection is to provide a process which prevents the emission of air contaminants. Failing this, controls should be provided (e.g., enclosure and local exhaust ventilation) which minimize worker exposure. Another part of the health protection program is adequate maintenance of potential emission sites such as pump shaft seals. If these preventive procedures are followed satisfactorily, air samples need only be taken initially to demonstrate the effectiveness of controls or when changes are made in process or controls. An exception is the need for some kind of alarm if there is the possibility of escape of immediately dangerous substances.

A common situation, which demonstrates the need for a sampling strategy, occurs when a process generates an air contaminant in the vicinity of one or more workers. The worker may generate the contaminant; a common example is spray painting. The customary way of estimating worker exposure to the air contaminant is by personal sampling. With some exceptions, such as long-term indicating tubes, the sample must then be analyzed in the laboratory. As noted by others in this symposium, the air contaminant concentration will vary with time, and estimated exposures will vary between workers doing the same or similar tasks. Although part of this variation will be systematic and part random, it will seldom be possible to distinguish these by air sample results.

The variation of these estimated exposures can usually be approximated by a lognormal distribution. The geometric standard deviation (GSD) of this distribution will customarily be from 1.3 to about 3. These GSDs might apply to typical samples of two or three hours, although few sets of short samples will generate distributions with a GSD as low as 1.3. For those operations where the exposures are negligible except for occasional releases, the GSD may be as high as 5.

The objective of the National Institute for Occupational Safety and Health (NIOSH), OSHA, most industries, and most unions in respect to worker health is to prevent workers from sustaining significant health damage because of their work. However, the viewpoints of research vs. enforcement for the two governmental agencies and economic factors for the two adversarial industrial groups lead to major disagreements on the application of sampling strategies. Complete ignorance of exposures, by eliminating air sampling, is not a viable option. Although initial expense may be less, all are held responsible, legally and ethically, for what they could reasonably be expected to know as well as what they actually do know. The legal implications alone indicate that some air sampling is necessary.

The purpose of this chapter is to look at the needs for air sampling on a cost-benefit basis. The generality of the chapter and the lack of hard data, especially on any benefits derived from air sampling, will prevent quantitative cost-benefit estimates.

Whether sampling strategies are being specified in regulations, designed for the enforcement of regulations, or designed by the employer for employee health protection, the general principles are the same. Workers potentially exposed are identified. Those with only trivial exposures are eliminated from the sampling program (except, perhaps, for epidemiologic studies). Those workers who apparently have the highest exposures are sampled first. Necessary actions for health protection are taken if some guideline or mandatory limit is exceeded. Sampling is repeated when processes or controls are changed, and sampling is repeated periodically for workers above some level.

EXPOSURE GUIDELINES

There are major disagreements as to sampling strategies. One most important factor is the selection of the exposure guideline to be used. Without some guideline, it is not possible to define "exposed" nor to decide what is a "trivial" exposure so that a reasonable sampling strategy may be applied. Most people are exposed, at some concentration other than zero, to the majority of industrial air contaminants of greatest concern. These include naturally occurring minerals, such as quartz and asbestos, as well as many halogenated hydrocarbons which are widespread and persistent, and can be analyzed at the levels of microliters per cubic meter. This must also include carcinogens in spite of the oft-stated contention that "there is no safe level for a carcinogen." In actual fact, people are all exposed at some level other than zero to a host of environmental carcinogens including the essential carcinogens of sunlight and potassium. If the linear hypothesis is accepted, safety becomes a matter of accepting some calculated, hypothetical degree of risk. In the generally accepted words of Lowrance, "Safety is that risk which is judged to be acceptable." Whether explicit, as in OSHA regulations, or implicit, as in NIOSH recommendations, limits for carcinogens exist and will continue to be used. As individuals and organizations disagree on quantitative risk assessments and degrees of acceptability, it is to be expected that major disagreements will continue on concentration limits in general and carcinogen limits in particular.

At this American Conference of Governmental Industrial Hygienists symposium, it is appropriate to use the recommendations of the Chemical Substances Threshold Limit Values (TLV) Committee as guidelines. In applying these guidelines, the TLV Committee criteria should govern, i.e., excursions above the TLV concentrations are permissible, provided that these are not extreme and that the time-weighted average is below the TLV. Partly for administrative convenience, and partly because many of the limits depend upon overnight metabolism and excretion of the substance, the single-shift, eight-hour average has come to be the basis for all but the ceiling TLVs and those for pneumoconiosis-producing dusts. In using the limits, an attempt is made to prevent workers from exceeding these eight-hour average exposures.

In addition to situations in which accepted guidelines, i.e., the TLVs, are used to decide whether further air sampling is necessary, there are other occasions in which the TLV is more or less irrelevant. For example, when workers are experiencing symptoms, e.g., nausea, eye irritation, and/or upper respiratory irritation, air sampling as well as corrective action will be required. Confirmed recurrence of symptoms well below accepted guidelines should call for notification of NIOSH and the TLV Committee as an indication that the current TLV may be inadequate for worker protection.

There are also many substances in use for which there are no TLVs, but judgment as to potential health hazards must be made. In such cases, conservative judgments as to acceptable concentrations must be made on the basis of available toxicological data. Fortunately, the reason for lack of a TLV is most often either that the material has low toxicity or that the exposures are expected to be exceedingly low.

If, as in most occupational exposure situations, a lognormal distribution of concentrations exists, the probability of a worker exceeding the eight-hour average concentration can be calculated when the geometric mean and geometric standard deviation are known. Even when these parameters are known within reasonable limits, it is almost impossible to estimate very small probabilities reliably. The data are seldom adequate even for a five percent probability, and any estimate below that is pure guesswork.

Public health professionals should not ask for air samples unless they are going to contribute to health protection. If previous sampling has demonstrated reasonable safety and there have been no changes, further sampling is not going to contribute to health protection.

A major defect in some proposed sampling strategies is that they suggest the same strategy for all situations. The strategies proposed tend to fit a common manufacturing situation where a routine operation is performed at the same location for most of the shift. Other common manufacturing situations include the roving worker who works on any part of a long line, such as in a paper mill, or the worker who works at a process which is ordinarily closed but in which spills are not an uncommon occurrence. With the routine operation, the exposure is relatively continuous, if variable, throughout the day. With the roving worker, the exposure may occur at only one point in the line or at only one time of day. With the enclosed operation, the exposure is negligible except for the times when there is a spill or release. These examples by no means exhaust the variations in the types of exposure experienced by workers.

For cost-effectiveness in air sampling, general principle questions are needed which address the main points rather than just a sampling strategy using the statistical parameters of geometric mean and geometric standard deviation and procedures to give confidence limits, relative certainties and percentile bands. These general principle questions include:

1. Will an air sample give information which will enable the prediction of exposures in the future, assuming no major changes in the operations?
2. Will an air sample give information which will assist in controlling the operation?
3. If the only purpose of the air sample is to estimate the exposure of the worker on the day and at the time sampled, will this information be of value in protection of his or her health?

It is necessary to take air samples to determine whether there is any potential hazard, whether there is an immediate hazard and whether some change has created a hazard where none existed before. The type of sample listed in point 3 above will then truly be of some value in protecting worker health. Furthermore, it will often be necessary to take more than one personal sample on a worker, or to sample more than one worker, in order to estimate the degree of hazard, if any. In these cases, statistics describing estimated exposures will indeed be of value. Experienced industrial hygienists have participated in the writing of OSHA regulations regarding exposure monitoring, and the suggestions in this chapter will bear many resemblances to these regulations. However, the necessity of attempting to write regulations to cover any possible contingency means that many more samples may be required than can be justified purely on a worker protection standpoint for most of the operations covered.

Any sampling strategy will explicitly or implicitly divide workers into categories. One such categorization is the following:

1. Unexposed, i.e., "trivial" exposure (Unexposed).
2. Readily measurable but "safe" (Measurable).
3. Needs watching (Significant).
4. Improvements desirable (Marginal).
5. Corrections necessary (Overexposed).
6. Immediate corrective action required (Critical).

The category within which the exposure falls will determine whether air samples should be taken, the necessary accuracy and precision of the exposure estimate, and frequency of sampling if repeated sampling is required. In the typical OSHA complete health standard, for example, the above exposure categories are combined as follows:

Safe/trivial and safe/significant—Below action level or unexposed.
Needs watching—Above action level but no exposure found above the Permissible Exposure Limit (PEL).
Improvements desirable—Some exposures at the PEL.
Corrections necessary—Some exposures above PEL.
Immediate correction required—All exposure estimates well above PEL.

In fact, an OSHA regulation will not distinguish between an exposure estimate just above the PEL and one several times the PEL, treating the PEL as that value which all recognize does not exist, a fine line between "safe" and "hazardous." However, in the actual application of the standards, there is considerable latitude given to the OSHA area directors as to penalties proposed and the length of time to come into compliance by engineering and administrative controls.

CATEGORIES WHICH REQUIRE SAMPLING

Workplace inspection can eliminate the exposure of many (perhaps most) of the workers in the "trivial" category and eliminate the necessity for sampling. In an OSHA standard, there might be a provision for an "Initial Determination" in which a determination is documented without any air sampling. Similarly, for many of the workers in the "Immediate correction required" category, there is no need for sampling. These exposures often represent conditions where the need for control measures is, or should be, recognized, but controls have not been applied. Air sampling may even be undesirable because conditions on the day the estimate is performed might give an exposure estimate below the PEL or TLV. OSHA standards, however, will not only call for air sampling at the operation but will require repeat sampling; this may be considered an additional penalty for not installing the necessary engineering controls.

For these two categories, "unexposed" and "immediate corrective action required," the requirements for accuracy and precision of individual exposure measurements are least stringent. Certainly accuracy with a factor of two would be adequate. On the low end, a "none detected" and on the high end a "greater than" can provide adequate information. In the first case, the result is below any level of interest; on the other end, it is more than is acceptable, whatever that is.

For the intermediate categories, the recommendations of a joint committee on indicating tube accuracy, subsequently endorsed by NIOSH, of ± 25 percent accuracy at the TLV and ± 35 percent at one-half the TLV are certainly adequate, even if not currently met by sampling and analytical methods for asbestos and silica. For even these less precise methods, the imprecision of the method is usually far overshadowed by environmental variability.

The exposure limits, i.e., PELs and TLVs for most materials, will be in terms of full-shift average exposures. The current NIOSH and OSHA approach to sampling is to estimate full-shift exposures for one worker at a time. For some exposures, e.g., mineral dust, full-shift sampling is a requirement of the sampling method. In other cases, such as the use of long-term passive indicating tubes, full-shift sampling may be the easiest sampling method. Where sampling time is limited because of the method, e.g., organic vapors poorly retained on solid sorbents, or where there are only a few sampling devices available for a large group of workers, there is a choice between direct full-shift estimates on a few workers or partial-shift samples on a much larger group. For example, if the sampling period must be limited to two hours, four consecutive samples must be taken on the same worker for a full-shift measurement. For maintenance workers and other workers with extremely variable and unpredictable exposures, this may be the only way to estimate the exposure on the day sampled. However, for such workers without routine exposures, rather than attempting to estimate averages, air samples should be taken at tasks which may be performed relatively frequently and which present the possibility of contaminant exposure. For those tasks which are done too infrequently and unpredictably to make air sampling possible, exposures should be minimized by mandatory controls, work practices, and personal protection. Where a number of workers are involved in the same operation, however, it is obviously more effective to estimate the population exposure by spreading the samples out over more workers. Even when only one worker is exposed at a continuous operation, it is questionable whether a full shift sample is necessary to obtain information adequate for health protection.

There is increasing evidence that the peak, or short-term, exposures are important in their own right and in their contributions to the overall full-shift exposure average. This is certainly true when sensory irritation is experienced, even though the TLV is not in terms of a ceiling. It may also be true even for the pneumoconioses, where all recognize that it is long-term retention of the dust which is the cause of the occupational disease.

More samples, even if partial rather than full shift, will surely give a better estimate of the group exposure. Direct full-shift exposure estimates of workers can show that none of the workers' samples exceeded the exposure guideline on a given shift; these data can also estimate group geometric means and variances for the full-shift exposures. Partial-shift samples cannot demonstrate conclusively that any worker did not exceed the full-shift guideline, but for the same total sample time, they will give as good and possibly a better estimate of group exposure since they will include more interworker variability. For the days not sampled, a larger number of partial-shift samples will give at least as good an estimate of the number of workers above some criterion concentrations as the same sampling time in full-shift samples.

If the exposure mean and variance have been estimated by air sampling, then there can be quantitative estimates of the probability of the group or some individual worker exceeding an exposure guideline. In general, TLVs were set on the basis of population estimates, whether of animal or human populations. Thus, it is most appropriate to consider exposures in terms of exposed populations rather than exposed individuals. Having considered the population, one can then look for individuals who, because of special circumstances, may be at greater risk. The first concern when statistical considerations are introduced is to be reasonably certain that the group mean for any population is below the exposure guideline. For example, if ten air samples were taken for a group of workers with apparently similar exposures to a solvent with a TLV of 200 ppm, the geometric mean concentration was 100 ppm, and the GSD was 2.5, the calculated mean concentration for the group would be 152 ppm and the probability that a concentration exceeded 200 ppm would be 0.33. If one were to use the common statistical criteria of a probability < 0.05 that a concentration was less than the TLV, the average concentration would have to be less than 125 ppm and the geometric mean concentration < 82 ppm. In considering the exposure of a group of workers in this way, the exposures of individual workers are not ignored. Industrial hygienists have always been concerned with individuals, but data are more readily accumulated from groups. In the above example, with the necessary assumptions about the exposure of other workers and other days being represented by this group of full-shift samples, the calculated probability that any worker will be exposed above the TLV is 0.33. Most industrial hygienists of any generation would have assumed that exposures were unacceptable if one in three samples were above their guideline. Further investigation to determine the cause of the high exposures would be undertaken. Controls to reduce the high exposure concentrations or the duration of such concentrations would then be recommended. Only then would additional air samples be taken.

If, as before, the GSD was 2.5, the mean concentration was 95 ppm, and the geometric mean (GM) was 60 ppm, there would be a 0.1 probability that an individual exposure estimate would exceed 200 ppm. In this case, it is considerably less likely that an industrial hygienist would have recommended controls. Without calculating that the 0.05 probability would be 270 ppm, she/he would have recognized that excursions above the guideline would be unlikely to exceed twice the guideline and, with averages well down in the safe range, would have felt conditions to be satisfactory. Further sampling would not have been recommended unless conditions changed.

AUTOCORRELATION

From a statistical standpoint, it would be desirable to have samples taken on randomly selected individuals in a given work category at random times. As a practical matter, samples must be taken when the industrial hygiene technician is in the vicinity, and nearby operations are sampled at almost the same time. This leads to the phenomenon known as autocorrelation. Each successive sample in the same area is influenced by the concentration which existed for the previous sample. Likewise, the concentration in a second, adjacent area being sampled at the same time may be influenced by contaminant generation in the first area. In terms of sampling strategy, this means that better estimates of actual average concentrations will be obtained if the air sampling is spread over several trips to the space rather than all done successively and/or simultaneously. Even though the old winter/summer differences do not exist in most modern manufacturing facilities, it is still better to conduct the sampling over the seasons.

INITIAL SAMPLING

In their previous work, Busch and Leidel (also see next chapter in this volume) have implicitly assumed that any full-shift average exposure in excess of a limit is harmful. The more conventional view is that exposures in excess of guidelines are to be avoided, but any useful limit must have enough built-in safety so that an occasional minor excursion above the limit will not cause any harm. A statistical calculation will demonstrate that average exposures must be less than half the guideline if relatively frequent full-shift excursions are to be avoided. The OSHA action levels of one-half the PEL are explicit recognitions of this fact. These combinations of field experience and statistical calculations enable listing of those factors which should be considered if a sampling strategy is to justify its cost in potential health benefits. These factors include:

1. Are materials used which could constitute a health hazard if generated into the air?
2. If so, are they used in such a way that they actually are generated into the air in the form of a gas, vapor, or aerosol?
3. Is such generation at points which are normally controlled by enclosure, e.g., at tank discharges?
4. Does the generation regularly or routinely occur? (This excludes rare, unanticipated, accidental releases.)

If number 3 is affirmative, periodic area samples at the potential release point may be necessary as well as a program of personal protection for workers involved. Only when question 4 is answered affirmatively is an air sampling program for determination of worker exposure necessary. If such a program is necessary, then the program may be divided into initial sampling and repeat sampling. Initial air sampling should be done:

1. Where required by regulation.
2. Where workers may be trivially exposed and documentation for legal purposes is desirable.
3. Where inspection suggests a potential inhalation hazard.

If there does appear to be a potential inhalation hazard, then the results of the initial sampling should be compared with exposure guidelines to make decisions as to the necessity for future sampling.

For exposures of interest, i.e., neither trivial nor critical, how many samples should be taken at an operation to classify it? The response must consider how many people are potentially exposed. If it is only one

TABLE I. Action Criteria Used After a Sample

Percent of Guideline	Category—Action
< 10	*Unexposed*—no further sampling
10–40	*Measureable*—no further sampling
40–70	*Significant*—include in routine sampling program
70–130	*Marginal*—improvements necessary—further sampling required
130–200	*Overexposed*—initiate or improve control program, consider respiratory protection, continue sampling
> 200	*Critical*—take immediate corrective action including respiratory protection

worker, more than three full-shift estimates are unlikely to be made, just because of the logistics involved. If there are 40 workers (an extremely large number for any homogeneous exposure group), ten or more samples might not be considered excessive. Ideally, the number of samples should be sufficient to calculate that the probability of any one full-shift exposure estimate exceeding the TLV would be less than 0.1. Without prior data, a fairly typical GSD of 2.0 might be assumed, in which case a single sample would have to be less than 40 percent of the guideline. As data on the operation were accumulated, a sample-based estimate of GSD could be used. For example, although a single sample of 70 percent of the guideline would not give any great confidence that the guideline would not be exceeded more frequently than allowed, neither would it call for any control action. In the same way, a single sample 20 percent over the guideline would indicate the possible need for control; by itself, it would not cause controls to be installed. Carrying this analogy a step further, one might hypothesize a set of action criteria to be used after one sample. In fact, this sort of a guideline has always been used implicitly and is explicit in some OSHA standards. One such criteria set is shown in Table I.

REPEAT SAMPLING

The frequency and number of repeat samples may be determined by regulation which must be followed. Where regulations give flexibility, then on principle, air samples should only be taken when the results are to be used for worker health protection. In the absence of a defined air sampling strategy, samples were probably taken most often where there was a perceived problem. The program suggested above would systematize this procedure and include samples representative of all those workers with significant exposure. For workers at or near the TLV, the sampling program would be combined with a program of control monitoring and improvement to reduce exposure of the workers. For those workers above the TLV, the emphasis should be on control improvement, with air sampling only as an adjunct to determine the progress of the control program.

If, as is often the case, the group doing the air sampling is not directly responsible for designing and installing controls, repeat sampling at least quarterly would be desirable for those operations at or above the TLV. This would ensure that improvements were made in a timely manner.

SAMPLING OF MIXTURES

It may not be necessary to analyze all compounds of a mixture, particularly when the mixture has been previously sampled and the approximate proportions of the components are known. If the proportion of silica in a dust is known, then a simple respirable mass determination is all that is required. Likewise, in a complex solvent mixture, it is not necessary to analyze more than one or two of the components once their proportions in the workplace air are known. In reporting and interpreting the results, the total air contamination is noted. In initial sampling, it may well be necessary to request analysis of all or most of the components of a mixture.

GENERAL

The foregoing, although suggesting some numerical guidelines for when air samples should be taken, has avoided categorical statements as to when samples must be taken. The principles stated are those which have been regularly followed by experienced, capable industrial hygienists. It has been stated as a fact that full-shift sampling is often not the most effective method of worker protection or the most cost-effective way to determine worker exposure. It is actually only required for enforcement of OSHA regulations. If air sampling is often required to determine the necessity for, or adequacy of, exposure control methods, it is also true that air sampling is not a substitute for control measures. A cost-effective air sampling program will provide enough information so that an employer and the workers can know that the workers are protected from health damage and the employer is protected from unjustified claims of such damage. It will be part of a total health and safety program which emphasizes prevention of harmful exposures to air contaminants while recognizing that zero exposure to such contaminants is neither necessary nor desirable.

DIFFERENCES

The viewpoints expressed herein often differ from sampling strategy viewpoints expressed by others. These differences are summarized as follows:

1. No single statistically-based sampling strategy can provide the most effective protection for all categories of workers.
2. When a work situation is obviously unsatisfactory in terms of potential air contaminant exposure, air sampling is not necessary to demonstrate the fact and may actually contribute to a delay in installation of necessary controls.
3. Undesirable effects on workers may occur below the TLV (or PEL); in this case, action is necessary, including further air sampling, control measures, and prompt notice to the TLV Committee and NIOSH.
4. If a lognormal distribution of worker exposures exists, no reasonable increment of the geometric mean exposure below the TLV will prevent all excursions above it, even on an eight-hour basis. Such inevitable excursions, if infrequent and minor, will cause no harm.
5. A sampling program should be examined to eliminate sampling unless it will contribute to worker protection, is required by governmental regulation, or is necessary to document "nonexposure" for legal purposes.

CHAPTER 22

Statistical Models for Occupational Exposure Measurements and Decision Making

KENNETH A. BUSCH[A] **and NELSON A. LEIDEL**[B]

[A]Senior Advisor for Statistics, NIOSH, Cincinnati, Ohio; [B]Senior Science Advisor, NIOSH, Atlanta, Georgia

GENERAL

Distributional models for occupational exposure measurements are presented as a basis for statistical procedures that have been developed for this type of data. The models include terms for several sources of random error that contribute to the total variability of exposure measurements. Models are discussed in terms of their theoretical basis, their relationship to the development of usable sampling strategies, and the effects of model choice on subsequent statistical analyses performed on the data.

Applications of statistical methods for parameter estimation and hypothesis testing are useful in several areas of occupational exposure monitoring by employers as well as in regulatory enforcement of occupational health standards. These include:

1. *Point Estimates*—of exposure distribution parameters, such as a) estimation of a time-weighted average (TWA) concentration based on the measured concentrations in one or more air samples; b) estimation of the daily TWA exposure level which is exceeded in five percent (or another chosen fraction) of an employee's work shifts; and c) estimation of the long-term percentage of an employee's work

shifts in which TWA exposure levels exceed a designated value, e.g., the Occupational Safety and Health Administration (OSHA) permissible exposure limit (PEL).

2. *Interval Estimates*—of exposure distribution parameters. Uncertainty limits (reflecting limits of error) for any of the three types of point estimates in point 1 are strongly recommended. Such limits can be in the form of one-sided or two-sided confidence intervals.
3. *Statistical Significance Tests*—of null hypotheses concerning true values of statistical parameters. For example, a compliance officer would properly test the null hypothesis of compliance (H_0: TWA < PEL) against an alternative hypothesis of noncompliance (H_1: TWA > PEL). Or, an employer would make a similar test except that he would properly test the null hypothesis of noncompliance (H_0: TWA > PEL) against an alternative hypothesis of compliance (H_1: TWA $\leq$ PEL).

Details of statistical methods which can be used for the above applications are given in previous publications[1,2] and are not the subject of this paper. The topic here is the *statistical distributional models* which underlie such data analysis procedures. In addition to the above-mentioned applications, the statistical models have been a theoretical vehicle for comparing efficiencies of alternative occupational exposure sampling strategies as well as the operating characteristics of significance tests performed with the resulting data. Limited discussion will be given to the nature of the disturbance in perceived statistical operating characteristics of significance tests caused by inexact agreement between actual and assumed statistical distributions of occupational exposure measurements.

Data models will be presented in this paper that are more detailed than the models used in practice. Separate distributional models will first be given for the different components of the total (observed) variability in exposure measurements: air sampling device error, chemical analytical error, and changes in true exposure level variations over time and space. Then, the separate models will be combined into a distributional model for the total (net) errors in exposure measurements. It will then be shown that this more detailed model "collapses" (simplifies) into a conventional lognormal model such as is usually employed to analyze actual exposure measurements or exposure monitoring data for repeated work shifts.

The terminology for different types of single-shift statistical sampling strategies was introduced earlier by Leidel, Busch, and Lynch.[1] The same terminology will be used here: full-period single sample, full-period consecutive samples, partial-period consecutive samples, and grab samples. Figure 1 illustrates the meanings of these terms. In addition, a data set

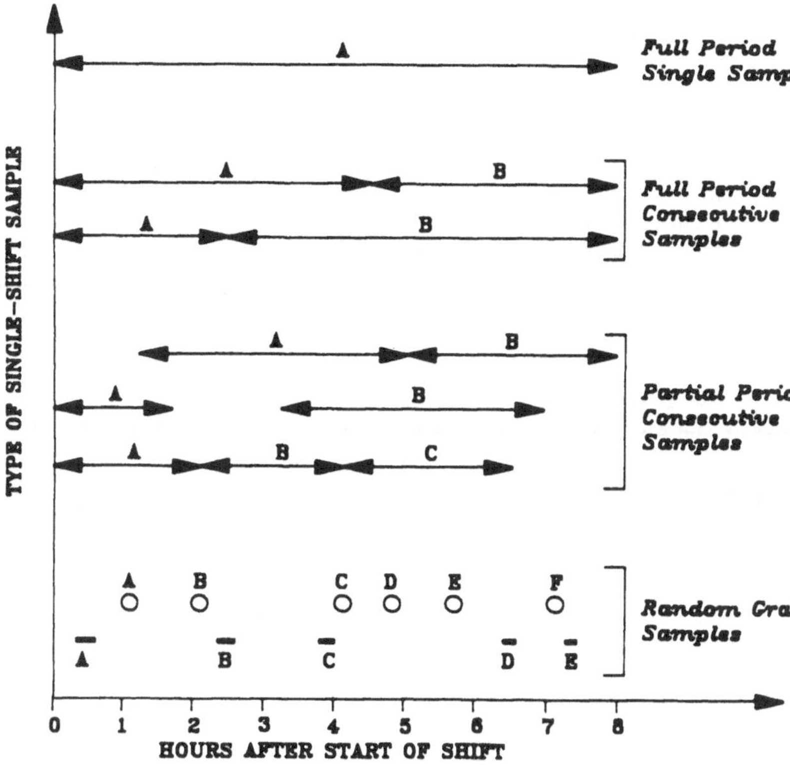

Figure 1. NIOSH terminology for different types of single-shift sampling strategies.

consisting of eight-hour TWA exposures of a single worker, measured during several shifts, is referred to as a *one-worker, multiple-TWA data set*. Finally, a data set consisting of several workers' eight-hour TWA exposures, measured during the same shift, is referred to as a *multi-worker, single-TWA data set*. Distributional models for each of these sampling strategies will be discussed under separate headings in succeeding sections of this report.

FULL-PERIOD SINGLE SAMPLE

A statistical distributional model is given in this section for replicate full-period single-sample exposure measurements of a single employee during a given work shift. (A single measurement at hand is considered to be a random sample from this hypothetical distribution.) Since air sampling and chemical analytical methods are assumed to be unbiased, air concentration measurement errors are considered to be random, not systematic.

Notation:

- x = measured eight-hour TWA concentration based on a single eight-hour air sample for an employee.
- μ = true TWA concentration for the employee during the sampled work shift.
- σ = standard deviation for total (net) random measurement errors of a measurement method.

$CV_T = \sigma/\mu$ = coefficient of variation of total (net) error in x, due to: e_S = error of the air sampling step, with coefficient of variation CV_S, not including pump error; e_p = error due to pump flow rate variations, with coefficient of variation CV_P; and e_A = error of chemical analysis, with coefficient of variation CV_A, including error in the determination of a desorption efficiency factor if one is used.

At this point, normal distributional models for the three types of errors will be combined into a normal model for the total (net) errors in exposure measurements. The total error (e_T) is an algebraic sum of the three separate errors. That is, the data model is:

$$x = \mu + e_S + e_p + e_A = \mu + e_T$$

Since three types of component errors affecting x are expected to be stochastically independent, the coefficient of variation for total (net) error (e_T) in x is given by:

$$CV_T = (CV_P^2 + CV_S^2\ CV_A^2)^{1/2}$$

In the National Institute for Occupational Safety and Health (NIOSH) Methods Validation Tests Program,[3,4] the sizes of CV_S, CV_A, and CV_T were separately estimated for over 300 measurement methods. However, only the CV_T value is needed for an exposure measurements distribution model as long as analyses are performed by the same NIOSH method that was used in the methods evaluation tests. Results from the NIOSH tests are not inconsistent with the following mathematical model for the probability density function of total (net) measurement errors in full-period single samples.

$$f(x) = [1/[((2\pi)^{1/2}\ CV_T\mu)]\exp[(-1/2)(x - \mu)^2/(CV_T\mu)^2]$$

This is the probability density function for a normal distribution with standard deviation proportional to the mean, i.e., $\sigma = CV_T\mu$. The CV_T is verified as being constant, at least over the fourfold range of the three concentrations tested by NIOSH: 0.5 PEL, 1.0 PEL, and 2.0 PEL.[3] Only infrequent significant departures from this normal distributional model have occurred, although the limited numbers of tests performed (6 per each of three concentrations) would not permit detection of small differences (i.e., those not having practical significance) among the three CV_T's. One chemical analysis method showed a higher CV_T for the lowest concentration, which bordered on the limit of detection for the method. Another method indicated non-normality due to an apparent "outlier" among the six samples taken at the highest concentration.

It is clear that the *standard deviation for total measurement errors of a measurement method is not constant.* Rather, the standard deviation (in concentration units) is approximately proportional to the concentration analyzed. This fact suggests use of a lognormal distribution with constant geometric standard deviation (GSD) as an approximation to a normal distribution of measurement errors with constant CV_T (or vice-versa). Hald[5] states that the lognormal distribution is close enough to the normal distribution for practical purposes when the CV is small, such as below 0.33 or 33 percent. Hald gives no quantitative basis for this statement, but theoretical data are presented in Table I which provide an index of comparability of normal and lognormal distributions in the context of noncompliance testing. The index used is the ratio between

TABLE I. Comparison of 95% Lower Confidence Limits for Arithmetic Means of Normal and Lognormal Distributions with the Same Known Coefficient of Variation (CV_T)

CV_T	LCL_{LN}/LCL_N	LCL_N/LCL_{LN}
0.381	0.950	1.053
0.364	0.952	1.050
0.333	0.957	1.045
0.200	0.978	1.022
0.184	0.981	1.019
0.128	0.990	1.010
0.100	0.993	1.007
0.090	0.994	1.006
0.080	0.995	1.005

LCL_N = 95% lower confidence limit for the arithmetic mean, based on a single eight-hour measurement and a normal distributional model.

LCL_{LN} = Corresponding 95% LCL for the arithmetic mean, based on the same measurement and a lognormal distributional model.

normal and lognormal lower confidence limit (LCLs) for the arithmetic mean, computed from a single eight-hour TWA exposure measurement and assuming that CV_T is known. The table shows that, for $CV_T = 1/3$, a 95 percent LCL based on a normal distribution is only 4.5 percent higher than a 95 percent LCL based on a lognormal distribution with the same CV_T. For any analytical method meeting the NIOSH accuracy standard (i.e., for $CV_T \leq 0.128$), the difference between normal and lognormal 95 percent LCLs is less than 1.1 percent. Since CV_T's for industrial hygiene measurements are usually below 0.10,[4] for practical purposes either a normal or a lognormal model could usually be used for measurement errors of full-period single sample data. Given this choice, however, the normal model is more convenient for compliance testing because its location parameter (μ, the arithmetic mean of x) is directly comparable to exposure standards based on health effects (e.g., OSHA PELs).[A] Under a normal model with known coefficient of variation (CV_T), exact statistical confidence limits for the arithmetic mean (μ) can be calculated by simple numerical methods, whereas approximate methods (using graphical decision charts) would be required under a lognormal model. References 1 and 2 give required statistical protocols, based on the normal distribution assumption, for making significance tests of compliance and for placing confidence limits on single-shift TWA exposure levels.

The normal distribution, as a model for measurement errors, can be rationalized theoretically in terms of a multistep measurement procedure which has independent, random, *additive* variations at each step.[5,6] Given many such steps, each with a small (or at least finite) error variance, the central limit theorem of mathematical statistics guarantees near-normality of the final measurements that result from the entire sampling and chemical measurement process.

Similarly, if the stepwise errors were independent, *proportional* (multiplicative) errors (as opposed to additive errors), the central limit theorem would suggest a lognormal distributional model of the x-values (concentrations).[5,7] Bennet and Franklin[6] note that the logarithmic normal model is the natural one to use for spectrographic analyses since the measured density of a spectral line is proportional to the logarithm of the amount of the given material present. However, even for this case, the normal

[A]The arithmetic mean of instantaneous exposure levels during an eight-hour period is the proper measure of central tendency to be compared to an eight-hour PEL. A TWA of consecutive full-period samples is an estimate of this arithmetic mean, which also corresponds to an air concentration based on a single full-period sample. The arithmetic mean air concentration is proportional to the inspired mass of the material in question during the sampled period. (The proportionality factor for a given worker is that worker's minute volume, i.e., respiratory rate × tidal volume.)

model is preferable for use in compliance/noncompliance hypothesis testing, provided that the GSD of the lognormal model (and hence the CV_T of the approximating normal model) are relatively small (e.g., GSD < 1.2, CV < 0.18, based on the authors' experience).

The relationship between the CV and the GSD of a lognormal distribution is given by Aitchison and Brown[7] as:

$$CV = [\exp(\ln^2 GSD) - 1]^{1/2}$$

Figure 2 shows the close agreement to be expected between a lognormal curve with GSD = 1.20 (CV = 0.184) and a normal curve with the same mean (μ = 10) and coefficient of variation (CV = 0.184). The agreement would be even better for actual exposure measurements since NIOSH-recommended measurement methods must have CV_T's below 0.128 in order to meet the NIOSH accuracy standard. Busch and Taylor[4] state that 90 percent of 310 methods evaluated by NIOSH were found to have CV_T's below 0.09. Leidel et al.[1] give an earlier listing of CV_T's for 189 methods, which shows only 16 methods above 0.09 and only one method (asbestos) above 0.13. Thus, the normal distributional model seems a useful and adequate choice for statistical analysis of individual occupational exposure estimates derived from full-period single samples.

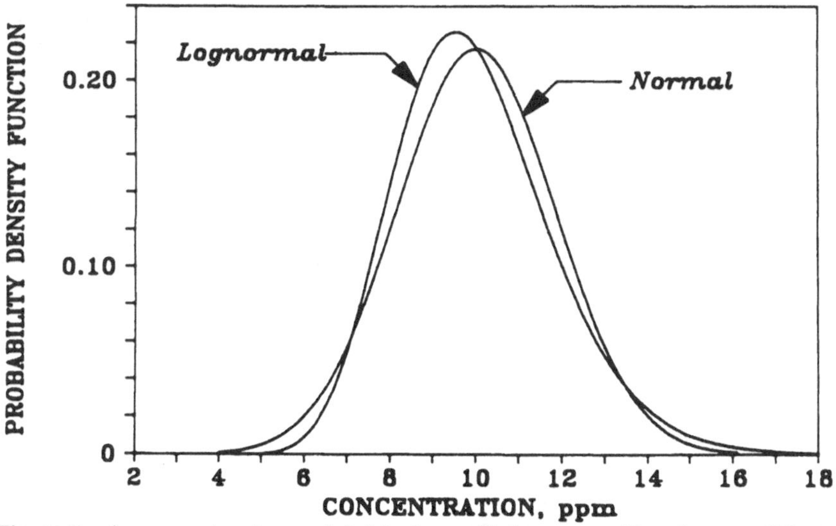

Figure 2. Lognormal and normal distributions with the same arithmetic mean (10 ppm) and coefficient of variation (CV = 0.184).

FULL-PERIOD CONSECUTIVE SAMPLES AND PARTIAL-PERIOD CONSECUTIVE SAMPLES

Means ($\bar{x}$) of consecutive samples taken over equal intervals are used as the eight-hour TWA estimates for these types of samples. The distributional model for averages of several consecutive exposure estimates during the same work shift is similar to the normal distributional model given for full-period single samples except that the coefficient of variation of the average ($CV_{\bar{x}} = CV_T/(n)^{1/2}$ replaces CV_T. (See Reference 1 for a more general formula applying to a TWA of consecutive samples with unequal intervals.) The normal curve is likely to be a good model for these types of TWA estimates because: 1) $CV_{\bar{x}}$ is smaller than CV_T (making for an even better normal approximation when x itself is lognormal), and 2) the central limit theorem applies more strongly since additional averaging has occurred in the (TWA) assay process. (Note: Reference 1 explains that use of partial-period sampling requires making an additional strong assumption about the nature of the production process itself, namely that the TWA exposure for the unsampled portion of the work shift is equal to the TWA exposure for the entire work shift).

ONE-WORKER, MULTIPLE-TWA DATA SET

If it were practical to implement random sampling of work shifts, this method would provide theoretical advantages over systematic sampling for estimation of parameters of an employee's long-term exposure distribution (i.e., over many months or years). (An example of systematic sampling would be to measure exposure during the first shift of every fourth week). Cochran[8] points out two problems of systematic sampling. 1) It may be in phase with periodicities in the exposure pattern. This could yield a biased estimate of the long-term mean, unless the day of the week for the starting shift were changed every so often. 2) There is no trustworthy method for using the sample data to estimate the precision of the mean. Cochran finds systematic sampling satisfactory if: a) the population is randomly ordered (i.e., shift-to-shift variability is nonperiodic, so that a systematic sample is effectively a random sample), or b) systematic samples are taken within numerous strata, artfully selected in such a way that hidden periodicities are canceled.

For some epidemiological studies, the population of TWA exposure levels that are of interest for each worker is all work shifts within a working lifetime. For such an exposure population, random sampling is only a theoretical concept, but it may be possible to sample randomly from a "pseudo-population" that is assumed to be representative of the entire population. For example, if a certain three-month period could be assumed to be fully representative of the exposure distribution for the entire working lifetime, a random sample of 10 shifts from the 65 or so shifts during that quarter would be interpreted as a random sample of the working lifetime. But a more realistic interpretation of such a sample is that is represents the *hypothetical* exposure which would be expected during a working lifetime *if* the worker continued to do the same work, under the same conditions that existed during the quarter actually sampled. Relevant conditions include such determinant factors[2] as production levels, seasonal environmental conditions, and level of expertise of the employee insofar as it affects his/her ability to avoid contamination of the working environment and/or ability to avoid exposure to a contaminated environment.

The preceding paragraph restricts the ability to extrapolate exposure data from a *simple random sample* of work shifts (during a limited period) to a working lifetime of eight-hour TWA exposure levels. It applies as well to other types of repeated-shifts sampling strategies (such as systematic sampling and stratified random sampling). No further discussion will be given to comparative properties of sampling strategies *per se* because the topic of this paper is statistical models for the resulting sampling data, not selection of the sampling strategy itself. Further information about the general statistical properties of survey sampling strategies has been available for decades.[8,9] Recent publications are also available dealing specifically with occupational health exposure sampling of repeated work shifts.[2,10]

A lognormal statistical data model is given below in this section which accounts for shift-to-shift exposure changes. The model has enough generality so that it can be adapted to sampling data from any of the aforementioned specific sampling strategies. The model has been used by Leidel et al.[1,2] as a basis for analytical statistical methodology to obtain the types of point estimates and interval estimates discussed in the *General* section of this chapter. A detailed lognormal model for shift-to-shift variability that was developed by Leidel, Busch, and Crouse[11] is the basis for the model to be given below. It is a mixture of lognormally distributed interday and intraday proportional variations of true exposure levels, along with normally distributed, additive measurement errors. The latter errors of measurement are assumed to follow the model given in the Full-Period Single Sample section of this chapter. Figure 3 depicts the general nature of the model. The asymmetrical curve is an example of a lognormal distribution of interday variations of true eight-hour TWA exposure levels, with GM = 8.41 ppm and GSD = 1.8. For the measure-

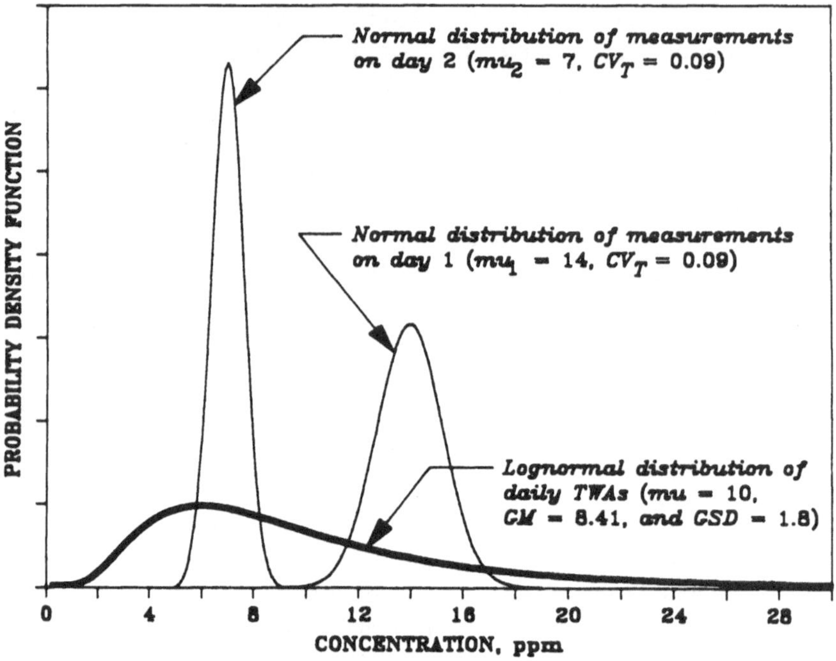

Figure 3. Examples of a lognormal distribution for true eight-hour TWA exposures and normal distributions of TWA exposure measurements on two particular days.

ment method, $CV_T = 0.09$ was used as an example, and related normal distributions of measurement variations are shown for two days (with randomly selected true TWA exposure levels of $\mu_1 = 14$ ppm ($\sigma_1 = 1.26$ ppm) and $\mu_2 = 7$ ppm ($\sigma_2 = 0.63$ ppm)).

The possible use of a lognormal distribution to approximate a normal distribution of measurement errors was discussed in the Full-Period Single Sample section, and such an approximation will also be used in the present model. This is even more defensible here than for full-period single-shift samples since measurement errors now contribute a minimal fraction of the total day-to-day variability of the exposure results. It was stated previously that measurement errors usually have CVs below 0.09 (equivalent to GSDs below 1.094 for approximately lognormally distributed assay errors). Such measurement errors are quite small relative to shift-to-shift variability of the exposure levels themselves. Ayer and Burg[12] compiled 105 GSD estimates for monitoring data on particulates and the median GSD was 1.65. Leidel et al.[11] compiled 59 GSD estimates for gases and vapors monitored in NIOSH health hazard evaluation studies and these had a median value of 1.55. For both data sets, over 20 percent of GSDs exceeded 2.0. A GSD of 2.0 is equivalent to a CV of 0.79, which far exceeds CVs for measurement error (0.09 or less). For such large GSDs the normal distribution cannot be used as an adequate model. TWA exposure measurements during successive shifts usually have obviously skewed-right distributions and numerous examples are given in references 2 and 11 which show applicability of a lognormal model.

The data model sought in this section, for eight-hour TWA exposure measurements on a given employee during randomly selected shifts, can be derived from the following more detailed model given in reference 11:

$$\begin{aligned} x_{ijk} &= \mu d_i e_{(i)j} + a_{(ij)k} \\ &= \mu d_i e_{(i)j} v_{(ij)k}, \text{ where} \\ v_{(ij)k} &= 1 + [a_{(ij)k}/(\mu d_i e_{(i)j})] \end{aligned}$$

where:

μ = long-term mean of eight-hour TWAs. The process is assumed to be stationary so that exposure levels exhibit no trends or cycles over time.

d_i = factor for the ith randomly selected day. The true mean for the ith day is denoted by $\mu_i = \mu d_i$.

$e_{(i)j}$ = factor for the jth randomly selected sampling period within the ith day. The true mean for the jth period of the ith day is denoted by $\mu_{ij} = \mu d_i e_{(i)j}$.

$a_{(ij)k}$ = differential total error (net) of sampling and analysis for the kth sample taken during the jth period of the ith day; $a_{(ij)k}$ has a coefficient of variance CV_T (discussed above).

The multiplicative form of the x_{ijl} model contains three error factors (d_i, $e_{(i)j}$, and $v_{(ij)k}$) which are treated as independently distributed lognormal variables.[B] Geometric means (GM) and geometric standard deviations (GSD) of these variables bear the following relationships to each other.

$$\begin{aligned} GM_d &= \exp[(-1/2)\ln^2 GSD_d] \\ GM_e &= \exp[(-1/2)\ln^2 GSD_e] \\ GM_v &= \exp[(-1/2)\ln(1 + CV_T^2)] \\ GSD_v &= \exp[(\ln(1 + CV_T^2))^{1/2}] \end{aligned}$$

[B]An unbracketed subscript is considered to be cross-classified with other unbracketed subscripts, but is not cross-classified with bracketed subscripts. That is, unbracketed subscripts vary *within* fixed levels of bracketed subscripts. Thus, the denominator of the fractional part of $v_{(ij)k}$ is fixed and the distribution relates only to the single subscript k.

Given this lognormal model for x_{ijk}, it is easy to derive a corresponding lognormal model for single assays of eight-hour samples, i.e., for assays $x_{i.1}$ ($i = 1,2 \ldots$). The data model is:

$$x_{i.1} = \mu d_i + a_{(i)1} = \mu d_i v_{(i)1}, \text{ where}$$
$$x_{i.1} = 1 + (a_{(i)1}/\mu d_i)$$

Terms μ and d_i are defined as before. The factor for relative assay error ($v_{(i)1}$) is approximately lognormally distributed with the same GM and GSD as given earlier for $v_{(ij)k}$.

Since the product of the two independently distributed lognormal variables ($d_i\ v_{(i)k}$) is itself lognormally distributed, the *measured* daily TWA exposures ($x_{i.1}$) are lognormally distributed with parameters:

$$GM = \mu \exp\left[-{}^1\!/_2\ (\ln^2 GSD_d + \ln(1 + CV_T^2))\right]$$
$$GSD = \exp\left[(\ln^2 GSD_d + \ln(1 + CV_T^2))^{1/2}\right]$$

Thus, the joint distribution of independently distributed variations have been "collapsed" (due to assay errors and day-to-day true exposure variations) into a conventional (univariate) lognormal distribution of measured daily TWA exposures. As an aside concerning proper statistical analytical methods under this model, note that Student's t distribution can be used to obtain standard confidence limits for the GM. However, it is the arithmetic mean (μ), not the GM, which should be compared to exposure limits in significance tests for compliance or noncompliance. To do this, the special decision charts supplied in references 1 and 2 can be used.

For the case of n, full-period consecutive samples taken during a randomly selected shift, the mean for day i, $x_i \ldots = (1/n)(x_{i11} + x_{i21} + \ldots + x_{in1})$, is a daily TWA exposure estimate which is approximately lognormally distributed (over shifts i) with parameters:

$$GM = \mu \exp\left[(-{}^1\!/_2)(\ln^2 GSD_d + \ln(1 + CV_T^2/n))\right]$$
$$GSD = \exp\left[(\ln^2 GSD_d + \ln(1 + CV_T^2/n))^{1/2}\right]$$

Note: These parameters are similar to those for full-period single samples taken on repeated shifts, except that CV_T^2/n has replaced CV_T^2.

GRAB SAMPLES DURING A SINGLE SHIFT FOR A GIVEN EMPLOYEE

A statistical data model for grab samples taken at random intervals during a given shift (i) can be obtained by replacing μd_i by μ_i in the previous model for x_{ijk}. The term μ_i is now a parameter (i.e., constant) for day i, not a random variable, and inferences are limited to exposure levels within day i. That is, the resulting lognormal model is relevant to compliance/noncompliance testing of single-shift PELs. The data model for grab samples is:

$$x_{(i)jl} = \mu_i e_{(i)j} + a_{(ij)l} = \mu_i e_{(i)j} v_{(ij)l}, \text{ where}$$
$$v_{(ij)l} = 1 + (a_{(ij)l}/\mu_i e_{(i)j})$$

$x_{(i)jl}$ is then (approximately) lognormally distributed with parameters:

$$GM = \mu_i \exp [(-1/2)(\ln^2 GSD_e + \ln(1 + CV_T^2))]$$
$$GSD = \exp [(\ln^2 GSD_e + \ln(1 + CV_T^2))^{1/2}]$$

As was the case for significance tests relative to μ, the special decision charts given in references 1 and 2 are needed for significance tests relative to μ_i.

MULTIWORKER, SINGLE-TWA DATA SET

A lognormal distribution is the preferred model for exposure measurements on different employees doing similar work under similar conditions. The data model can be written in terms of proportional exposure factors $w_{(i)j}$ for different workers. The $w_{(i)j}$ factor for the jth worker during shift i multiplies the average area concentration (μ_i) which exists during that shift. The data model is:

$$x_{(i)jl} = \mu_i w_{(i)j} + a_{(ij)l} = \mu_i w_{(i)j} v_{(ij)l}, \text{ where}$$
$$= 1 + (a_{(ij)l}/\mu_i w_{(i)j})$$

The product $w_{(i)j} v_{(ij)l}$ is lognormally distributed so that $x_{(i)jl}$ is lognormally distributed with parameters:

$$GM = \mu_i \exp [(-1/2)(\ln^2 GSD_w + \ln(1 + CV_T^2))]$$
$$GSD = \exp [(\ln^2 GSD_w + \ln(1 + CV_T 2))^{1/2}]$$

Such a lognormal distribution can be rationalized in terms of a multiplicity of factors having independent proportional effects on a given employee's exposure levels. Such effects would cause a given worker's exposure to have a characteristic percentage difference from the general average exposure level at any given time. For example, an individual's typical work habits and the location of his work station (distance from emission source, air flow patterns, etc.) may induce exposure differences (ppm) which are proportional to pollution source emissions.

PROPORTION OF WORKER-SHIFTS WITH HIGH TWA EXPOSURES

The following statistical question is sometimes asked: "What is the proportion (P) of worker-shifts with exposures above the PEL?" Given appropriate exposure monitoring data (discussed further below), the proportion P can be estimated. However, such a proportion cannot necessarily be interpreted as any given worker's personal probability of being overexposed. Each worker has his/her own distribution of shift-to-shift exposure levels, and a particular worker's probability of being overexposed (i.e., the proportion of that worker's TWA exposure levels above the PEL) could be above or below the long-term proportion (of worker-shifts with TWAs above the PEL) for the entire workforce. The data models given earlier will now be manipulated to provide an estimate of P, but caution is advised to interpret this proportion carefully in the context of the appropriate exposure environment and worker population to which it applies.

By omitting assay error from a combination of the data models given in the One-Worker, Multiple TWA Data Set and the Multiworker, Single TWA Data Set sections, a model for the *true* exposure level of a randomly selected worker (i) assigned to a randomly selected workshift (j) can be obtained:

$$\mu_{ij} = \mu d_i w_j$$

Exposure levels μ_{ij} are then lognormally distributed with parameters:

$$GM = \mu \exp[(-1/2)(\ln^2 GSD_d + \ln^2 GSD_w)]$$
$$GSD = \exp[(\ln^2 GSD_d + \ln^2 GSD_w)^{1/2}]$$

The desired proportion P is the area under this lognormal distribution to the right of μ_{ij} = PEL. To find this area, first compute the standard normal deviate (z-value) which corresponds to PEL:

$$z = [\ln(PEL) - \ln GM]/(\ln GSD)$$

The proportion P can then be obtained from a published table of right tail areas under the standard normal distribution (P = area to the right of z).

In order to *estimate* P from a random sample of n worker-shift exposure *measurements*, $x_{ij} = \mu d_i w_j + a_{(ij)l}$, it will be necessary to obtain estimates of the lognormal parameters, denoted (GM)^ and (GSD)^ by "deducting measurements error" (see below) from values (GM_{meas})^ and (GSD_{meas})^ calculated from the n exposure measurements.

$$(GM_{meas})\hat{} = \exp[(1/n) \sum (\ln x_{ijl})]$$
$$(GSD_{meas})\hat{} = \exp[((1/(n-1)) \sum (\ln x_{ijl} - \ln (GM_{meas})\hat{}\ )^2)^{1/2}]$$

This adjustment is needed because errors of sampling rates and chemical analysis cause the "spread" (GSD) of the exposure-level measurements to be greater than the actual "spread" of the true exposure levels. The summations, denoted by operator Σ, extend over the n combinations of i and j which occurred in the random sample.

Adjusted parameters are then obtained as follows:

$$(GM)\hat{} = \exp\ [\tfrac{1}{2} \ln\ (1 + CV_T^2)] GM_{meas}$$
$$(GSD)\hat{} = \exp\ [(\ln^2 GSD_{meas} - \ln\ (1 + CV_T^2))^{1/2}]$$

As before, CV_T denotes the coefficient of variation of measurement errors.

An estimate of P is obtained by calculating an estimated z value:

$$\hat{z} = [\ln(PEL) - \ln GM]/(\ln GSD)$$

and obtaining P as the area under a standard normal curve to the right of $\hat{z}$.

Statistical methods are given in references 1 and 2 for obtaining confidence limits for P or tolerance limits for μ_{ij}. The methods depend on having a random sample of n worker-shifts in which only one worker is sampled per shift. Each x_{ijl} exposure measurement in the sample then reflects independent sampling variations from both the d_i and w_j lognormal distributions. If the same workers were sampled repeatedly, measurements of a given worker would be intercorrelated since their variability would reflect only d_i (shift-to-shift) sampling variations. Similarly, the variability of exposures for several workers sampled during the same shift would reflect only w_j (worker-to-worker) sampling variations. Such data can be treated by methods of variance components analysis in order to construct estimates of GM and GSD parameters which reflect the desired net sampling variations ($d_i w_j$), but to apply such methods would usually require assistance from a professional statistician.

AUTOCORRELATION BETWEEN SUCCESSIVE SAMPLES

When successive samples are taken at the same location, samples taken close together in time may yield results more similar than samples taken farther apart. In this case, there would be a nonzero correlation coefficient between successive samples, referred to as a first order autocorrelation coefficient. The distributional models presented so far in this chapter have not included any provision for autocorrelation. In fact, no autocorrelation exists for full-period single samples or for full-period consecutive samples. In these cases, variability in the true concentrations during partial periods need not be included in the data model for eight-hour TWAs because such variability does not affect the precision of the TWA exposure estimate. This is true whether or not the true partial-period concentrations are autocorrelated.

However, for repeated-shifts samples, ignoring any substantial autocorrelation can compromise the accuracy of statistical confidence limits computed for the long-term mean. Confidence limits would be computed from an empirical variance of samples taken at different times, and ignoring autocorrelation causes negative bias in this estimated variance. Therefore, the result of ignoring autocorrelation would be to produce too-narrow confidence limits and a resulting increased chance of false positive results of significance tests. Snedecor and Cochran[9] point out that the most effective precaution against inferential mistakes due to unknown autocorrelation is the skillful use of randomization and "blocking" while collecting the samples. Even though autocorrelation would be evident if evenly spaced systematic samples were taken, it would probably have little effect on the results of data analyses if sampling had been done at random intervals. If periodic (cyclical) changes occur in exposure levels, sampling performed at evenly spaced intervals that are in phase with the exposure cycles would yield data showing *no* autocorrelation. In this case, the sample mean would be biased and the data would not allow the precision of the mean to be determined.

When systematic sampling must be used, a method of time series analysis can be employed wherein the autocorrelation coefficient is estimated and used to adjust the empirical variance estimate. Box, Hunter, and Hunter[13] give the following formula for the variance of the mean of n values of x taken from a time series with a first-lag autocorrelation coefficient equal to w.

$$V(\bar{x}) = (\sigma^2/n)[1 + (2\,w1(n - 1)/n]$$

As an example, for n = 10 the formula shows that an autocorrelation coefficient of only 0.3 would cause the true standard error of $\bar{x}$ to be 24 percent higher than the value $\sigma/(n^{1/2})$ which would apply if autocorrelation were ignored. Bennett and Franklin[6] point out that observations spaced more widely in time tend to give a more accurate estimate of the true variance because they have lower autocorrelation.

Thus, a long-term average exposure estimate calculated from widely spaced daily TWA exposure measurements can probably be reliably statistically analyzed using methods based on an assumption of independent random variations. However, if trends or cycles exist in data from systematic samples, a statistical model can be used which includes linear or periodic regression effects. Alternatively, "blocking" could be used to compute separate exposure averages within periods when trends or cycles are absent. A weighted average of block exposure levels can then be used to estimate the long-term average, and an appropriate weighted variance estimate can be used to compute confidence limits or significance tests. For these more complicated data sets, the assistance of a professional statistician would probably be needed.

ACKNOWLEDGMENTS

The authors gratefully acknowledge the excellent comments and suggestions made by three NIOSH reviewers: Dennis D. Zaebst, Randall J. Smith, and Peter M. Eller, Ph.D.

REFERENCES

1. Leidel, N.A., K.A. Busch and J.R. Lynch: *NIOSH Occupational Exposure Sampling Strategy Manual.* DHEW (NIOSH) Pub. No. 77-173 (1977).
2. Leidel, N.A. and K.A. Busch: *Statistical Design and Data Analysis Requirements*, Chapter 8, pp. 395-507. *Patty's Industrial Hygiene and Toxicology*, 2nd ed., Vol. 3A, *Theory and Rationale of Industrial Hygiene Practice—The Work Environment.* L.J. Cralley and L.V. Cralley, Eds. John Wiley and Sons, Inc., New York (1985).
3. Busch, K.A.: Appendix A. SCP Statistical Protocol. *Documentation of the NIOSH Validation Tests*, pp. 7-12. D.G. Taylor et al., Eds. DHEW (NIOSH) Pub. No. 77-185 (April 1977).
4. Busch, K.A. and D.G. Taylor: Statistical Protocol for the NIOSH Validation Tests, Chapter 31. *Chemical Hazards in the Workplace—Measurement and Control.* G. Choudhary, Ed. ACS Symposium Series No. 149. American Chemical Society, New York (March 1981).

5. Hald, A.: *Statistical Theory with Engineering Applications*, p. 164. John Wiley and Sons, Inc., New York (1952).
6. Bennett, C.A. and N.L. Franklin: *Statistical Analysis in Chemistry and the Chemical Industry*. John Wiley and Sons, Inc., New York (1954).
7. Aitchison, J. and J.A.C. Brown: *The Lognormal Distribution*. Cambridge at the University Press, Cambridge, Great Britain (1957).
8. Cochran, W.G.: *Sampling Techniques*. John Wiley and Sons, Inc., New York (1953).
9. Snedecor, G.W. and W.G. Cochran: *Statistical Methods*, 6th ed. The Iowa State University Press, Ames, Iowa (1967).
10. Esmen, N.: A Problem of Exposure Assessment: Is Random Sampling Appropriate? (Draft). Presented at the Workshop on Strategies for Measuring Exposure, Dec. 9–10, 1986, Washington D.C. Sponsored by Chemical Manufacturers Association, American Industrial Hygiene Association and American Petroleum Institute.
11. Leidel, N.A., K.A. Busch and W.E. Crouse: *Exposure Measurement Action Level and Occupational Environmental Variability*. DHEW (NIOSH) Pub. No. 76-131 (December 1975).
12. Ayer, H. and J. Burg: Time-weighted Average vs. Maximum Personal Sample. Presented at the American Industrial Hygiene Conference, Boston (1973).
13. Box, G.E.P., W.G. Hunter and J.S. Hunter: *Statistics for Experimenters*. John Wiley and Sons, Inc., New York (1978).

CHAPTER 23

Biological Considerations for Designing Sampling Strategies

STEPHEN M. RAPPAPORT

School of Public Health, University of California, Berkeley, California 94720

INTRODUCTION

It is now well accepted that workers' exposures to airborne chemicals vary greatly over time. This variability obviously complicates the process of assessing exposures;[1] for example, it no longer appears appropriate for the industrial hygienist to base decisions concerning a working lifetime upon only a few measured air concentrations. Rather it seems that a premium is often placed upon gathering sufficient relevant data to characterize exposure distributions received over time. The parameters of these distributions can then be used in a variety of useful ways to quantify, classify, and limit exposures.

This process of characterizing variable exposures to a particular toxicant with reference to appropriate hygienic standards is often referred to as a "sampling strategy." Here, the assumed "sample" is considered in the statistical sense where the number of measurements as well as the timing and duration are selected *a priori* to allow appropriate inferences to be drawn. Indeed, most recent work in this area has focussed upon the many statistical issues which surround the estimation of parameters and exceedances, the testing of hypotheses, and the proper classification of worker groups.[2-10]

However, statistical methods are of dubious value if they do not relate to important biological phenomena which dictate the risks of disease. These biological factors and risks are often assumed to lie in the exclusive domain of toxicologists, epidemiologists, and physicians. Yet, it is probably the industrial hygienist, working at the interface between process and exposure, who is best able to evaluate and reduce health risks. Thus, it would seem that hygienists should weigh the physiological consequences of exposure at least as heavily as those of control and compliance when they select sampling strategies; this is particularly true regarding chronic toxicity where exposure is assessed over long periods. Unfortunately, the strategies most commonly employed by hygienists today cannot be clearly related to the risk of disease.[1,5]

The purpose of this chapter is to investigate the strategic implications of exposure variability as a potential determinant of health risk. The following questions will be addressed:

1. What are the mechanistic linkages between exposure and health risk, and how might they be influenced by exposure variability?
2. How might exposure variability dictate the evaluation of exposures to acutely toxic chemicals?
3. What is the likely impact of short-term (< 15 min) variation in exposure upon chronic health risk?
4. What is the likely impact of day-to-day variation in exposure upon chronic health risk?

This is only a partial list of relevant biological questions which could be discussed. It is hoped that the following analyses will close some of the gaps in current knowledge and suggest promising avenues for future inquiries.

EXPOSURE-RESPONSE

A conceptual model has been developed which organizes the functional relationships between the exposure distribution and an individual's risk of disease.[11] The model, shown in Figure 1, defines the processes of exposure, dose, damage, and risk as functions of time. The input is the air concentration inhaled by the worker, which can be considered in terms of either the continuous function, C(t), or the discrete series, (C_t), inhaled at regular intervals. (The latter function is perhaps more useful for considering particular occupational exposure regimens and will be used here.) It is assumed that (C_t) is a stationary process, i.e., its mean, variance, and autocovariance functions are all constant and independent of time over the period of interest. The time scale between exposure and

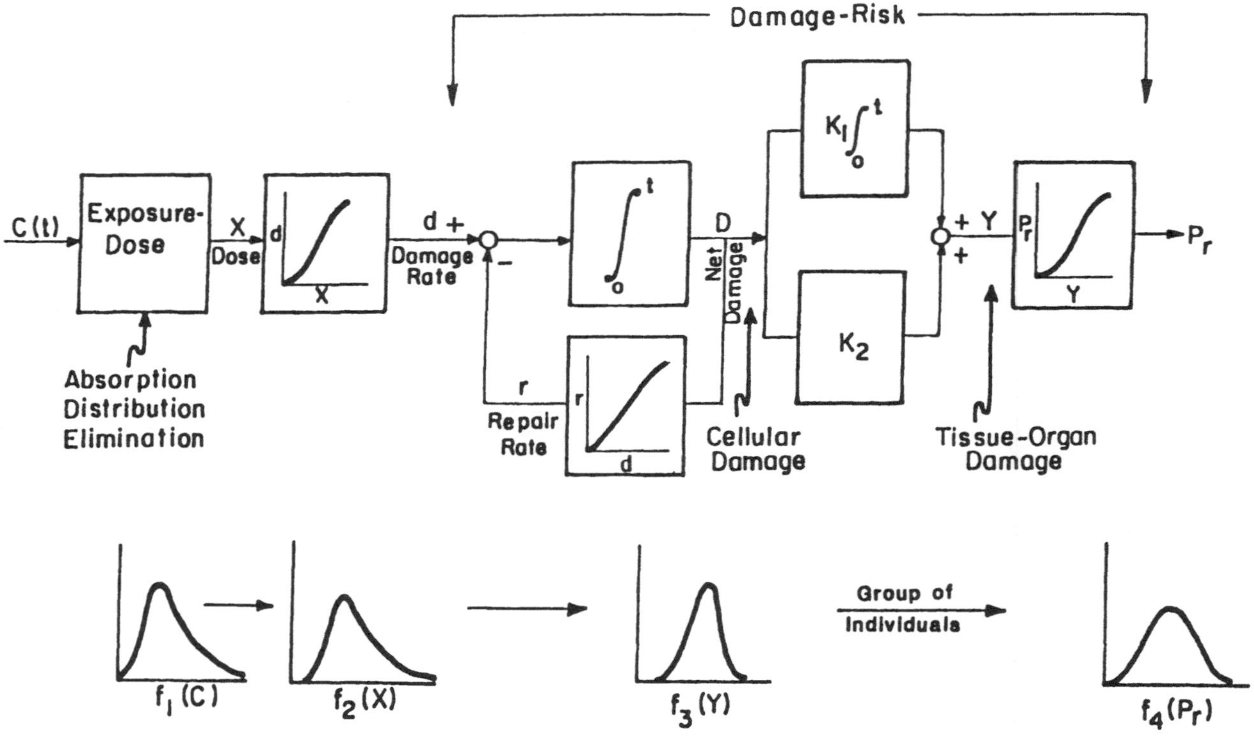

Figure 1. A conceptual model relating exposure, dose, damage, and risk of disease for an individual worker or uniformly-exposed group of workers. The distribution functions shown at bottom represent the following: exposure [$f_1(C)$], dose or burden [$f_2(X)$], tissue-organ damage [$f_3(Y)$], and risk to a group of uniformly-exposed workers [$f_4(P_r)$].

response can range from a few minutes for certain acute toxicants to many years in cases of chronic poisoning.

Some portion of the chemical or its toxic metabolite is distributed to target tissues; the result is a series of biologically-effective doses or burdens, (X_t). This time series approaches a stationary process after some necessary period has elapsed, the length of which depends upon the elimination rate of the toxic species. Then there are only transient deviations from the mean burden at the target site. X_t is the argument of a continuous dose-response function which describes the rate, d, at which molecular or cellular damage is generated.

The integral of d over time less that of the rate of repair, r, yields the net damage, D_t, at any time. After sufficient time elapses given the value of r, (D_t) also becomes stationary; then there are only transient deviations from the mean number of damaged cells. In some cases, the likelihood of a clinical response is directly proportional to this level of cellular damage, for example, when considering acute pulmonary edema or chronic fibrosis. In others, such as with carcinogens, the probability of response is related to the period of time that the mean level of damage is maintained. The model allows for both situations and, in either case, the variable, Y_t, is the amount of damage accumulated at the tissue-organ level.

Y_t is the argument of an analog of the quantal dose-response function the output of which is P_r, referred to as "individual risk." This represents the health risk borne by a worker given some genetic predisposition for the disease. A group of workers exposed to the same distribution of air levels would exhibit a distribution of P_r due to chance and because of differences in the various physiologic rates and genetic factors among the individuals.

ACUTE TOXICANTS

The above model clarifies some of the biological issues involved in selecting a sampling strategy. Consider first the situation involving acute toxicity or severe irritation. Here, the rate constants for biological elimination and damage are in the range of seconds to minutes; thus, the transfers between exposure and tissue damage proceed within the time scale of events during a single shift. This indicates that exposure variability during brief periods on a given day cannot be ignored; indeed, this variation is likely to be the primary determinant of acute response.

The industrial hygienist intuitively recognizes the importance of brief exposures to acute toxicants since problems typically arise from "unusual" events, such as a spill or leak. Unfortunately, rare events are difficult to measure since one must be present when they occur. Thus, in order to assess the likelihoods that acute episodes may occur, the hygienist is often forced to make inferences by extrapolation from the body of the exposure distribution into the right tail.[2,12] This extrapolation requires large sample sizes and depends critically upon assumptions concerning the independence of measurements, the stationarity and continuity of the exposure series, and any parametric models which might have been used. Thus, the hygienist is in a position where considerable resources must be expended merely to determine whether exposures might be excessive under some set of possibly unknown circumstances. Even after concluding that controls are not indicated (because the frequency of large transients appears to be small), one remains uncomfortable given the health consequences associated with accidental release.

Perhaps it would be more sensible as a general rule to forego the assessment of exposure in cases involving acute toxicity. It seems preferable to either invest monitoring resources in sensors which can continuously measure source concentration and thereby warn workers of excessive releases or automatically adopt control measures to contend with particular chemicals, processes, and work practices. In either case, the strategic focus moves away from exposure assessment *per se* to source assessment and control.

CHRONIC TOXICANTS

The model shown in Figure 1 also defines the strategic implications of long-term exposure to chemicals which are primarily associated with chronic diseases. The issue is to decide when the expected value of the exposure series (C_t) should be a sufficient predictor of individual risk P_r to justify ignoring the variance and higher moments of the exposure distribution.[13] Then exposure variability is only important insofar as it affects the uncertainty with which the mean level is estimated. This is a statistical rather than a biological issue and can be dealt with fairly easily.[6]

However, if P_r is not strictly proportional to the mean exposure, then evaluation and control of the variance as well as the mean level becomes necessary. This raises interesting biological questions (what physiological or biochemical factors affect transient exposure-damage relationships?) as well as queries pertaining to statistical methodology (how can the mean and variance be evaluated simultaneously?) and control measures (how can exposure variability be reduced?). The biological issue addressed by this chapter concerns *a priori* classification of toxicants into groups for which exposure variability is either likely or unlikely to be an important determinant of risk.

Logically, it would appear that variability in exposure can influence chronic health risk only to the extent that the following two conditions are met: 1) variability is transmitted from the air environment to the target site, and 2) nonlinear kinetics are involved in the transfers of exposure to damage. The first condition reflects the physiological dampening of exposure variability resulting at the target site from accumulation of either burden or damage over many exposure events.[11,14-18] Given sufficient dampening, then risk necessarily becomes insensitive to variation in exposure. This will be discussed more extensively later.

The second condition relates to the situation where exposure variability is not dampened at the target site. Then, in the context of Figure 1, variability becomes important insofar as damage D_t is proportional to C_t^a, where $a > 1$ due to saturable detoxification.[19-23] That is, the continuous dose-response relationship between X and d at a particular time is concave upwards. Unfortunately, relevant empiric evidence of such nonlinear kinetics associated with transient behavior of inhaled toxicants is extremely rare. Thus it is difficult at this time to determine the extent to which such behavior may occur.

PHYSIOLOGICAL DAMPENING OF EXPOSURE VARIABILITY

Since exposure variability must be transmitted through the subsequent dose and damage compartments in order to condition the risk of disease, it is useful to define quantitative relationships between the variance of (C_t) and those of succeeding series. Of the various linkages involved, perhaps the one which best lends itself to this type of analysis is that of the relationship between exposure C_t and burden X_t. Here, the series of exposures (C_t) becomes the input to a compartmental model of chemical disposition, the output of which is the resulting series of burdens (X_t).

The relationships between the means and variances of the stationary series (C_t) and (X_t) were derived by Roach for a single-compartment linear model.[14] Then, the series (X_t) is obtained from the equation:

$$X_t = (KC_t/k)(1 - e^{-kt_i}) = X_{t-1}e^{-kt_i} \quad (1)$$

where: K = the uptake constant (m^3/hr)
k = the first-order biological elimination constant (hr^{-1})
t_i = the interval between observations (hr)

(Biological elimination can also be described in terms of the elimination half-time, $T_{1/2} = 0.693/k$). Roach showed that if the series (C_t) is random and uncorrelated, with mean μ_c and variance σ_c^2, then the mean and variance of (X_t) are respectively:

$$\mu_x = \mu_c K/k \tag{2}$$

and

$$\sigma_x^2 = \sigma_c^2 \, [K(1 - e^{-kt_i})/k(1 + e^{-kt_i})]^2 \tag{3}$$

He noted that as k becomes small, reflecting slow biological elimination of toxicant, the variance of the series of burdens, σ_x^2, decreases relative to that of exposure, σ_c^2, due to accumulation of chemical in the body.

This dampening behavior is demonstrated in Figure 2, which depicts the fluctuations in C_t and X_t about their mean values when $\mu_c = 1$ mg/m³, $\mu_x = 2.9$ mg, K = 1 m³/hr, k = 0.347 (hr⁻¹) ($T_{1/2} = 2$ hr), and $t_i = 0.25$ hr. Values of C_t were derived from a lognormal distribution with $\sigma_c = 1$ mg/m³ and zero autocovariance; initial values of C_t and X_t were set equal to the respective means. The figure clearly illustrates that the relative variation in burden at intervals of 15 minutes is much less than that of the input concentration when $T_{1/2} = 2$ hr.

Roach characterized this physiological dampening of exposure variability in terms of the ratio of the coefficients of variation of burden ($CV_x = \sigma_x/\mu_x$) and of air concentration ($CV_c = \sigma_c/\mu_c$), respectively, which, from equations 2 and 3 is:

$$1/A(t_i) = CV_x/CV_c = [(1 - e^{-kt_i})/(1 + e^{-kt_i})]^{1/2} \tag{4}$$

The dimensionless factor $1/A(t_i)$ varies between zero, which indicates great dampening of exposure variability, and one, indicating no dampening in the body. The example shown in Figure 2 has a value of $1/A(0.25) = 0.21$, indicating that only about 21 percent of the purely random variation of instantaneous exposure concentrations, obtained at intervals of 15 minutes, is transmitted to the tissues (when the elimination half-time is 2 hours).

Since the dampening factor is a function of both the interval between observations and the biological elimination constant, it is useful to standardize the interval t_i so that dampening can be quantified in terms of the k (or $T_{1/2}$) values for individual chemicals. Two scenarios appear important in this context, i.e., one involving the dampening of variation in instantaneous air concentrations noted above when the interval between observations $t_i \leq 0.25$ hr (as regarding short-term exposure limits) and

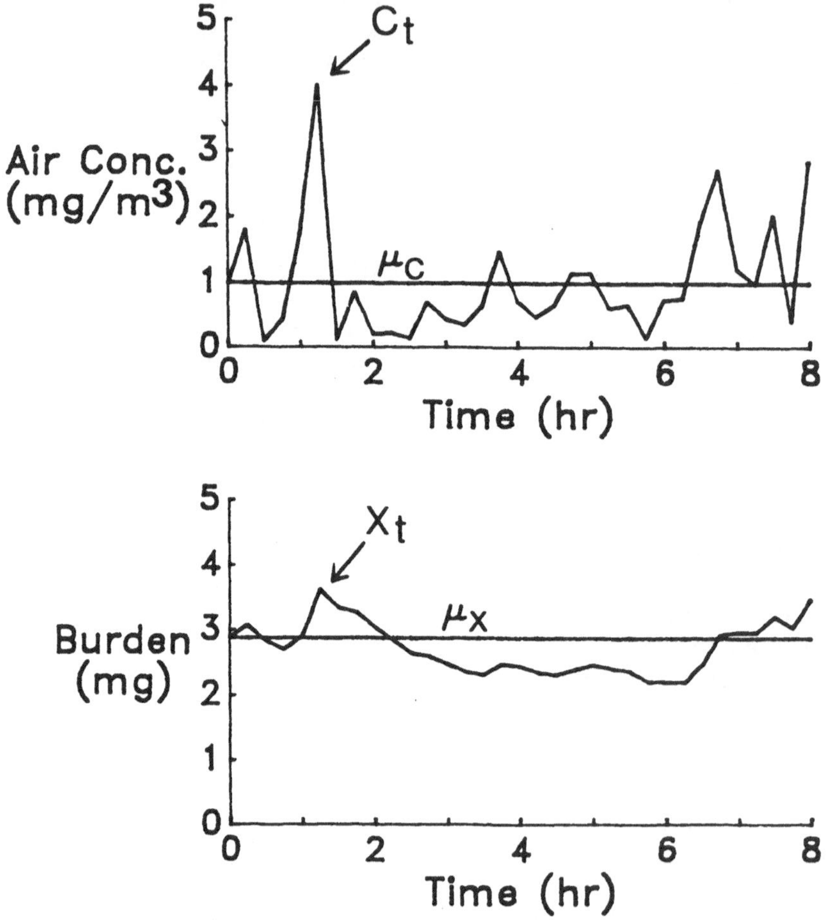

Figure 2. Series of 15-minute exposures (top) and burdens (bottom) resulting from 32 random uncorrelated values of C_t from a lognormal distribution with mean and standard deviation equal to 1 mg/m^3. The corresponding values of the burden X_t are derived from equations 1–3 in the text for a single-compartment linear toxicokinetic model when the elimination half-time $T_{1/2}$ is 2 hours and the uptake constant K is 1 m^3/hr.

the other involving variability in eight-hour time-weighted average (TWA) air concentrations from day to day.

Short-Term Variation

The relationship given in equation 4 for the dampening factor $1/A(t_i)$ was derived for the situation where short-term exposures are random and uncorrelated. Yet, it has been noted that serial air concentrations over time scales of minutes to hours tend not to be purely random but rather to be autocorrelated due to the episodic nature of events or practices which produce exposures.[7,16,24] This is more consistent with interpretation of (C_t) as a first-order autoregressive process where the current value of C_t is equal to a weighted fraction of the immediate past value plus a random shock,[25,26] i.e.:

$$C'_t = w(C'_{t-1}) + Z_t \tag{5}$$

where: C'_t and C'_{t-1} = the instantaneous deviations from the mean value μ_c at consecutive times
w = the first-lag autocorrelation coefficient
Z_t = derived from a purely random process with mean zero

The physical interpretation of (Z_t) is that of a series of random perturbations in the breathing-zone concentration arising from several variables including the source of the contaminant, the rate of ventilation, and the mobility of the worker.

Assuming that the transport of contaminant into and out of the breathing zone is described by a first-order process such as turbulent diffusion, then $w = e^{-bt_i}$, where b (hr^{-1}) is the air exchange rate constant. This exponential model is consistent with previous analyses which considered the constant b in the context of air exchange in the workroom as a whole.[7,16,24] By focussing here upon the air-exchange rate in the breathing zone, the problem of worker mobility is made more tractable.[26]

Given this diffusive-transport model, the dimensionless dampening factor becomes:

$$1/A(t_i) = \left[\frac{(1 - e^{-kt_i})(1 + e^{-(k+b)t_i})}{(1 - e^{-kt_i})(1 - e^{-(k+b)t_i})}\right]^{1/2} \tag{6}$$

a derivation of which is given in reference 26. As the air-exchange rate b becomes large, indicating efficient ventilation in the breathing zone, the quantity $e^{-(k+b)t_i}$ goes to zero and equation 6 reverts to the expression when the input is purely random (equation 4). Conversely, as b approaches zero (poor ventilation), $1/A(t_i) = 1$ and no dampening is indicated.

Figure 3 illustrates the expected range of dampening factors 1/A(0.25) for $T_{1/2} \leq 4$ hr. The three curves represent the autocorrelation coefficients expected for air exchange rates of 1, 3, and 10 hr^{-1}, i.e., w = 0.78, 0.47, and 0.08, respectively. One could envision these as high, average, and low rates of ventilation in the breathing zone. All curves display sharp reductions in 1/A(0.25) over the range of half-times $0 < T_{1/2} < 2$ hr with more gradual declines thereafter. This illustrates the increased physiological dampening associated with accumulation of toxicant at the target site. The lower curve in the figure, corresponding to an air exchange rate of 10 hr^{-1}, is essentially the same as that obtained from equation 4 for the case where exposures are purely random. The middle and upper curves illustrate the reduction in physiological dampening associated with autocorrelated exposure series.[26]

The relationships in Figure 3 suggest that less than half of the short-

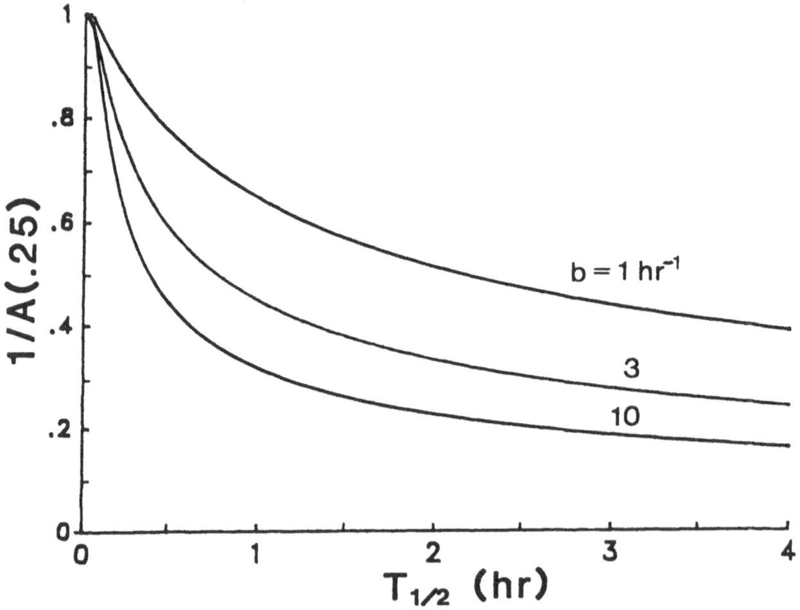

Figure 3. The short-term dampening factor 1/A(0.25) vs. the elimination half-time $T_{1/2}$. Assumes instantaneous exposures from a random autocorrelated series given by equation 5 in the text, observed at an interval of 0.25 hour. The first-lag autocorrelation coefficient is given by $w = e^{-25b}$ where b (hr^{-1}) represents the air-exchange rate in the breathing zone. Values of 1/A(0.25) are obtained from equation 6 in the text.

term exposure variability over intervals of 0.25 hour or less is transmitted to the tissues when the elimination half-time is more than about two hours. This value of $T_{1/2} = 2$ hr is, therefore, suggested as a strategic benchmark; it would appear to be unnecessary to evaluate short-term exposures to agents with half-times greater than two hours assuming that there is at least one air change per hour in the breathing zone.

If the elimination half-time of an agent is less than about two hours, short-term variation in exposure can influence individual risk to the extent that nonlinear dose-damage kinetics are involved as described above. If it is known or suspected that such nonlinear behavior occurs for a particular toxicant, then the sampling strategy should evaluate both the mean and the variance of the exposure distribution. Theoretically, this may be accomplished in several ways including the use of dual limits which specify eight-hour TWAs and short-term-exposure values.

Day-to-Day Variation

The assumption of random and uncorrelated exposures appears to be more appropriate when considering the dampening of exposure variability from day to day. However, the analysis is complicated by the irregular exposure schedule encountered in most work situations. Given the typical regimen of five 8-hour exposures per 168-hour week, the following useful approximation for the dampening factor was obtained:[13]

$$1/A(8\text{–}168) \cong \frac{(1 - e^{-24k})(1 - e^{-168k})}{(1 - e^{-120k})\sqrt{1 - e^{-48R}}} \qquad (7)$$

where 1/A(8–168) refers to the dampening factor associated with this regimen.

A graph of 1/A(8–168) vs. the elimination half-time is shown in Figure 4 as well as some data obtained from workers exposed to lead and mercury.[13] The figure indicates that the proportion of day-to-day variation in exposure which is transmitted to the tissues diminishes rapidly over the range $10 < T_{1/2} < 100$ hr. Indeed, less than half of the exposure variability is transmitted when $T_{1/2}$ is more than about 40 hours. This value of $T_{1/2} = 40$ hr is, therefore, suggested as another strategic benchmark; i.e., if $T_{1/2} > 40$ hr, it appears unnecessary to evaluate and control exposure variability from day-to-day in most circumstances. However, if $T_{1/2} < 40$ hr, it *may* be necessary to do so inasmuch as nonlinear dose-damage kinetics appear to be involved as noted previously.

Many notable toxic agents have half-times greater than 40 hours,

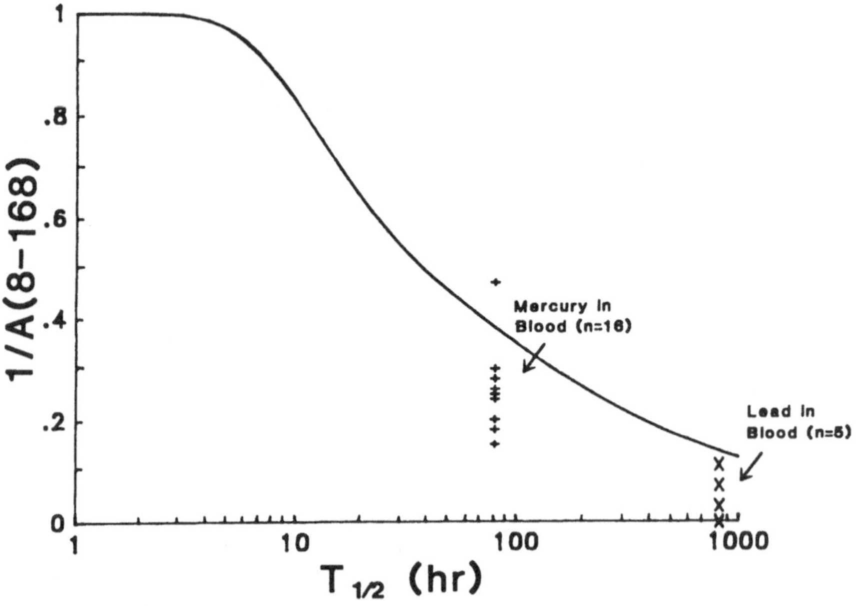

Figure 4. The day-to-day dampening factor 1/A(8–168) vs. the elimination half time $T_{1/2}$. Assumes a regimen of five eight-hour exposures per week. Values of 1/A(8–168) are obtained from equation 7 in the text.

including the following: fibrogenic dusts, heavy metals such as lead and cadmium, organo-metallic compounds such as methyl mercury, and highly lipophilic organic species such as 2,3,7,8-tetrachlorodibenzodioxin (TCDD).[13,27-28] It appears reasonable that monitoring strategies for these agents should focus upon the mean exposure received over a relatively long period of time without undue regard for the occasional high value.

One possible approach would be to ensure that $\mu_c <$ STANDARD or some fraction thereof at a significance level of 0.05.[6] This assumes that the exposure schedule is more or less continuous, i.e., that workers are exposed to the agent during most shifts. If the schedule is highly intermittent so that exposures occur only occasionally, then additional administrative safeguards may be required. A possible solution would be to define the mean exposure μ_c in terms of the population of shifts in which the worker is exposed instead of the population of all shifts worked.

CONCLUSIONS

This discussion of the impact of biological factors upon sampling strategies embraces issues concerning the relationship between physical and biological processes which come together in the work environment. The linkages between the various series of exposure, dose, and damage offer convenient avenues for investigation of the impacts of exposure variability upon health risk.

Tissue damage necessarily occurs very rapidly following transient exposures to acute toxicants. Thus, an effective sampling strategy for these agents would have to consider variation in exposure during brief periods. However, given the difficulties and uncertainties involved in assessing brief exposures, such a strategy would be more wisely used to monitor and control releases of these chemicals at the source.

By removing acute toxicants from consideration, debate concerning sampling strategies centers upon long-term exposures to chemicals which produce chronic diseases. Here, the linkage between the series of exposure and absorbed dose or tissue burden suggests that an essential piece of information is the rate at which the agent is eliminated from the target site in the body. If this rate constant is not known and cannot be estimated or bounded, then selection of a biologically-based sampling strategy is problematic. It is important, therefore, to develop a database of human elimination rate constants to assist in designing strategies. The hygienist of the future would refer to such a list in much the same way that one currently uses the compilations of the Occupational Safety and Health Administration (OSHA) standards and threshold limit values (TLVs).

With information concerning elimination rates, it is possible to develop strategies for toxic agents which share particular ranges of constants. Two strategic benchmarks are suggested here to illustrate this approach. The first concerns chemicals with elimination half-times greater than about two hours. These agents are likely to be accumulated in the tissues to such an extent that intraday variation in exposure at intervals of 15 minutes or less is efficiently dampened. It therefore seems unnecessary to consider such transient exposures to agents as, for example, when applying short-term exposure limits. It is noted, however, that the degree of dampening is also a function of the autocorrelation coefficient of the exposure series which, in this analysis, has been related to the air exchange rate in the breathing zone. Thus, the $T_{1/2} = 2$ hr benchmark would only be appropriate when the effective air exchange rate is at least 1 hr^{-1}.

The second benchmark concerns chemicals which have elimination half-times greater than about 40 hours. These agents are likely to be accumulated in the tissues to such an extent that the variation in eight-hour TWAs from day to day should be efficiently dampened in the body. It may be that the appropriate strategy for these agents is to evaluate the mean exposure received over a relatively long period of perhaps a year or more; statistical procedures are available which allow this to be done. If exposures are highly intermittent at intervals of a week or longer, it may be useful to define the mean exposure in terms of the average exposure received on days when the worker is exposed.

The above conclusions define classes of toxic agents for which variation in exposure over particular time periods is unlikely to be an important determinant of health risk. Agents with half-times less than 40 hours *may* require evaluation of the variability of exposure if it is known or suspected that the relationship between dose and damage at a particular time is concave upwards. Likewise, agents with half-times less than two hours *may* require evaluation of transient variability inasmuch as this same nonlinear behavior is expected during brief periods. Future investigations should focus upon this important issue.

These recommendations are based upon relatively simple models of chemical disposition which assume first-order behavior and a single physiological compartment. It is suspected that predicted dampening effects will be conservative in the sense that, if actual chemical disposition is better described by a multicompartment model or if the chemical is metabolically activated, the actual dampening will be greater than predicted by a single-compartment model. This is discussed further in references 13 and 26. In any event, it is essential that additional data be gathered for many toxicants to test the predicted dampening behavior over a wide range of elimination rates.

REFERENCES

1. Rappaport, S.M.: The Rules of the Game: An Analysis of OSHA's Enforcement Strategy. *Am. J. Ind. Med. 6*:291 (1984).
2. Leidel, N.A. and K.A. Busch: Statistical Design and Data Analysis Requirements. *Patty's Industrial Hygiene and Toxicology*, 2nd ed., Vol. 3A, *Theory and Rationale of Industrial Hygiene Practice: The Work Environment*, pp. 395–507. L.J. Cralley and L.V. Cralley, Eds. John Wiley and Sons, New York (1985).
3. Tuggle, R.M.: Assessment of Occupational Exposure Using One-Sided Tolerance Limits. *Am. Ind. Hyg. Assoc. J. 43*:338 (1982).
4. Rock, J.: A Comparison Between OSHA-Compliance Criteria and Action-Level Decision Criteria. *Am. Ind. Hyg. Assoc. J. 43*:297 (1982).

5. Selvin, S., S.M. Rappaport, R.C. Spear et al.: A Note on the Assessment of Exposure Using One-Sided Tolerance Limits. *Am. Ind. Hyg. Assoc. J.* (in press).
6. Rappaport, S. M. and S. Selvin: A Method for Evaluating the Mean Exposure from a Lognormal Distribution. *Am. Ind. Hyg. Assoc. J.* (in press).
7. Spear, R.C., S. Selvin and M. Francis: The Influence of Averaging Time on the Distribution of Exposures. *Am. Ind. Hyg. Assoc. J. 47*:365 (1986).
8. Esmen, N. and Y. Hammad: Lognormality of Environmental Sampling Data. *Environ. Sci. Health A12*:29 (1977).
9. Corn, M. and N.A. Esmen: Workplace Exposure Zones for Classification of Employee Exposures to Physical and Chemical Agents. *Am. Ind. Hyg. Assoc. J. 40*:47 (1979).
10. Spear, R.C., S. Selvin, J. Schulman and M. Francis: Benzene Exposure in the Petroleum Refining Industry. *Appl. Ind. Hyg. 2(4)*:155 (1987).
11. Rappaport, S.M., R.C. Spear and S. Selvin: The Influence of Exposure Variability on Dose-Response Relationships. *Ann. Occup. Hyg.* (in press).
12. Rappaport, S.M., R.C. Spear, S. Selvin and C. Keil: Air Sampling in the Assessment of Continuous Exposures to Acutely-Toxic Chemicals. Part 1: Strategy. *Am. Ind. Hyg. Assoc. J. 42*:831 (1981).
13. Rappaport, S.M.: Smoothing of Exposure Variability at the Receptor: Implications for Health Standards. *Ann. Occup. Hyg. 29*:201 (1985).
14. Roach, S.A.: A More Rational Basis for Air-Sampling Programs. *Am. Ind. Hyg. Assoc. J. 27*:1 (1966).
15. Saltzmann, B.E.: Significance of Sampling Time in Air Monitoring. *J. Air Poll. Control Assoc. 20*:660 (1970).
16. Roach, S.A.: A Most Rational Basis for Air-Sampling Programmes. *Ann. Occup. Hyg. 20*:65 (1977).
17. Koizumi, A., T. Sekiguchi, M. Konno and M. Ikeda: Evaluation of the Time Weighted Average of Air Contaminants with Special References to Concentration Fluctuation and Biological Half-time. *Am. Ind. Hyg. Assoc. J. 41*:693 (1980).
18. Saltzmann, B.E. and S.H. Fox: Biological Significance of Fluctuating Concentrations of Carbon Monoxide. *Environ. Sci. Technol. 20*:916 (1986).
19. Hoel, D.G., N.L. Kaplan and M.W. Anderson: Implication of Nonlinear Kinetics on Risk Estimation in Carcinogenesis. *Science 219*:1032 (1983).
20. Ehrenberg, L., E. Moustacchi and S. Osterman-Golkar: Dosimetry of Genotoxic Agents and Dose-Response Relationships of Their Effects. *Mut. Res. 123*:121 (1983).
21. Starr, T.B. and R.B. Buck: The Importance of Delivered Dose in Estimating Low-Dose Cancer Risk from Inhalation Exposure to Formaldehyde. *Fund. Appl. Tox. 4*:740 (1984).
22. Andersen, M.E., H.J. Clewell, III, M.L. Gargas et al.: Physiologically-Based Pharmacokinetics and the Risk Assessment for Methylene Chloride. Submitted for publication.
23. Krewski, D., D.J. Murdoch and J.R. Withey: The Application of Pharma-

cokinetic Data in Carcinogenic Risk Assessment. *Proceedings of Pharmacokinetics in Risk Assessment*. National Academy of Sciences, Washington, DC (in press).
24. Coenen, W.: Beschreibung des Zeitlichen Verhaltens von Schadstoffkonzentrationen durch einen steigen Markowprozess. *Staub Reinhalt. Luft 36*:240 (1976).
25. Chatfield, C.: *The Analysis of Time Series*, 3rd ed., pp. 44–50. Chapman and Hall, New York (1984).
26. Rappaport, S.M. and R.C. Spear: Physiological Dampening of Exposure Variability During Brief Periods. Submitted for publication.
27. Kershaw, T., P. Dhakir and T. Clarkson: The Relationship Between Blood Level and Dose of Methyl Mercury in Man. *Arch. Environ. Health. 35*:28 (1980).
28. Poiger, H. and C. Schlatter: Pharmacokinetics of 2,3,7,8-TCDD in Man. *Chemosphere 15*:1489 (1986).

CHAPTER 24

Sampling Strategies for Epidemiological Studies

THOMAS J. SMITH

Environmental Health Sciences Program, Department of Family and Community Medicine, University of Massachusetts Medical Center, Boston, Massachusetts

INTRODUCTION

The general approach to exposure assessment for epidemiological studies is significantly different from that generally used by industrial hygienists to assess the presence of a recognized hazard by the use of exposure limits. The principal source of this difference is that the epidemiologic study is performed to detect a hazard and determine the nature of any dose-response relationship, whereas compliance testing is performed with the presumption that the dose-response relationship is known and built into the exposure limit. These differences have important implications for sampling strategy.

The objective of this paper is to present a rationale for developing exposure assessments for epidemiology and to contrast this approach with that used for compliance testing. The application of the rationale will be illustrated in the design of a study to follow up on male reproductive effects observed among shipyard workers using 2-methoxyethanol-based (methyl Cellosolve) paints. Finally, the specific differences in epidemiologic and compliance sampling strategies will be highlighted.

EXPOSURE EVALUATION FOR EPIDEMIOLOGIC STUDIES

The objective of epidemiologic studies is to determine if exposure to a toxic material affects human risk of adverse health effects, such as loss of pulmonary function, or the risk of lung cancer. Epidemiologists study health risks in groups of workers with different exposures.[1] The relative risks of effects are compared among groups of workers who are as similar as possible in all respects except their exposure to a toxic agent. For example, Lloyd compared the cancer risk of all noncoke-oven steel workers to those who work around coke ovens and found that the latter had significantly higher lung cancer risks.[2] This comparison has two important assumptions: 1) all nonexposed steel workers have no contact with coke oven emissions or other occupational causes of lung cancer, and 2) coke oven workers have sufficient exposure to carcinogens to change their risk of lung cancer.

Classically, epidemiologic studies have made little use of quantitative data on exposures to assign subjects to exposure groups. Surrogates were used to indicate potential exposure to a given toxic agent. Checkoway has reviewed these surrogates and noted their advantages and disadvantages.[3] The most common are "current" job title, longest held job title, duration of work in an exposed job or work area, and total duration of work. These variables have been used successfully to detect gross differences in risk for many serious hazards including asbestos and benzene.

All exposure assignment approaches suffer from two problems: misclassification and confounding. Misclassification is putting workers in the wrong exposure category, i.e., exposed instead of not exposed or "high" instead of "low"; quantitative measures of exposure have similar problems. If misclassification is sufficiently severe, it will reduce the apparent relationship between exposure and the effect.[4] Confounding is observing an apparent positive relationship between an exposure and an effect, which is actually caused by an exposure to another unmeasured agent whose level is correlated with the agent of interest. For example, if cigarette smoking is not measured but is correlated with exposure to a suspected carcinogen, then the carcinogen may appear to have more influence on lung cancer risk than is appropriate because of the effects from cigarette smoking. If smoking is random across exposure groups, then the baseline rate of cancer may be increased, but there will be no confounding with the agent's effects. Because of practical limitations, it is frequently necessary to assume that unmeasured agents are randomly distributed relative to the exposure agent of interest; this may be a poor assumption. Measuring exposure to all agents that might cause an effect tends to reduce the risk of confounding. It is very important to measure the exposures of all subjects, including those that are not exposed to verify their levels of exposures.

Beyond the evident need to measure all potential agents of the effects, there is also a need to measure the variation of exposures with time because the level of the effects will to some degree reflect changes in exposure. Finally, there is the serious problem of how to process the exposure data and develop dose indices and other measures that can be related to the effects. Appropriate responses to these concerns depend strongly on the nature of the health effects and hypotheses about causal mechanisms. Consequently, the optimum approach for sampling strategy depends on the underlying pharmacological processes. The nature of these processes and their connection to sampling strategy will be discussed next.

PHARMACOLOGIC RATIONALE FOR EXPOSURE EVALUATION

This section will summarize an approach presented in more detail in a recent paper by Smith.[5] The pharmacologic processes that causally relate exposures to toxic materials to the adverse health effects are diagrammed in Figure 1. Exposure causes a toxic material to enter the body through the lungs, skin, or gastrointestinal tract, depending on the mode of exposure. Once inside the body, toxicokinetic processes transport the agent to the target tissues and are responsible for its metabolism, storage, and excretion. The presence of sufficient quantities of the agent or a metabolite within the target tissues leads to some tissue effect that may or may not be directly detectable. These tissue effects may accumulate, provoke other processes, or be sufficiently severe that eventually there is an observable effect which may also represent a clinical problem. The processes relating the tissue concentration of the agent to the effects are called its pharmacodynamics. Toxicologists and pharmacologists will commonly measure quantitative dose-response relationships to describe the pharmacodynamics of an agent. They have found that there is no single expression for the dose that will fit all toxic agents. Similarly, there is no single form of dose expression, such as exposure intensity times duration ("cumulative exposure"), which is the appropriate form for quantifying the risk for all occupational exposures.[6] It is clear that exposure is only indirectly related to the observed effects and that many important biological processes must be considered when attempting to quantify the relationship between exposure and effects.

The pharmacologic approach to sampling for epidemiologic studies emphasizes that the objective of exposure assessment is to characterize

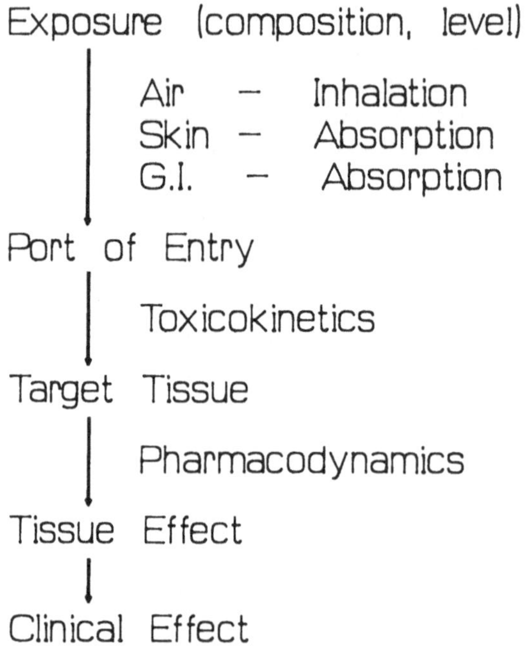

Figure 1. Pharmacologic sampling approach.

each subject's tissue dose of toxic agent, not just the exposure, because it is the tissue dose that causes the effect, as shown in Figure 1. The industrial hygienist measures the exposures but must do so with the parameters of the biological relationship as guides. The nature of the toxic materials and their potential effects will guide: 1) the selection of the agent(s) to be measured; the physical forms and modes of exposure of interest will guide 2) the choice of sampling methods; and the toxicokinetics and pharmacodynamics of the agent and its effects will guide 3) the selection of sampling averaging time and total period of interest.

There has been much interest in using toxicokinetic models of toxic materials behavior as a way to understand the complex interactions between temporal variations in occupational exposures and tissue concentrations, especially as an aid to interpreting biological monitoring data.[7] Roach[8] and Rappaport[9] have clearly shown that the averaging time of sampling must be matched to the toxicokinetics of the agent; otherwise, important variations in body burden may be obscured. The same consideration applies to the agent's concentration in the target tissues. Similarly, the total time period of interest for exposure characterization depends on the mechanism of the effects. For example, the total effect from agents whose effects are cumulative, such as crystalline silica in the lungs or cadmium in the kidneys, depends on the total period of exposure. However, the total effects from agents whose effects are reversible, such as airway irritation from sulfur dioxide, are related to relatively recent exposures, because the older effects have been repaired. In these cases, the relevant time period for exposure characterization is not apparent. Smith has shown that "effect indices" which model the damage-repair process can be used to both identify the time period for exposure characterization and provide a means of summarizing the time profile of exposure so it can be directly related to the measures of health effects.[5]

The toxicokinetics and mechanism of an effect are critical in defining a biologically important "peak exposure," i.e., one that is likely to cause tissue effects. The accumulation of an agent in the body will damp out some of the short-term variability in exposures, but the amount of damping depends on the location of the target tissues and the toxicokinetics. External surfaces, such as the eyes and the pulmonary airways, can show very rapid irritation responses to very brief parts per million peaks of airborne ammonia. On the other hand, a peak pulmonary exposure to quartz dust requires several months duration of milligram per cubic meter levels. Both types of exposures will produce "acute" responses but by very different mechanisms. The hourly and day-to-day variations in quartz dust exposure are not relevant because they do not produce a large change in the quartz content of the lungs relative to that needed to produce the effect. These points can be made somewhat clearer by considering their application in an example.

AN APPLICATION OF THE PHARMACOLOGIC APPROACH

A preliminary epidemiologic study was conducted by Drs. Mark Cullen and Laura Welch to determine if changes in sperm count were associated with exposures to ethylene glycol ethers (EGEs) (Cellosolves). The study compared semen analyses from a group of shipyard painters, who are exposed to solvents in paints, and a group of comparable workers drawn from other trades at the shipyard. The methods included semen samples; urine samples which were analyzed for methoxyacetic acid (MAA), the principal metabolite of methyl Cellosolve; and a questionnaire to obtain reproductive history, personal data, and a job history. There were two measures of exposure: urinary MAA to reflect current exposure level and duration of work as a painter to reflect long-term exposure.

A significant decrease in sperm count was seen in the painters relative to the other trades group. The painters also showed a significantly higher fraction of workers (17% versus 5%) with a sperm count less than 20 million per cc, which is an indication of infertility. Thus there were indications that the painters had reduced sperm production relative to the comparison group; however, the cause was uncertain because the painters' sperm counts were not correlated to either urinary MAA or duration of work as a painter. A follow-up study has been proposed which would more closely relate the exposures to EGEs and spermatogenic effects.

Using the pharmacologic approach, the first step is to identify the target tissue and the possible mechanisms of effects. A review of spermatogenesis and toxicology of male reproductive effects by Dixon[10] was used to understand the nature of the pathological processes governing the effects; this was supplemented by consultations with clinicians and toxicologists. The target tissue appears to be the stem cells (spermatogonia) in the seminiferous tubules of the testes, although it might also be the Sertoli cells. Toxic effects on these cells will reduce or prevent the production of sperm. The timing of a reduction in sperm count relative to reduced spermatogenesis has a lag of 80 days, depending on sexual activity, because there is a 76-day sperm production cycle. Animal studies indicate that toxic damage to the stem cells requires about 120 days to completely repair, depending on the degree of injury, which suggests a half-time of repair of about 20 days. It was assumed that humans would behave similarly. This information is important for sampling strategy in two ways: 1) the total time period for exposure characterization for each subject probably need not be any longer than 100 days, and 2) the relative timing of exposure and effects measurements is offset by 76 days.

The active agents of the toxic effects by EGEs and their toxicokinetics are not known. Studies in rats showed that MAA produced testicular toxicity; similar metabolites of other cellosolves are also suspected.[11] A reasonable alternative is a short-lived, reactive aldehyde intermediate that is produced when EGEs are oxidized to acids; the oxidation could be done by mixed function oxidases and cytochrome P-450 enzymes in testicular tissue. Thus there are two likely agents; both are metabolites of EGEs but have very different time courses in the body. It is necessary to have an approximate model of the toxicokinetics to understand the relationship between exposure and target tissue concentrations of the suspected agents.

A general single compartment model of EGE uptake, metabolism, and excretion can be constructed with three reasonable assumptions: 1) EGE and acid metabolites (all highly water soluble) are distributed uniformly throughout body fluids, 2) EGE only leaves the body by metabolism to an acid, and 3) alkoxyacetic acids only leave the body by urinary excretion. EGEs are not directly excreted to any significant degree; less than five percent are excreted by exhalation.[11] Alkoxyacetic acids are the main metabolic products; only small amounts of other materials are formed. The model is diagrammed in Figure 2. Data from recent chamber studies by Groeseneken and coworkers in Belgium were used to estimate the metabolic and excretion process rates for ethylene glycol monoethyl ether (EGEE).[12,13] Rates for other EGEs would probably be somewhat different but should follow the same pattern of relationships. Since exhaled breath levels of EGEE are proportional to serum levels, the rate of decline of breath levels can be used to estimate the rate of decline of serum levels. Groeseneken and coworkers observed that once exposure stopped, and a short period was allowed for the EGEE to redistribute into the less accessible body fluids, the breath level dropped with a half-time of 1.7 hours.[12] This can be used as an estimate of the metabolic rate in the model (given assumption 2). The same investigators also examined the time course of urinary ethoxyacetic acid (EAA) levels during and after a four-hour steady exposure to EGEE.[13] They found that the urinary EAA peaked about six hours after the exposure stopped, after

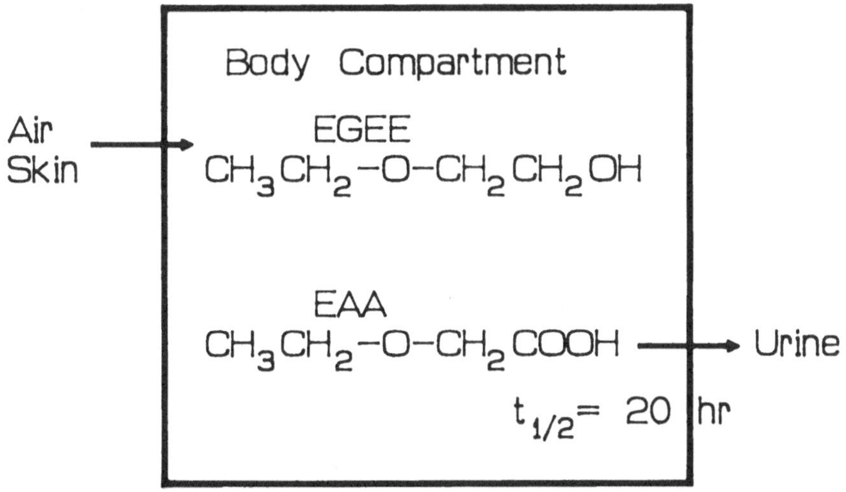

Figure 2. Metabolic pathways of ethylene glycol monoethyl ether.

which the EAA declined with a half-time of 20 hours. This rate can be used to estimate the EAA elimination rate in the model (based on assumption 3).

The applicability of the model was tested by determining how well it predicted urinary EAA given a four-hour steady inhalation exposure such as that studied by Groeseneken. Figure 3 shows that the model EAA follows the observed time course reasonably well; the timing of the peak in the EAA is very sensitive to errors in the metabolic rate. These data show that the simple model is a reasonable first approximation of the behavior of EGEE and its major metabolite, EAA, within the body.

The model was used in an exposure simulation to determine how the normal variations in inhalation exposure experienced by a shipyard painter might affect tissue levels of EGEE and EAA. Random, lognormal hourly exposures (geometric mean = 16 mg/m³, threshold limit value (TLV) = 19 mg/m³; geometric standard deviation = 2.0) were calculated with a random number generator so that typical daily exposures could be simulated. It was assumed that the time-weighted average (TWA) air level was constant within each hour. The differential equations that describe the model were solved for each hour by the Runge-

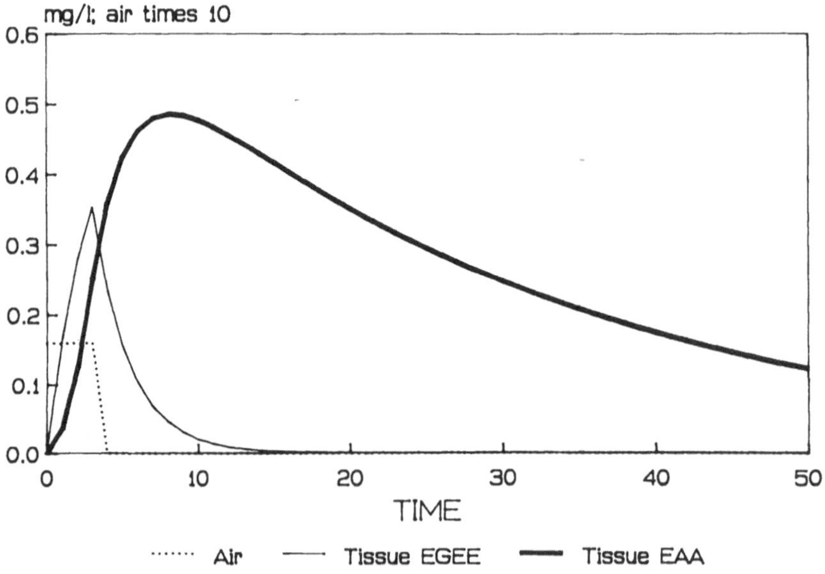

Figure 3. Ethoxyacetic acid observed time course.

Kutta method for a coupled system of first order differential equations using software from the Turbo Pascal Numerical Methods Toolbox (Borland International, Scotts Valley, CA). The estimated tissue EGEE and EAA levels derived from the two work days of random exposures are shown in Figure 4.

Several important and useful things can be seen from the simulation. First, the levels of tissue EGEE follow the hourly exposures fairly closely and are sensitive to high values. Consequently, the tissue levels of the aldehyde intermediate metabolite will also follow the hourly exposures, if the rate of aldehyde formation is proportional to the EGEE levels, which is very likely. If stem cell damage is caused by the tissue concentration of aldehyde, it may not be a linear relationship in which each unit of aldehyde in the tissue causes the same amount of damage. For example, the characteristic dose-response pattern for toxicity from irritant aldehydes is generally nonlinear: low concentrations produce little or no damage and high levels produce disproportionately large amounts of damage. Therefore, the occurrence of peak tissue concentrations of aldehyde may be very important; hence, peak exposures to EGEE may be very important.

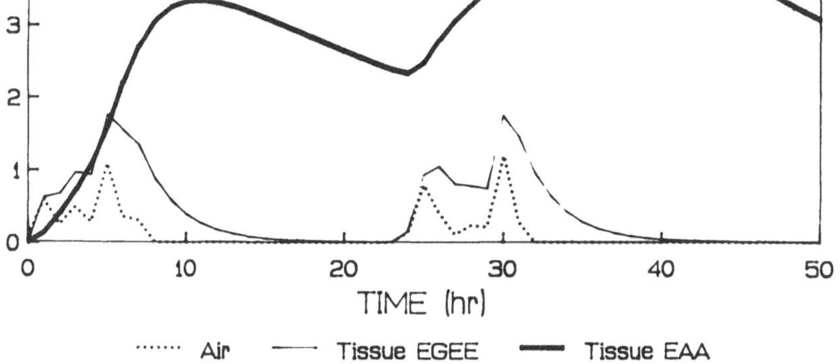

Figure 4. Estimated tissue levels of ethylene glycol monoethyl ether and ethoxyacetic acid.

The second useful observation was that the EAA levels are not sensitive to hourly peaks: the slope of the EAA increase during the day mostly reflects the daily TWA exposure level. The EAA levels in the tissues are substantially higher than the EGEE levels because the rate of elimination of EAA is much slower than its rate of formation. The morning level of EAA may contain some carryover from the previous day, but most of the tissue burden will be excreted over a weekend. Thus, morning EAA urinary level is an estimate of the TWA exposure from the previous day or two. As with the aldehyde intermediate, it is quite possible that the effects of the EAA may be nonlinear; thus, days with the highest exposures may produce disproportionately large amounts of tissue damage. Thus both EGEE and EAA in the tissues, the TWA, and frequency of peaks larger than some threshold must be characterized to examine possible linear and nonlinear mechanisms of effects. These findings can be used to guide the design of an exposure evaluation for a follow-up epidemiological study.

A damage-repair model can be very useful for showing the interrelationships between exposures, and the effects that might be caused by the hypotheses for mechanisms of effects, in the same way that the toxicokinetic model was useful above. However, it is beyond the scope of this paper to develop damage-repair models for the EGEE example (see reference 11 for a detailed discussion of the approach). A simple dose index based on an underlying damage-repair mechanism is given below in the discussion of the data analysis for the example.

DEVELOPING A SAMPLING STRATEGY USING PHARMACOLOGIC DATA

The industrial hygienist is faced with three tasks:

1. To identify a group(s) of workers with a range of glycol ether exposure who are without potential confounding exposure to other agents.
2. To estimate each subject's time profile of exposure to glycol ethers over the 100 days that occurred 76 days prior to the semen test, for all modes of exposure.
3. To develop dose indices that can be calculated from each subject's exposure estimates.

Site visits and walk-through surveys can identify painters with exposures to glycol ether vapors and aerosols as well as those job activities that are most likely to cause exposure. All painters using significant amounts of paints with other solvent bases should be excluded from the study group. This reduces the risk of confounding; the sampling demands that the only environmental agents to be measured will be glycol ethers. The absence of these or other toxic exposures in the "unexposed" comparison group must be verified.

Given a suitable group of exposed workers for the study, the sampling strategy will depend on the details of the epidemiological design. The ability of a study to detect and quantify a dose-response relationship depends on the precision of both the exposure and effect measurements. The optimum strategy for exposure measurements to estimate dose is to measure every subject's exposure over the whole time period relevant to the effect. However, if a large number of subjects must be measured over a long time period, then the optimum is not feasible and alternative strategies must be used. As the strategy moves away from personal measurements to group estimates, the potential for misclassification increases and the strength of the apparent dose-response relationships may decrease unless the range in exposures among exposure groups is very wide.

The number of subjects that must be measured will depend on the precision of the effect measurement technique; a low-precision method requires that many workers be tested at each dose level to detect changes in effects associated with changes in dose. Some epidemiological studies try to deal with poor precision in effects measurements by adding large numbers of subjects whose exposures are poorly characterized. This misguided approach will increase the cost of the study with little improvement in overall resolution of the dose-response relationship.

The specifics of sampling strategy are controlled by the number of workers to be evaluated for effects, the time period for the effects, the time scale of the toxicokinetics, and the hypothesized mechanism of effects (especially linear versus nonlinear mechanisms). Since sperm count is highly variable, it is necessary to measure many workers to detect an effect. Since the tissue concentration during a 100-day period may contribute to the effect, it is necessary to characterize the exposures of many workers over a long time period; therefore, it is not feasible to measure each subject's personal exposure. Alternatively, the average exposure of tasks performed by the painters may be measured and coupled with a personal record of each worker's job activities during the 100-day exposure period, 76 days prior to his semen sample. A painter's job activities may be subdivided into several well-defined tasks with different levels of exposure, such as set up, spray painting, roller painting, and cleanup, and his daily routine will vary widely depending on the project needs. The exposures during each of the tasks can be characterized by

TWA sampling. Those tasks with probable high exposures, e.g., spray painting, can be sampled to detect peaks by obtaining consecutive one hour, or shorter, interval samples. Collection of daily diaries of job activities during the 100-day study period for each subject can permit reasonably specific estimates of personal TWA and peak exposures. This assumes that there is relatively little variation between workers in the distribution exposure for a given task. This may not be a good assumption if work habits are important in determining exposure, such as skin washing with solvents during cleanup. First void in the morning urine sampling can be conducted to estimate MAA tissue levels.

Highly sensitive methods are available to measure air and breath levels for glycol ethers and urinary metabolites.[12-14] Consecutive, personal one-hour samples are feasible to measure airborne exposure peaks. Airborne exposure can be measured by lapel sampling with a combined aerosol and vapor collector, e.g., a glass fiber filter followed by a silica gel adsorbent tube. The worker's use of respirators may necessitate sampling inside the mask. If there is appreciable skin exposure, then matched breath and air sampling should be performed so that the amount of skin absorption can be estimated; a series of breath samples could be collected, one at the end of each hour of air sampling. Commonly, several alternative sampling strategies are available. The choice of the optimum approach depends on the hypothetical mechanism of effects; the sources of exposure variability; availability of sampling and analytical resources; and the costs of sample collection, analysis, and data processing.

Once the time profiles of exposure have been developed for each subject during the relevant time periods, then they must be processed into summary statistics for use in the epidemiologic data analysis. Traditionally, indices such as cumulative exposure have been calculated assuming that the damage mechanism is linear, i.e., each unit of exposure contributes equally to the effect. It is not obvious how to process the time profile into a more appropriate dose index if there is repair of the damage or the mechanism of effects is nonlinear.

A simple index can be developed for the testicular tissue effects to illustrate derivation of an effect index. Since testicular damage is repaired, an increment in damage will contribute to the observed reduction in sperm count inversely with the time since the damage occurred. The half-time of repair gives the time scale, which is probably about 20 days for testicular damage. Counting backward from the time of the semen sample, allowing for the 76-day lag in effects on sperm counts, the tissue concentrations during the first 20 days (77 to 97 days prior to the semen sample) will contribute the most to the observed effect, the second 20 days will contribute half as much, the third 20 days will contribute one quarter as much, and so on. A reasonable dose index can be obtained by calculating a weighted average of tissue concentrations of either EGEE or EAA for the exposure period prior to the semen sample for each subject. A similar dose index could be developed for the peak exposures to EGEE and EAA.

COMPARISON OF EPIDEMIOLOGIC AND COMPLIANCE TESTING

The principal differences between the epidemiologic and compliance testing are shown in Table I. It can be seen that there are major differences for all of the major aspects of the evaluations. Perhaps the most evident differences are who to sample, the choice of measurement methods, and the interest in past exposures. An industrial hygiene survey conducted for one purpose can provide data for the other, but it is not uncommon that there are limits in their cross-application. Historical surveys of compliance status are an important source of data on past

TABLE I. Comparison of Epidemiologic and Compliance Types of Exposure Evaluations

Feature	Epidemiologic Testing	Compliance Testing
Persons to be sampled	All exposed and controls	Most highly exposed
Agents/methods	Must measure all suspected agents; frequently must develop and validate methods	Only measure regulated agents; methods NIOSH-approved or equivalent
Sampling strategy	Combination of personal and area with time/motion data; both day-to-day and worker-to-worker variation should be examined	Personal sampling; area sampling rarely acceptable; variation across days and workers assumed equivalent
Air sample duration	Defined by toxicokinetics of agent in target tissue	Defined by OSHA; 8-hour (full shift) or 15 minutes
Statistics	Arithmetic mean estimated for exposure of individuals or groups; use lognormal statistics to describe distributions	Probability of exceeding PEL estimated using lognormal statistics
Number of samples	Defined by precision desired for mean exposure; related to size of differences among exposure group means	Defined by desired probability for detecting values over PEL
Past exposures	Estimated from variety of data	Irrelevant

exposures for epidemiologic studies. Ulfvarson reviewed the limitations of this use of compliance data and found that there is a tendency to overestimate average exposures because of the bias toward sampling those conditions leading to high exposures.[15] This concern is particularly important for the use of compliance surveillance data as part of a medical monitoring system which is intended to provide the raw data for future epidemiologic studies. This application can work, but it works best if the differences in epidemiologic and compliance testing are recognized and adjustments made in the surveillance system to meet the probable needs of future studies.

CONCLUSION

It is important to recognize that sampling for epidemiologic studies has a set of special requirements that are different from compliance testing. Second, evaluations for epidemiologic studies must be firmly grounded in the toxicokinetics of the agents being measured and must be related to the biological hypotheses about the mechanisms of effects that are to be investigated.

REFERENCES

1. Monson, R.R.: *Occupational Epidemiology*. CRC Press Inc.; Boca Raton, FL (1980).
2. Lloyd, J.W.: Respiratory Cancer in Coke Plant Workers. *J. Occup. Med. 13*:53 (1971).
3. Checkoway, H.: Methods of Treatment of Exposure Data in Occupational Health. *Med. Lav. 77*:48 (1986).
4. Armstrong, B.G. and D. Oakes: Effects of Approximation in Exposure Assessments on Estimates of Exposure-Response Relationships. *Scand. J. Work Environ. Health 8 (Suppl.)*:20 (1982).
5. Smith, T.J. Exposure Assessment for Occupational Epidemiology. *Am. J. Ind. Med.* (in press).
6. Atherley, G.: A Critical Review of Time-Weighted Average as an Index of Exposure and Dose, and of its Key Elements. *Am. Ind. Hyg. Assoc. J. 46*:481 (1985).
7. Fiserova-Bergerova, V.: *Modeling of Inhalation Exposure to Vapors: Uptake, Distribution, and Elimination*. CRC Press, Inc., Boca Raton, FL (1983).
8. Roach, S.A.: A Most Rational Basis for Air Sampling Programmes. *Ann. Occup. Hyg. 20*:65 (1977).
9. Rappaport, S.M.: Smoothing of Exposure Variability at the Receptor: Implications for Health Standards. *Ann. Occup. Hyg. 29*:201–214 (1985).

10. Dixon, R.L.: Toxic Responses of the Reproductive System. *Casarett and Doull's Toxicology*, 3rd ed., pp. 432–477. C.D. Klaassen, M.O. Amdur and J. Doull, Eds. Macmillan Publishing Co., New York (1986).
11. Andrews, L.S. and R. Snyder: Toxic Effects of Solvents and Vapors. *Casarett and Doull's Toxicology*, 3rd ed., pp. 636–668. C.D. Klaassen, M.O. Amdur and J. Doull, Eds. Macmillan Publishing Co., New York (1986).
12. Groeseneken, D., H. Veulemans and R. Masschelein: Respiratory Uptake and Elimination of Ethylene Glycol Monoethyl Ether in Experimental Human Exposure. *Br. J. Ind. Med. 43*:544 (1986).
13. Groeseneken, D., H. Veulemans, and R. Masschelein: Urinary Excretion of Ethoxyacetic Acid in Experimental Human Exposure to Ethylene Glycol Monoethyl Ether. *Br. J. Ind. Med.* (in press).
14. Smallwood, A.W., K.E. DeBord, and L.K. Lowry: Analyses of Ethylene Glycol Monoalkyl Ethers and Their Proposed Metabolites in Blood and Urine. *Environ. Health Persp. 57*:249 (1984).
15. Ulfvarson, U.: Limitations to the Use of Employee Exposure Data on Air Contaminants in Epidemiologic Studies. *Int. Arch. Occup. Environ. Health 52*:285 (1983).

CHAPTER 25

Community Air Sampling Strategies

PAUL J. LIOY

Division of Exposure Measurement and Assessment, Department of Environmental and Community Medicine, UMDNJ-Robert Wood Johnson Medical School, 675 Hoes Lane, Piscataway, New Jersey 08854–5635

INTRODUCTION

Studies of community air pollution problems can use one of a number of different approaches and each has evolved substantially over the past 25 years. Early activities in community air sampling used very simple tools, including the dustfall bucket, and collected data primarily in specific geographic-population centers. The location studied could include: a rural center, a small town or a city, a suburb of a large urban center, or a selected portion of a city. In each case one or more fixed monitoring stations were usually placed at selected locations and these comprised a sampling program. This form of community air sampling grew from the use of simple manual monitoring techniques, such as the high volume sampler and spot samplers, into complex monitoring networks.[1] Today, a number of pollutants are measured continuously at a site; for large networks, the monitoring data are usually sent by telemetry to a central data acquisition and validation center.

Since the promulgation of the Clean Air Act in 1970,[2] one of the basic objectives of community air sampling has been primarily to measure the concentrations of one or more pollutants originating from any number of sources. For most state and local agencies, these are, at a minimum, the criteria pollutants: carbon monoxide, ozone, nitrogen oxides, lead, total suspended particulate matter, and sulfur dioxide; however, other pollutants and pollutant indicators are measured at specific sites. Data from these types of routine monitoring programs can be used in epidemiological or long-term trend studies; however, when a specific problem is addressed, they are usually augumented by other community air sampling programs. Overall community air sampling programs have been designed to examine the following: the potential for human health effects; the damage to vegetation, materials, etc.; the compliance with the National Ambient Air Quality Standards and emissions standards; and pollutant formation, transport and deposition. Part of the reason for conducting these types of studies is that for criteria pollutants the primary ambient air quality standards are based upon the potential for human effects and the secondary standards on ecological and degradation effects. In addition, many community air pollution studies are designed and conducted to investigate basic chemical and physical processes in the atmosphere which will assist scientists in attempts to reduce the intensity of air pollution episodes and to develop more effective control strategies.

TYPES OF COMMUNITY STUDIES

Some of the air sampling programs developed to address the above issues are categorized as special short-term studies. The original special survey studies have, in some cases, evolved into the present National Air Monitoring Station (NAMS) network or the State Long-Term Air Monitoring Station (SLAMS) network for criteria pollutants monitored by regulatory agencies. Both of these networks are important because they measure indicators of the variety of emission sources, including fossil fuels, combustion and industrial processes, and those sources with national scope of the emissions.

At the present time, research investigations on community air pollution use short-term studies to assess acute health effects exposures and the chemical characteristics of the atmosphere, and long-term studies to investigate the nature of acid rain or pollutant trends and chronic health effects. Depending upon the objectives of any particular study and the resources available for a study, variations from program to program will be noted in the size of the area, the site locations, the number of samplers, the pollutants measured, the frequency and length of sampling, and the duration of the samples. The area covered by community air sampling programs can be defined by the meteorological influence regions: microscale, mesoscale and synoptic scale, which translate into community air pollution problems confined to radial distances of < 10 km, 100 km, and < 3000 km.[3] The microscale investigations can be subcategorized since problems may exist in a specific neighborhood, in a section of a city, around a group of small sources (< 50 tons/year emission), or downwind of a single point source. The mesoscale influence can involve emission from a number of point or line sources which ultimately combine to produce the urban plume and its downwind impacts. Synoptic scale events are associated with high- or low-pressure weather systems and the contributions from secondary pollutants such as ozone, sulfate, and nitrates.

The preceding discussion has emphasized a more traditional approach to community air sampling. In recent years, however, concerns about community air pollution have extended to the indoor environment[4,5] and, in some cases, the total exposure of an individual to specific pollutants.[6-8] These new avenues of study have developed because of the potential for the accumulation of high pollutant concentrations in indoor environments. This has been partly a result of the desire to reduce energy costs by sealing up homes and the construction of public and commercial buildings with windows that do not open. Since there are no standards at the present time, the community studies are directed toward source identification, pollutant characterization, indoor-outdoor relationships, and health effects.

The purpose of the remainder of this chapter is to examine the features of various types of community air sampling programs and the parameters that must be considered in the design of individual programs. A discussion of representative examples of different community air sampling studies is also presented.

COMMUNITY AIR SAMPLING

Fixed Outdoor Sampling

The selection of different protocols and methods for measuring air pollution is mainly dependent upon the specific goals of the investigation. The earliest efforts in air monitoring focused on a central monitoring station; for instance, in New York City it was the 121st Street Laboratory.[9] Various pollutants were measured. Originally, mechanized sampling techniques, such as bubblers and high volume samplers, integrated ambient concentrations of pollutants for periods of 24 hours or longer. This was done on a daily basis or on a statistically selected number of days (e.g., every sixth day) at regular intervals throughout the year. Air quality data from these types of monitoring networks provided valuable information on long-term trends and eventually provided a basis for assessing compliance with local and national standards. The original network was called the National Air Surveillance Network (NASN) and was a volunteer effort conducted at selected locations throughout the United States.[10] This network has been superseded by the NAMS and SLAMS network which selects sites and pollutants to be measured using specific siting criteria. The siting documents have been developed by the U.S. Environmental Protection Agency for ambient air criteria pollutants such as photochemical oxidants.[11] In addition, Ott[12] recommended six types of sites, or monitoring categories, that could assist in identifying situations where a variety of human exposures could be measured (Table I).

Aside from the established air monitoring network, the approach to sampling is different for specific types of studies. For example, a state or local control agency may initially conduct a short-term, multiple station intensive survey for particular ambient air pollutants. In recent times this has been done for pollutant or pollutant classes such as volatile organic compounds, ozone, acid sulfates, and nitric acid. After assessment of the measured concentrations, the network can be reduced in size to two strategically located stations. One would monitor an area that receives maximum ambient concentrations while the other would be located in an area that receives minimum ambient concentrations. Any monitoring strategy can be modified as required to examine impacts from emission increases (or decreases) as new control technology is placed on a source, process characteristics are altered, or fuel conversions are implemented. Sometimes monitoring strategies are designed to measure the amount of a particular pollutant transported into a state, country, or province from another jurisdiction.

Industry at times may take a different approach to ensure compliance with existing air quality standards. Managers of an industrial plant responsible for the control of a single pollutant may be most interested in averaging concentrations over specific sampling times from their own source when it is operating at different production levels. Therefore, the industry may design short-term studies. In fact, a plant's total monitoring effort may be focused on relating pollution concentrations to production emissions as a basis for choosing the correct control devices. Consequently, these monitoring efforts could evolve into long-term studies that measure levels both before and after the implementation of a control strategy. The sites would be located primarily at the plant fence line and at particular locations either upwind or downwind of the facility.

In the past ten years, fixed outdoor sampling studies have been

TABLE I. Recommended Criteria for Siting Monitoring Stations

Station Type	Description
TYPE A	Downtown Pedestrian Exposure Station. Locate station in the central business district (CBD) of the urban area on a congested, downtown street surrounded by buildings (i.e., a "canyon" type street) and having many pedestrians. Average daily travel (ADT) on the street must exceed 10,000 vehicles/day, with average speeds less than 15 miles/hour. Monitoring probe is to be located 0.5 meter from the curb at a height of 3 ± 0.5 meters
TYPE B	Downtown Background Exposure Station. Locate station in the central business district (CBD) of the urban area but not close to any major streets. Specifically, no street with average daily travel (ADT) exceeding 500 vehicles/day can be less than 100 meters from the monitoring station. Typical locations are parks, malls, or landscaped areas having no traffic. Probe height is to be 3 ± 0.5 meters above the ground surface.
TYPE C	Residential Population Exposure Station. Locate station in the midst of a residential or suburban area but not in the central business district (CBD). Station must not be less than 100 meters from any street having a traffic volume in excess of 500 vehicles/day. Station probe height must be 3 ± 0.5 meters.
TYPE D	Mesoscale Meteorological Station. Locate station in the urban area at appropriate height to gather meteorological data and air quality data at upper elevations. The purpose of this station is not to monitor human exposure but to gather trend data and meteorological data at various heights. Typical locations are tall buildings and broadcasting towers. The height of the probe, along with the nature of the station location, must be carefully specified along with the data.
TYPE E	Nonurban Background Station. Locate station in a remote, nonurban area having no traffic and no industrial activity. The purpose of this station is to monitor for trend analyses, for nondegradation assessments, and large-scale geographical surveys. The location or height must not be changed during the period over which the trend is examined. The height of the probe must be specified.
TYPE F	Specialized Source Survey Station. Locate station very near a particular air pollution source under scrutiny. The purpose of the station is to determine the impact on air quality, at specified locations, of a particular emission source of interest. Station probe height should be 3 ± 0.5 meters unless special consideration of the survey require a nonuniform height.

From Ott.[12]

designed to investigate the nature of pollutants deposited in acid rain and their relationship to potential ecological effects in lakes and on forest vegetation.[13] In addition, visibility degradation in the western vistas of the U.S. and in the urban-rural areas of the eastern U.S. have been the subject of multipollutant fixed site sampling studies.[14] Epidemiological and field health effects studies require an understanding of the origin and general distribution of a pollutant. However, each health investigation requires sampling in time periods and locations which are appropriate for relating the species or pollutant to a potential effect. At a minimum, the information derived should be appropriate for estimating the dose (concentration × time) presented to an individual for inhalation in a particular environment.

Indoor and Personal Sampling

Unfortunately, in many cases a pollutant or pollutants may have both outdoor and indoor sources. This situation may require a thorough evaluation of a person's total exposure or, at a minimum, the important microenvironments. This will ensure that the major source of exposure can be accurately identified and the result compared to a health effect. Confounding factors, such as occupation, weather, etc., must also be explored to ensure that any potential effect is associated with air pollution.

For some pollutants, the use of a fixed ambient air site in a population center may not accurately assign the exposures for a given individual or population.[15] Prior to major initiatives on emission controls, this was probably more reasonable for compounds such as sulfur dioxide, benzo(a)pyrene, or particulate matter. However, in most areas of the U.S. the strict pollution control regulations have reduced the levels of a number of pollutants (e.g., nitrogen dioxide, volatile organic compounds) to the point where there may be equivalent or higher concentrations indoors.[5] In developing or third world nations, the outdoor levels for some pollutants may still be well above indoor concentrations for these pollutants; however, for situations in which open habachi cooking occurs, the indoor concentration of criteria and non-criteria pollutants can be excessively high.[15] Therefore, for some pollutants indoor and personal air sampling will be required to ensure high concentration situations are not ignored (e.g., volatile organic compounds and nitrogen dioxide).

Chronic health effects studies would require identifying areas or subpopulations that would be subjected to conditions conducive to high, medium, and low pollution exposure. Acute effects studies may be designed in one location where significant temporal changes are possible and are on a scale comparable to the potential effect. Personal monitoring of individuals is very desirable in many such situations. For some pollutants the monitors have become very inexpensive to make and inconspicuous to wear (e.g., passive diffusion monitors).[16]

Pollutant Characterization Sampling

Other common types of community air sampling studies are classified as applied research and are directed toward understanding the physical, chemical, and biologically active nature of the atmosphere (both indoor and outdoor). For outdoor studies the focus or foci can be on: 1) the formation or decay processes of individual compounds, 2) the transport and transformation of pollutants in industrial or urban plumes, 3) the wet or dry deposition of pollutants, 4) the dynamics of pollution accumulation and removal in urban and rural locales, 5) photochemical smog and other types of episodes, and 6) source tracer measurements. These can involve fixed site sampling, mobile vans and trailers, and aircraft sampling platforms. For indoor studies, the emphasis will be on characterization of indoor sources, outdoor pollution penetration, adsorption, absorption, and transformation and transformation rates. In these cases, samplers will be located in one or more rooms throughout a house or, more commonly, a group of houses.

GENERAL FEATURES OF COMMUNITY STUDIES

Using the previous section as a guide, it is immediately apparent that a number of factors must be considered when designing a community air pollution sampling program. Key articles or books can be useful in designing the details of a specific type of study and representative examples are found in the reference list. There are some basic or fundamental steps that must be considered prior to any community air sampling study, and these are outlined in the following sections.

Sampler Location

Selection of an outdoor air monitoring site requires addressing a number of considerations which will affect pollution values recorded at any given time. These include the effects of 1) point sources, 2) obstructions or changes to airflow caused by tall buildings, trees, etc., 3) abrupt changes in terrain, and 4) height above ground for a sampler or sampler probe.

Beyond considerations of the physical location of the monitor, each site must be representative of the design questions being asked. All too often, the investigator can select a location which seems practical in terms of availability and electrical needs, but would severely compromise the intent of the study. At times, practical problems cannot easily be resolved, but the integrity of the site for answering pollutant-related questions is the prime consideration.

For example, if the objective is to monitor the concentrations of carbon monoxide from automobiles within the center of a city, the concentrations inhaled would be found in the breathing zone approximately two meters above the street and the highest values between tall buildings forming a street canyon. In addition, personal exposure may be enhanced by the driving habits of an individual and time spent outdoors. Thus, the maximum concentration for a specific sub-population may be associated with exposure from a specific microenvironment like the cabin area of a car or bus or parking garage.[17]

In contrast, ozone monitors are normally placed at some distance from the primary sources of their precursors, nitrogen oxides and hydrocarbons. In the early 1970s, Stasiuk and Coffey made a major observation in a rural area of New York State.[18] Their results showed the presence of ozone concentrations in a rural area at or above levels found in major urban areas. This finding required the scientific and regulatory community to re-evaluate where and when high ozone would occur in the Eastern U.S. and where population exposures could be significant.[19] Today ambient ozone monitors are located in rural and suburban areas throughout the United States and other countries. During the summer, the eight-hour averages of ozone can be above the occupational threshold limit value (TLV) of 100 ppb which creates a situation ripe for the study of potential health effects.[20] Since this occurs in rural and suburban areas, a study would not be confounded by the presence of many locally generated pollutants.

Sampler Location for Health Studies

Generally, air sampling conducted in support of health effects studies requires the data to be representative of population exposures and should be conducted coincidentally with any measurements of a health outcome. Once a population or populations are defined and the need for personal and indoor sampling has been assessed, a number of central sampling stations, personal monitors, and/or indoor air samplers must be positioned to obtain adequate representation of where the selected population lives. The size of the population to be studied as well as the number of homes to be used can be determined from epidemiologic principles[21,22] and must take into account a variety of typical personal habits and lifestyles. The positioning of the outdoor monitors requires a full understanding of the nature of pollutant accumulation under ambient conditions. From our previous examples, adequate measurement of the distribution of ambient carbon monoxide in an urban environment requires many more samplers than are required to measure exposure to ozone. The basic reason is that carbon monoxide accumulates in confined spaces, and produces a large range in concentrations over a short spatial range. In contrast, ambient ozone concentrations will vary rather uniformly over a large area, although some local differences will be observed because of high local concentrations of nitrogen oxide.

The mobility of the study population is important in defining the outdoor sampling situations. Consideration must be given to the variations in exposure to a pollutant emitted near a residence, a place of employment, transportation routes, school, and recreational activities (Figure 1).

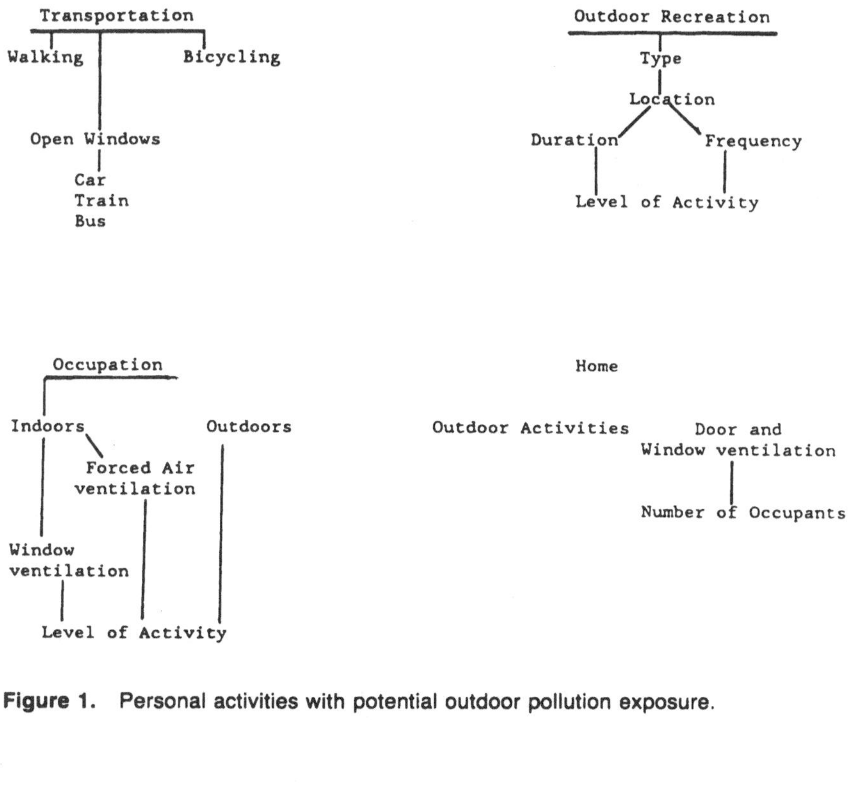

Figure 1. Personal activities with potential outdoor pollution exposure.

Sampling Frequency and Duration

In any community air pollution study the length of the sampling program is shaped by a number of factors, not the least of which is the amount of resources available to conduct the study. However, this perennial problem aside, the purpose of a community air pollution program will have a major influence on the duration of the sampling activities. For example, long-term trend studies should be conducted for multiple years to obtain data on the range of concentrations, overall influence of meteorology, and changes in emission strength.[23] Examination of peak concentrations, diurnal variations, and episode conditions requires a study design which attempts to obtain a sufficient number of sampling days or hours to include a representative number and range of events. Unfortunately, in many instances only one or two major events will occur. For example, the study of photochemical smog requires sampling to be conducted for an extensive period during the summer (approximately 30 days). This would provide a sufficient number of sampling days for measuring the impact of one or more episodes.[24] If continuous samplers are used for hydrocarbon, ozone, organic species, and nitrogen oxide measurements, the kinetic processes associated with the accumulation of oxidant species can be examined and subsequently modeled.[25]

Another example would be the examination of local source impacts on the surrounding neighborhood. In this case, an intensive community air sampling program would have an array of sampling sites placed around the facility. The size of the emission source would dictate the extent of the array, e.g., an industrial source vs. a gasoline station vs. wood burning, but the number should be sufficient to detect concentration variability due to changes in wind direction (Figure 2). This type of spatial coverage provides an investigator with the opportunity to determine the background concentrations. The duration of the study would have to be sufficient to identify the meteorological conditions conducive to maximum plume impact.[1] The study could be at least a year in duration, but the approach could include an intensive period of study for gathering baseline information and then a second phase which is only activated when specific meteorological conditions are predicted to occur. For individual pollutants present in a plume, it may be important to consider the use of indoor and personal samplers. The true picture of the impact a plant is having on a community is better assessed by understanding the significance of the indoor penetration of outdoor accumulated pollutant(s). In some cases, however, the deposition of particulate matter over time on the soil or in water may potentially be the most important source of a pollutant to man, plant, or animal. Obviously, this situation will

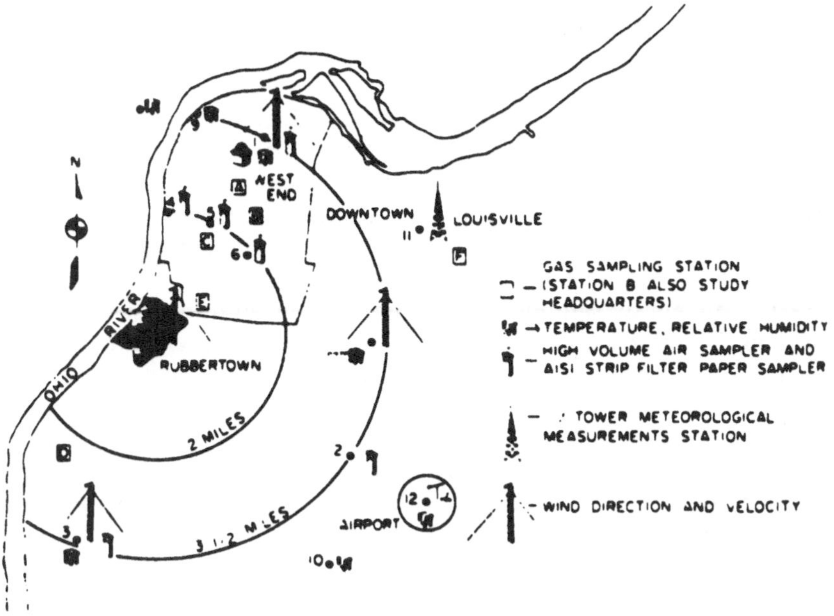

Figure 2. Site locations for air samplers and meteorology around a major source.[1]

require determining a pollutant's concentration in the soil or the groundwater and exposure estimation from ingestion as well as inhalation.

Averaging Time

Sample averaging time is dependent upon the instrumentation available to conduct a study, the detection limits for a particular compound, and the time resolution required to discriminate particular events or effects. For many of the gaseous criteria pollutants, this will not pose a problem since most devices are continuous samplers.[26]

For volatile organic compounds (VOC), the devices primarily integrate a sample over an interval that is usually 24 hours in duration.[27] Unfortunately, there is a wide range of artifact, breakthrough, and equilibrium problems associated with VOC which limit the number of compounds that can be detected reliably. At the present time, some state agencies have set up routine air monitoring networks for VOC. The EPA has initiated a Toxic Air Monitoring System (TAMS) which measures approximately 15 VOC using a steel canister or a set of four distributed volume Tenax samplers for 24 hours on an every-sixth-day sampling cycle.[28] However, the results of the Total Exposure Assessment Methodology (TEAM) studies have indicated that the major route of population exposure to VOC is indoor air, suggesting a different focus for future investigations.[6] Further development of VOC sampling techniques, such as continuous monitors, will occur as more researchers attempt to measure more of the volatile compounds present in the outdoor and indoor environment, especially around specific toxic pollutant sources, landfills and hazardous waste sites.

Particulate matter sampling has normally been conducted with devices that integrate samples. For most routine EPA and state regulatory monitoring efforts through the early 1980s, Total Suspended Particulate (TSP) samples are collected on a statistically based every-sixth-day schedule and are 24 hours in duration. This approach is adequate for the determination of mass and selected inorganic and organic constituents. For other material, such as semi-volatile organics, and hydrogen ion (which represents aerosol acidity), other samplers with much shorter ($<$ 6 hours) and more frequent samples (4/day) should be used during a community air sampling program. For some compounds, denuders are required prior to a filter; for others, additional samplers are required after the filter to reduce artifacts due to filter absorption or desorption of compounds during a specific sampling period.

In Chapter H of *Air Sampling Instruments*,[29] the need for conducting size-selective particle sampling was discussed, and the types of instruments and methods were identified. Presently, a number of size-selective devices are used in ambient, personal, and indoor studies. These are designed to collect particles presented to various regions of the lung (Figure 3). The most common are the thoracic samplers ($d_{50} = 10\ \mu m$), respirable samplers ($d_{50} = 3.5\ \mu m$), and fine particle samplers ($d_{50} = 2.5\ \mu m$). Devices have been developed which can be used to detect particle size fractionated mass and many inorganic and organic constituents in time intervals of four hours or less. In addition, for the dichotomous filter sampler,[30] automated models are available which provide the opportunity to obtain up to 36 consecutive samples over various time durations. The minimum or maximum duration of any particular sample would be dependent upon the detection limits for the compounds measured, and the range of pollution levels present in a particular area.

The EPA is considering the adoption of a PM-10 (particulate mass collected with a 50% cut size of 10 μm), and has published the rationale for this change.[29] Community air pollution sampling will be conducted for this particulate fraction, and it is anticipated that the sampling schedule will be every day. This will be a very labor intensive program which

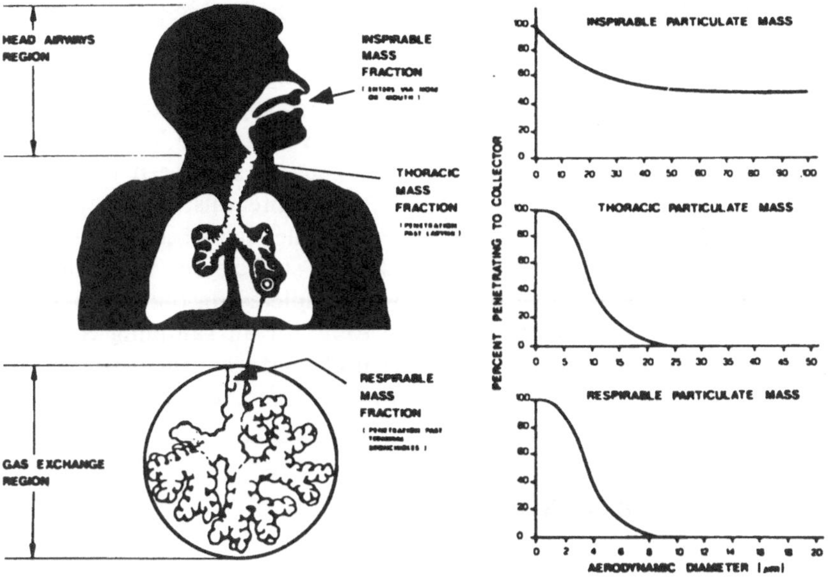

Figure 3. The three aerosol mass fractions recommended for particle size-selective sampling.

will ultimately lead to the development of reliable continuous particle samplers. Personal samplers with respirable particle inlets are presently used in indoor air pollution studies.[30] At the present time, a personal impactor is available, and it is anticipated that a PM-10 Personal Impactor will be available in the near future.

Chemical Analyses

Presently, community air sampling surveys for particulate matter usually require analyses beyond a traditional mass determination. Some fairly routine items include numerous trace elements (Pb, Cd, Cs, Fe, Se, As, Br, etc.), water soluble sulfate, ammonium ion, chloride, and nitrate.[32] More sophisticated studies measure the acidity of sulfate particles,[33] the organic mass fractions,[34] specific organic species,[35] and elemental carbon.[36]

Samplers have been developed for outdoor studies that can routinely collect particulate organics on a filter and have a vapor trap that contains polyurethane foam. The trap collects semi-volatile species that evaporate from the filter during sampling. These devices are housed in a Hi-volume sampler shell and operate at 40 CFM.

Indoor air pollution area particulate matter samplers have been developed by a number of investigators. The most sophisticated are the Marple-Spengler-Turner samplers, which collect mass on a Teflon filter and can have single or multiple aerodynamic cut sizes between 10 and 1 μm.[37] The samples are collected for various time intervals: $<$ 1 day to 4 days. Most analyses performed for ambient samples can be conducted on indoor impactor filters, if the mass loading is sufficiently high, and a limited number of destructive analyses are completed on the samples.

A careful review of modern analytical tools for air sampling can be found in an APCA critical review.[32] Other useful information is found in the works of a number of authors listed in the reference section.

Biological Assay of Air Samples

The identification of potentially carcinogenic compounds in the atmosphere has led to the development and application of techniques for the measurement of the biological activity of particulate and gaseous samples. In principle, actual animal bioassays of carcinogenic air pollution are the most direct measure of potential effects;[38] however, these are difficult to conduct on ambient air. As an alternative to actual animal bioassays, short-term *in vitro* bioassays have been employed to examine the mutagenic properties of organic materials. Of the known carcinogens, over 80 percent have been shown to be mutagenic in the Ames Assay,[39] which has been the method of choice for application to air pollution samples.[40] The concentration results of such assays have been reported as revertant colonies per m^3 of air; for mutagenic potency, these have been reported as revertant colonies per μg of sample. Through 1985 over 50 studies have used the Ames Assay on air pollution samples, and the results with references have been summarized by Louis et al.[41] At present, further research is being directed to the identification of the actual compounds contributing to the mutagenic activity of the air pollution samples.[42]

The field of biological measures or markers of exposure is just beginning to evolve. It is anticipated that techniques will continue to be developed which will permit the precise measurement of a biologically effective exposure and dose to individuals in a community. This could include the use of cytotoxicity tests as well as the more frequent application of breath analyses.[16]

Data Retrieval and Analysis

For the state monitoring networks, most data are sent by telemetry to a central station for computerized data recording, validation, and processing of many of the continuously measured pollutants. At a minimum the criteria pollutants NO_x, O_3, SO_2, and CO are monitored. Eventually, these data are formatted with identifying parameters and sent to the National Aerometric Data Bank. The format used for data archived by the EPA is called SAROAD.

Manually collected particulate pollution data are eventually entered by tape or by hand into a computer after a number of chemical analyses have been performed on a series of samples.

Intensive field studies conducted at a given location have become much more sophisticated in both study design, monitoring equipment, and data gathering practices. The former will be discussed in a separate section; however, improvements in the use and application of sampling and data retrieval equipment has been significant. Beyond the construction of fixed monitoring sites with telemetry systems, a number of groups have developed and utilized fully equipped mobile vans or trailers. One of the most sophisticated mobile units is a trailer designed by General Motors Research Laboratories, which has a complete computerized system for continuous samplers, calibration facilities, and integrated samplers.[43] The trailer has been used in field studies on atmospheric chemistry and air pollution throughout the United States. Another mobile unit has been developed cooperatively by the New York University and the Environmental Protection Agency. This van houses both monitoring equipment and health measurement equipment in separate sections of the same vehicle. The continuous environmental exposure monitors (O_3, NO_x, SO_2, H_2SO_4) and Total Sulfur Analyzer use a datalogger which feeds a cassette recorder. Designed for acute health effects studies, this van is equipped to make measurements of the pulmonary function parameters (Force Vital Capacity, Force Expiratory Volume in one second, Peak Expiratory Flow Rate) from a spirometry system. The spirometry is conducted on two computerized systems with memory to store the lung function tracings and calculate function parameter values for an individual. The information is displayed on a monitor for visual observation during actual tests.[44]

Air Pollution Modeling

When a community air pollution program or study is designed, one of the primary activities to be considered for the collected data is model development. Models play an important role in the examination of air pollution since they can be used to: 1) determine the effectiveness of control strategies, 2) predict pollutant concentrations downwind of sources, 3) examine chemical processes, 4) examine regional transport questions, 5) establish source-receptor relationships, and 6) estimate human exposure. In all cases, though, an important final step is the validation of a model using community sampling data. The acquisition of these types of pollution data would require carefully selected protocols since application of a model would pose certain constraints on the selection of variables, study duration, sampling frequency, number of samples, and site selection.

The most common form of model applied to air pollution is the *dispersion model*. It estimates the contributions of a pollutant emitted by a source and the impact at a downwind receptor site for a variety of meteorological conditions. The technique has been used in community air pollution for many years and is available for specific source and terrain applications as off-the-shelf-models. Turner[45] in 1979 published a review of dispersion modeling which covers the general features of the technique and the source and meteorological information necessary to apply the various types of models. In 1984, Hidy[46] completed a follow-up review of air pollution modeling issues, but focused on regional models and their application to the source-receptor relationships of acid deposition.

Another modeling approach which has been used extensively since the late 1970s is *receptor-source* apportionment. The general technique involves constructing a model that determines the sources that contribute to pollution levels observed at a receptor location. In contrast to dispersion modeling, it derives information from composition data collected at the receptor (sampler). The significance of the source is obtained by measuring the concentration of a source tracer at the receptor and other pollutants emitted by the source by using the principal of conservation of mass.

A number of different approaches are available for receptor models and include Chemical Element Balance,[47] Factor-Analysis Multiple Regression,[48] and Target Transformation.[49] Applications of these models are summarized by Hopke[50] and a review of their use in transport modeling studies has been published by Thurston and Lioy.[51]

A more recent approach to modeling that examines the frequency and distribution of a pollutant among a population is human exposure modeling. This involves the acquisition of pollutant concentration data, identification of the activities of an individual, and the estimation of time spent by an individual in a particular microenvironment. From these data, a model can be constructed that links the presence of an individual at a location in a microenvironment to the concentration of a pollutant present at that location for a certain length of time. The exposures estimated are usually the average or integrated values for each microenvironment. A recent application of this technique was completed by Ott for carbon monoxide.[7]

EXAMPLES OF COMMUNITY AIR POLLUTION STUDIES SINCE 1970

The California Aerosol Characterization Experiment (ACHEX)

One of the most comprehensive investigations of outdoor air pollution was ACHEX[52] (Table II). The major objectives of the study focused on describing the physical and chemical characteristics of photochemical smog aerosols and their relationship to the reduction of visibility in Southern California. As an adjunct to this study, the three-dimensional distribution and transport of a number of pollutants were examined in the region.

The approach involved very intensive case study sampling protocols within a particular 24-hour (midnight to midnight) period. Actual experiments were conducted based upon the results of a meteorological forecast one day before an experiment and the potential for high ozone levels. Depending upon the instrumentation used for a particular pollutant, the sample durations were from 10 minutes to two hours. This provided an opportunity to obtain detailed information on the origin and evolution of the smog aerosol. The basic ACHEX study lasted from 1972 through 1973, with most of the field activity concentrated in the summer and the early fall. In contrast to other studies, the measurement systems were located in a mobile van which took samples at a number of locations. These sites were areas with either high source emission densities, or receptors of reacted species, or non-urban areas. The measurements in the van were supplemented by a number of fixed sites in the Southern California Basin.

In addition to the particulate mass, a number of species were measured including sulfates, organic compounds, and trace elements. Particle size distributions were obtained from a series of analyzers that covered the

TABLE II. Site Locations for the Mobile Van During the ACHEX Study

Occupancy Date	Site	Location	Environment
A. 1972			
July 15–30	Berkeley	College of Agriculture tract, 2175 Hearsat Avenue Berkeley, California	Acceptance tests and preliminary checkout
Aug. 1–13	Richmond	San Pablo Water Pollution Control Plant, 2377 Garden Tract Road, San Pablo, California	Urban-Industrial (downwind from chemical complex)
Aug. 15–27	San Francisco airport	On the airport property at Bayshore Drive and end of north-south runway	Aircraft-enriched aerosol; receptor of aerosol from west San Francisco Bay area
Aug. 28–Sept. 11	Fresno	Fresno County Fairgrounds, 1121 Chance Avenue, Fresno, California	Photochemical-agricultural
Sept. 12–18	Hunter-Liggett Military Reservation	Meadow areas, 1 mile south of the public road connecting Hunter-Liggett with Route 1	Vegetation-enriched (expected photochemical aerosol production from vegetation organic emissions)
Sept. 19–Oct. 2	Freeway Loop	May Company parking lot, 33rd Street and Hope Street, Los Angeles, California	Auto-enriched (100 ft. east of Harbor Freeway)
Oct. 3–30	Pomona	Los Angeles County Fairgrounds, Pomona, California	Los Angeles photochemical (receptor)
Oct. 30–Nov. 4	Goldstone	Goldstone Tracking Station Barstow, California	Desert background
Nov. 6–10	Point Arguello	U.S. Coast Guard LORAN Station	Marine-enriched
B. 1973			
July 11–Aug. 9	West Covina	Hospital on Sunset Avenue, West Covina, California	Los Angeles photochemical (receptor)
Aug. 11–25	Pomona	Los Angeles County Fairgrounds, Pomona, California	Los Angeles photochemical (receptor)
Aug. 27–Sept. 26	Rubidoux (Riverside)	Rubidoux Water Treatment Plant, Rubidoux, California	Los Angeles photochemical (receptor)
Sept. 29–Oct. 11	Dominguez Hills	California State College, Dominguez Hills, California	Source-dominated (refineries, chemical industry)

From Hidy et al.[53]

nuclei through the coarse particle size ranges[29] and were processed as 10-minute averages. Continuous measurements were obtained for a number of gases including total hydrocarbons and ozone. A complete listing of the chemical constituents measured and the techniques used is found in Hidy et al.[53]

Some major findings in the study include detailed information on the duration and multi-modal nature of the particle size distribution, the chemical composition of different size fractions, the episode intensity, the source apportionment of the particulate mass in the atmosphere, and the particle size fractions and sources causing visibility reduction.

The Denver "Brown Cloud" Study

The "Brown Cloud" study[54] was a major study which characterized the ambient atmosphere, but this time during the winter. This mesoscale ($<$ 100 km) pollution phenomenon is topographically induced and affects the metropolitan Denver area. A major feature of this investigation, which was conducted in November and December of 1978, was the measurement of both the organic and elemental carbon content of size-fractionated particulate mass. Chemical analyses similar to those conducted in ACHEX provided the opportunity to conduct source apportionment studies. In addition, the species contributing to visibility reduction in Denver were estimated.

A very elaborate and well-designed community sampling program was established for this intensive investigation (Figure 4). It included surface-based sites to measure 1) the maximum impact of the pollution contained in the cloud, 2) the background concentrations, and 3) the characteristics of the cloud throughout various parts of the city. Aircraft measurements were made in the vertical and horizontal direction to examine cloud dynamics.

Measurements were made of the fine and coarse particle mass; sulfate, nitrate, and ammonium ions; trace elements; and the carbon fractions. Source apportionment studies were conducted using the chemical element balance technique, and regression models were developed to estimate the contributors to visibility reduction.

The Airborne Toxic Element and Organic Substance (ATEOS) Project

The ATEOS was another community air pollution characterization study but was designed to investigate not only the atmospheric dynamics

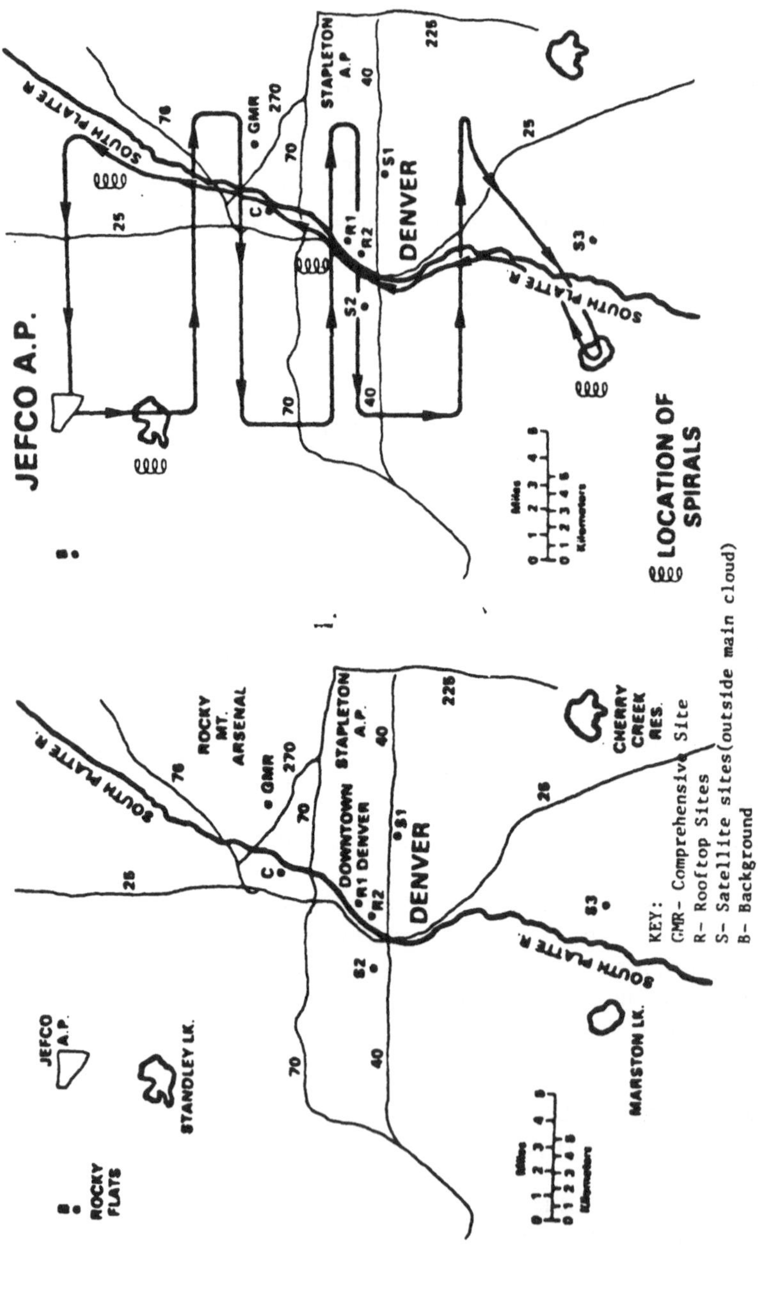

Figure 4. The Denver "Brown Cloud" study.[54]

and distribution of pollutants, but also the potential risks to the human health from biologically active materials present in the outdoor air.[55] More than 50 pollutants were measured simultaneously at fixed monitoring sites within three urban areas in New Jersey and at a rural location. Each site was selected to reflect a different type of industrial-commercial-residential interface that was representative of a different type of exposure situation rather than the entire city. For example, the Newark site was within a residential area surrounded by small and moderate size industrial facilities, e.g., body shops, chemical manufacturers (Figure 5). The site was a surrogate for the exposures to ambient pollutants in the local community rather than the central business district and its residents.

Each day measurements were completed in Newark, Elizabeth, Camden, and Ringwood (the latter is the rural site) over the course of six weeks in the summers of 1981 and 1982 and the winters of 1982 and 1983. Most samples were 24 hours in duration, and these included Inhalable Particulate Mass (d_{50} = 15 μm) and volatile organic compounds (VOC). The mass was analyzed for a number of components including: nonpolar through polar organics, polycyclic aromatic hydrocarbons, sul-

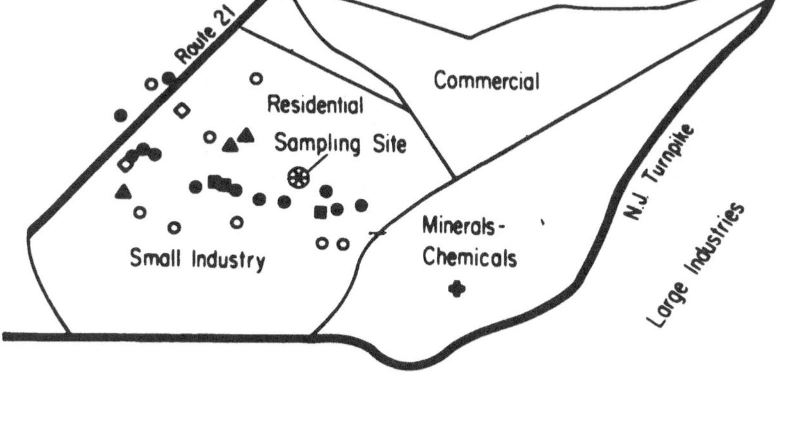

Figure 5. The distribution of sources around the subpopulation studied in ATEOS.[55]

fates, trace elements, mutagenicity, and alkylating agents. A total of 26 VOC were measured including benzene, toluene, chlorinated hydrocarbons, and vinyl chloride

The results of the study have produced useful information on potential human exposures to biologically active compounds in different types of community settings and identified target populations for future epidemiological studies. Other results have defined seasonal and interurban variations, the intensity of summer and winter episodes, and the sources of the mass and its organic fractions.[56]

The Harvard Air Pollution Health Study

The community air sampling being conducted in the Harvard study, which is familiar to many as the Harvard Six Cities Study, includes outdoor, indoor, and personal air monitoring.[57] The air sampling is in support of the ten-year prospective examination of respiratory symptoms and pulmonary function of children and adults living in the six communities of Topeka, KA, Portage, WI, Watertown, MA, Kingston, TN, St. Louis, MO, and Steubenville, OH. Indices of acute and chronic respiratory effects and pulmonary function performance are being examined in relation to any adverse effects of ambient and indoor air pollutants.

Fixed outdoor air sampling sites are located in each community. However, since exposure to a number of air pollutants, such as nitrogen dioxide, respirable particles, and formaldehyde, can be associated with a number of microenvironments, indoor and personal samples are required for individuals participating in the study. The use of each of the above types of samples provides microenvironment and personal pollution data for the development of exposure models as well as estimation of the influence of various activity patterns. The most extensive indoor data base developed in the Harvard study is for nitrogen dioxide and respirable particles. As can be seen in Figure 6, the respirable particles measured or monitored indoors contribute the dominant proportion of the personal exposure.

The fixed monitoring sites measure total suspended particulates, respirable particles, trace elements, sulfate, acid sulfates, ozone, and other pollutant gases. The indoor particulate samples are analyzed for a number of the above and in special studies they are analyzed for tracers specific to tobacco smoke. The Harvard study has provided an opportunity to conduct a number of supplementary studies and add pollutants as the measurement technology; the potential for pollutants to affect public

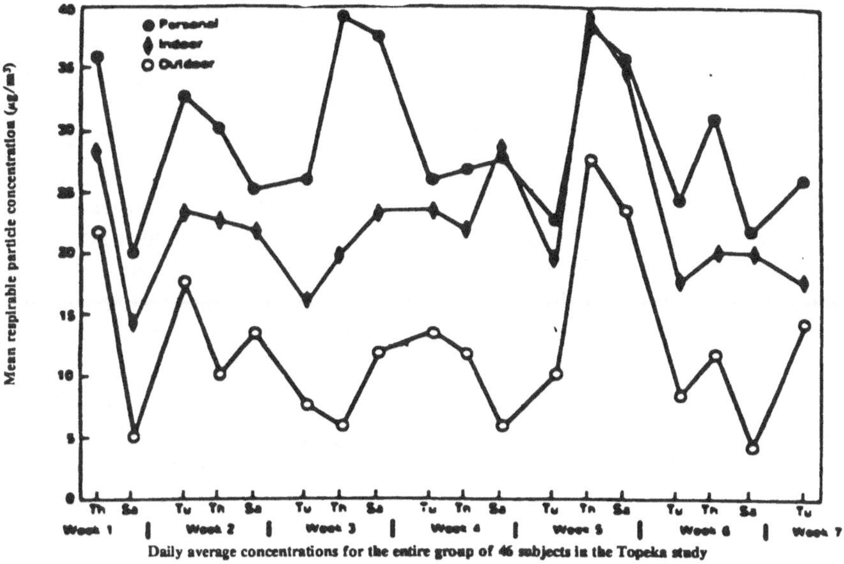

Figure 6. Personal, indoor, and outdoor levels of respirable particles in Topeka, KN during the Harvard Health Study.[57]

health is defined. For instance, measurement of acid sulfates has been added after investigations conducted by a number of researchers determined that significant concentrations of acid-sulfate species are present at times in the outdoor atmosphere. In fact, the interurban differences in the pulmonary function measured for the children in each city appear to be linked to the variation in acidic species from city to city.

The Total Exposure Assessment Methodology (TEAM)

The last study that will be examined in this chapter, although there are many other examples, is the TEAM.[6] This was another unique approach to the study of community pollution. The investigation was actually a series of studies conducted in ten cities in the U.S. from 1980 through 1984. Since the TEAM was a statistically designed study, inferences could be drawn about the general population living in certain areas of Elizabeth/Bayonne, NJ, the South Bay of Los Angeles, CA, and Antioch/Pittsburgh, PA.

The investigation primarily involved measuring the personal exposures of 700 individuals to 20 volatile organic compounds and the corresponding body burden. Fixed site outdoor sampling was conducted next to the homes of the participants. Personal monitoring was completed on each individual, and the levels of the compounds were measured in exhaled breath as an indication of levels which could be found in an individual's blood. Drinking water and beverage concentrations of the VOC were determined, and detailed questionnaires were administered concerning occupation, hobby, and home VOC sources. Other sub-studies conducted in the TEAM included indoor microenvironmental sampling.[58]

A major result of the TEAM has been that the outdoor environment is not the primary contributor to personal VOC exposures (Figure 7). In many cases the concentrations were 10 to 100 times higher indoors than those observed outdoors. Also significant variations in the VOC concentrations can occur in a small geographic area during the day. Therefore, when planning epidemiological studies, investigators will have to consider the possibility of not having uniform exposures for VOC.

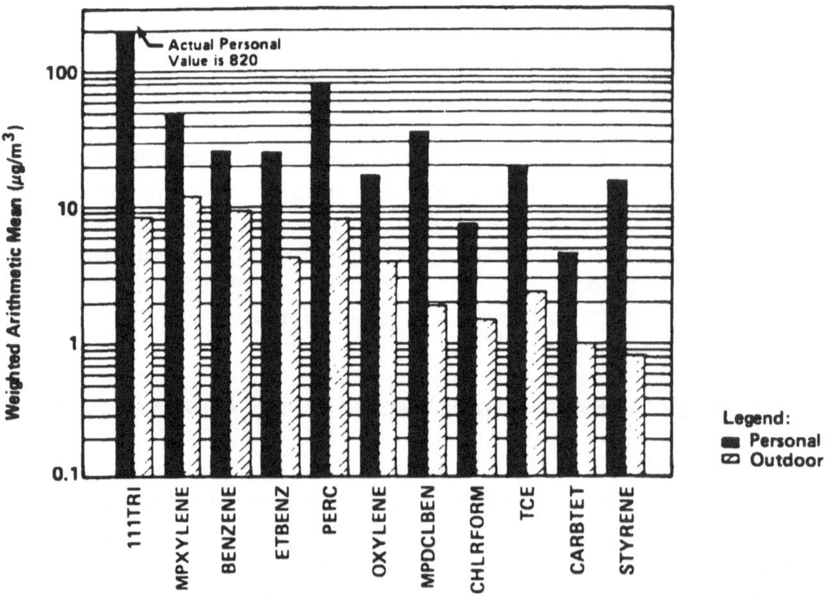

Figure 7. Estimated arithmetic means of 11 toxic compounds in daytime (6:00 am to 6:00 pm) air samples for the target population (128,000) of Elizabeth and Bayonne New Jersey, between September and November 1981. Personal air estimates based on 340 samples, outdoor air estimates based on 88 samples.[6]

Summary

All of the above illustrate different types of research studies that have been conducted with regard to community air pollution. While these are by no means inclusive of all the types of investigations that can be designed, they do demonstrate: 1) the breadth and depth to which studies of these types can examine the nature of the outdoor or indoor community atmosphere, 2) the advances in instrumentation and analyses for measuring pollutants, and 3) the flexibility available for finding appropriate sampling locations and for conducting experiments of adequate duration to obtain meaningful exposures.

COMMUNITY AIR SAMPLING AT HAZARDOUS WASTE AND LANDFILL SITES

Chemical waste disposal in the United States has been a significant problem for many years. Until recently, the methods of disposal have not been responsibly developed and the disposal has usually been uncontrolled by industry or insufficiently regulated by governments.[59] Numerous examples of chemical disposal sites can be identified across the nation, with a few of the more well known being the Love Canal, NY; Hyde Park Land Fill, Niagara Falls, NY; Rollins Landfill, Baton Rouge, LA; Chemical Control, Elizabeth, NJ; and Jersey City Landfill, NJ.[60] The individuals exposed to the emissions from the many types of disposal sites include persons living and working in communities adjacent to the area. Their air pollution exposures may result from the inhalation of particles (fugitive dusts), fumes, or vapors dispersed from a dump in the active stage, inactive stage, or mitigation stage. They may also occur as a result of leachate in moving groundwater which contaminates a well and is used for showers and tap water in a home.

The documentation of the air pollution exposure will require a number of activities similar to those described in previous sections. These include a screening stage as well as an intensive study and a follow-up long-term investigation. The study design would, of necessity, have a component that characterizes the emissions from the site. The typical approach would include bore hole testing for potential volatile organic emissions and groundwater sampling for compounds that could be dispersed to wells. Further sampling could be made of the soil in the dump to determine the content of organics and trace elements that could be dispersed as fugitive emissions to the surrounding neighborhood.

In contrast to general community air sampling studies, the nature of the emissions may be quite variable since the waste will generally not be homogeneous within a dump. Also, any active depositing of wastes or any activities associated with removal could change the emission rates. Thus, a series of emissions tests may be required throughout the period encompassed by the study. Other screening activities would involve the use of hydrogeologists and meteorologists to model the movement of contaminants through the water and air, respectively.

The second stage of a study would document the nature and extent of population exposure to the pollution and would involve both area and personal sampling. A study would include: the identification of the most probable downwind directions for outdoor air and indoor air contamination, the downstream direction for groundwater contamination and the identification of a control area. After a series of samples are taken, the most exposed area would be subject to follow-up investigation of the ambient environment and persons living in the area. The approach would eventually provide the basis for the development of epidemiological studies on specific health end points.

The analytical approaches used would probably follow those presently available for both air pollution and industrial hygiene investigations. However, the specific protocols would be dependent upon the scope of the investigation. The sensitivity of the techniques used would probably be that required for ambient air sampling, since ambient air usually contains the lowest concentrations of contaminants such as volatile organic compounds.

ACKNOWLEGMENTS

The author wishes to thank Dr. Jed Waldman of UMDNJ for reviewing the content and Mrs. Arlene Bicknell for typing and editing the manuscript. Support for this work was derived from the State of New Jersey, Division of Environmental Quality and Office of Science and Research, #DEP C29529.

REFERENCES

1. Sterns, A.C., Ed.: *Air Pollution*, Vol. I-V. Academic Press, New York (1977).
2. *Federal Register 36*:8186 (1971).
3. Wolff, G.T.: Mesoscale and Synoptic Scale Transport of Aerosols. *Aerosols: Anthropogenic and Natural—Sources and Transport*, pp. 338, 379–388. T.J. Kneip and P.J. Lioy, Eds. Annals of NY Acad. of Sciences (1980).

4. National Research Council, Committee on Indoor Pollutants: *Indoor Pollutants*. National Academy Press, Washington, DC (1981)
5. Yocum, J.: Indoor-Outdoor Air Quality Relationships—A Critical Review. *JAPCA 32*:500–519 (1982).
6. Wallace, L., E. Pellizzari, T. Hartwell et al.: Personal Exposures. Indoor-Outdoor Relationships and Breath Levels of Toxic Pollutants Measured for 355 Persons in New Jersey. *Atmos. Environ. 19*:1651–1661 (1985).
7. Ott, W.: Exposure Estimates Based Upon Computer Generated Activity Patterns. *J. Toxicol.—Clin. Toxicol. 21*:97–128 (1983–84).
8. Lioy, P., J. Waldman, A. Greenberg et al.: The Total Human Environmental Exposure Study (THEES). Prepared for presentation at the 80th Annual Meeting of the APCA, New York, New York (1987).
9. Eisenbud, M.: Levels of Exposure to Sulfur Oxides and Particulates in New York City and their Sources. *Bull. New York Acad. Med. 54*:991–1011 (1978).
10. U.S. Environmental Protection Agency: *Air Quality Criteria for Particulate Matter and Sulfur Oxides*, Vol. I-IV. EPA-600/8–82–029a. ECAO. Research Triangle Park, NC (December 1982).
11. U.S. Environmental Protection Agency: *Site Selection for the Monitoring of Photochemical Air Pollutants*. EPA-450/3–78–013. OAQPS (April 1978)
12. Ott, W.: Development of Criteria for Siting of Air Monitoring Stations. *JAPCA 27*:543–547 (1977).
13. U.S. Environmental Protection Agency: *The Acid Deposition Phenomenon and Its Effects*. EPA-600/8–83–016BF. OAQPS (July 1984).
14. White, W.H., Ed.: Plumes and Visibility: Measurements and Model Components. *Atmos. Environ. 15*:1785–2406 (1981).
15. National Research Council, Committee on Air Pollution Epidemiology: *Epidemiology and Air Pollution*, 224 pp. National Academy Press, Washington, DC (1985).
16. Wallace, L. and W.R. Ott: Personal Monitors: A State of the Art Survey. *JAPCA 32*:601–610 (1982).
17. Akland, G.G., T.D. Hartwell, T.R. Johnson and R.W. Whitmore: Measuring Human Exposure to Carbon Monoxide in Washington, DC, and Denver, CO, during the Winter of 1982–1983. *Environ. Sci. Tech. 19*:911–918 (1985).
18. Coffey, P.E. and W.N. Stasiuk: Evidence of Atmospheric Transport of Ozone into Urban Areas. *Environ. Sci. Tech. 9*:59–62 (1975).
19. U.S. Environmental Protection Agency: *Review of the NAAQS for Ozone: Preliminary Assessment of Scientific and Technical Information*. OAQPS, Research Triangle Park, NC (March 1986).
20. Rombout, P.J.A., P.J. Lioy and B. Goldstein: Rationale for an Eight-hour Ozone Standard. *JAPCA 36*:913–919 (1986).
21. Morris, J: *Uses of Epidemiology*, 262 pp. Churchill Livingston, New York (1975).
22. Lilenfeld, A.M. and D.E. Lilenfeld: *Foundations of Epidemiology*, 2nd ed., 375 pp. Oxford University Press, New York (1980).

23. Lioy, P.J., R.P. Mallon and T.J. Kneip: Long-term Trends in Total Suspended Particulates, Vanadium, Manganese, and Lead at a Near Street Level and Elevated Sites in New York City. *JAPCA 30*:153–156 (1980).
24. Lioy, P.J. and P.J. Samson: Ozone Concentration Patterns Observed During the 1976–1977 Long Range Transport Study. *Environ. Intl. 2*:77–83 (1979).
25. Seinfeld, J.H.: *Atmospheric Chemistry and Physics of Air Pollution*, 738 pp. Wiley Interscience, John Wiley & Sons, New York (1986).
26. Lioy, P.J. and M.J. Lioy, Eds.: *Air Sampling Instruments for the Evaluation of Atmospheric Contaminants*, 6th ed., Chaps. A-V. American Conference of Governmental Industrial Hygienists, Cincinnati, OH (1983).
27. Thompson, R.: Air Monitoring for Organic Constituents. *Air Sampling Instruments for the Evaluation of Atmospheric Contaminants*, 6th ed, Chap. D. P.J Lioy and M.J. Lioy, Eds. American Conference of Governmental Industrial Hygienists, Cincinnati, OH (1983).
28. Walling, J.F.: The Utility of Distributed Air Volume Sets When Sampling Ambient Air Using Solid Adsorbants. *Atmos. Environ. 18*:855–859 (1984).
29. Lippmann, M.: Size-Selective Health Hazard Sampling. *Air Sampling Instruments for the Evaluation of Atmospheric Contaminants*, 6th ed., Chap. H. P.J. Lioy and M.J. Lioy, Eds. American Conference of Governmental Industrial Hygienists, Cincinnati, OH (1983).
30. Lou, B.W., J.M. Jaklevic and F.S. Goulding: Dichotomous Vertical Impacters for Large Scale Monitoring of Airborne Particulate Matter. *Fine Particles: Aerosol Generation Meaurement, Sampling and Analysis*, pp. 311–350. H. Liu, Ed. Academic Press, New York (1976).
31. Spengler, J.D., R.D. Treitman, T.D. Losteson et al.: Personal Exposures to Respirable Particulates and Implications For Air Pollution Epidemiology. *Environ. Sci. Tech. 19*:700–706 (1985).
32. Katz, M.: Advances in the Analysis of Air Contaminants: A Critical Review. *JAPCA 30*:528–559 (1980).
33. Lioy, P.J. and M. Lippmann: Measurement of Exposure to Acidic Sulfur Aerosols. *Aerosols*, S.D. Lee, Ed. Lewis Publishers, Chelsea, MI (1986).
34. Daisey, J.M.: Organic Compounds in Urban Aerosols. Aerosols: Anthropogenic and Natural, Sources and Transport. T.J. Kneip and P. J. Lioy, Eds. *Ann. N.Y. Acad. Sci. 338*:50–69 (1980).
35. Lee, M.L., S.R. Goates, K.E. Markides and S.A. Wise: Frontiers in Analytical Techniques for Polycyclic Aromatic Compounds. *Polynuclear Aromatic Hydrocarbons: Chemistry Characterization and Carcinogenesis*, 9th International Symposium, pp. 13–40. M. Cooke and A.J. Dennis, Eds. Battelle Press, Columbus, OH (1986).
36. Cadle, S.H. and P.J. Groblicki: An Evaluation of Methods for the Determination of Organic and Elemental Carbon in Particulate Samples. *Particulate Carbon: Atmospheric Life Cycle*, pp. 89–110. G.T. Wolff and R.L. Klimesch, Eds. Plenum Press, New York (1982).
37. Turner, W.: Personal communication (1986).
38. U.S. Environmental Protection Agency: *Review and Evaluation of the Evi-*

dence for Cancer Associated with Air Pollution. EPA-450/5-83-006R. QAQPS (1984).
39. Maron, D.M. and B.N. Ames: Revised Methods for the Salmonella Mutagenicity *Test. Mutat. Res. 113*:173–215 (1983).
40. Ames, B.N., J. McCann, and E. Yamasaki: Methods for Detecting Carcinogens and Mutagens with the Salmonella/microsomal Mutagenicity Test. *Mutat. Res. 113*:347–364 (1974).
41. Louis, J.B., L.J. McGeorge, T.B. Atherholt et al.: Mutagenicity of Inhalable Particulate Matter at Four Sites in New Jersey. *Toxic Air Pollution*. P.J. Lioy and J.M. Daisey, Eds. Lewis Publishers, Chelsea, MI (1986).
42. Butler, J.P., T.J. Kneip and J.M. Daisey: *Atmos. Environ.* (in press).
43. Wolff, G.T.: Personal communication. General Motors Research Laboratories, Warren, MI (1984).
44. Lioy, P.J., D. Spektor, G. Thurston et al.: The Design Consideration for Ozone and Acid Aerosol Exposure and Health Investigations: The Fairview Lake Summer Camp. Photochemical Smog Case Study. *Environ. Internat.* (in press).
45. Turner, D.B.: Atmospheric Dispersion Modeling: A Critical Review. *JAPCA 29*:502–519 (1979).
46. Hidy, G.M.: Source-Receptor Relationships for Acid Deposition: Pure and Simple. *JAPCA 34*:518–531 (1984).
47. Miller, M.S., S.K. Fiedlander and G.M. Hidy: A Chemical Balance for the Pasadena Aerosol. *J. Coll. Interface Sci. 37*:165–176 (1972).
48. Kleinman, M.T., B. Pasternack, M. Eisenbud and T.J. Kneip: Identifying and Estimating the Relative Importance of Sources of Airborne Particles. *Env. Sci. Tech. 14*:62–65 (1980).
49. Hopke, P.E., D.J. Alpert and B.A. Roscoe: Fantasia—A Program For Target Transformation Factor Analysis to Apportion Sources in Environmental Samples. *Computers in Chem. 7*:149–155 (1983).
50. Hopke, P.E.: *Receptor Modeling in Environmental Chemistry*, 319 pp. J. Wiley & Sons, New York (1985).
51. Thurston, G.D. and P.J. Lioy: Receptor Modeling and Aerosol Transport. *Atmos. Environ.* (in press).
52. Hidy, G.M. et al.: Summary of the California Aerosol Charaterization Experiment. *JAPCA 25*:1106–1114 (1975).
53. Hidy, G.M., P.K. Mueller, D. Grosjean et al.: *The Character and Origin of Smog Aerosols: A Digest of Results from the California Aerosol Characterization Experiment (ACHEX)*, 761 pp. J. Wiley & Sons, New York (1980).
54. Countess, R.J., G.T. Wolff and S.H. Cadle: The Denver Winter Aerosol: A Comprehensive Chemical Characterization. *JAPCA* 30:1194–1200 (1980).
55. Lioy, P.J. and J.M. Daisey: Airborne Toxic Elements and Organic Substances. *Env. Sci. Tech. 20*:8-14 (1986).
56. Lioy, P.J. and J.M. Daisey, Eds.: *Toxic Air Pollutants: Study of Non-Criteria Pollutants*, 283 pp. Lewis Publishers, Chelsea, MI (1986).

57. Ferris, B.G. and J.D. Spengler: Harvard Air Pollution Health Study in Six Cities in the USA. *Tokai J. Exp. Clin. Med. 10*:263–286 (1985).
58. Pellizzari, E., L. Sheldon, C. Sparcino et al.: Volatile Organic Levels in Indoor Air. *Indoor Air*, Vol. 4, *Chemical Characterization and Personal Exposure*, pp. 303–308. Swedish Council for Building Research, Stockholm, Sweden (1984).
59. Landrigan, P.J.: Epidemiologic Approaches to Persons with Exposures to Waste Chemicals. *Environ. Health Persp. 48*:93–97 (1983).
60. Melius, J.M., R.J. Costello and J.R. Kominsky: Facility Siting and Health Questions: The Burden of Health Risk Uncertainty. *Natural Resources Lawyer 17*:467–472 (1985).

INDEX